11

12

ENGINEERING HYDROLOGY TECHNIQUES IN PRACTICE

ELLIS HORWOOD SERIES IN CIVIL ENGINEERING

Series Editors
Structures: Professor R. EVANS, Department of Civil Engineering, University College, Cardiff
Hydraulic Engineering and Hydrology: Dr R. SELLIN, Department of Civil Engineering, University of Bristol
Geotechnics: Professor D. WOOD, Department of Civil Engineering, University of Glasgow
North American Editor: Professor N. G. SHRIVE, Department of Civil Engineering, University of Calgary, Canada

Addis, W.	**Structural Engineering: The Nature of Theory and Design**
Allwood, R.	**Techniques and Applications of Expert Systems in the Construction Industry**
Bhatt, P	**Programming the Matrix Analysis of Skeletal Structures**
Blockley, D.I.,	**The Nature of Structural Design and Safety**
Britto, A.M. & Gunn, M.J.	**Critical State Soil Mechanics via Finite Elements**
Bljuger, F.	**Design of Precast Concrete Structures**
Calladine, C.R.	**Plasticity for Engineers**
Carmichael, D.G.	**Structural Modelling and Optimization**
Carmichael, D.G.	**Engineering Queues in Construction and Mining**
Cyras, A.A.	**Mathematical Models for the Analysis and Optimisation of Elastoplastic Systems**
Dowling, A.P. & Ffowcs-Williams, J.E.	**Sound and Sources of Sound**
Edwards, A.D. & Baker, G.	**Prestressed Concrete**
Fox, J.	**Transient Flow in Pipes and Open Channels**
Graves-Smith, T.R.	**Linear Analysis of Frameworks**
Gronow, J.R., Schofield, A.N. & Jain, R.K.	**Land Disposal of Hazardous Waste**
Hendry, A.W., Sinha, B.A. & Davies, S.R.	**Loadbearing Brickwork Design, Second Edition**
Heyman, J.	**The Masonry Arch**
Holmes, M. & Martin, L. H.	**Analysis and Design of Structural Connections: Reinforced Concrete and Steel**
Iyengar, N.G.R.	**Structural Stability of Columns and Plates**
Jordaan, I.J.	**Probability for Engineering Decisions: A Bayesian Approach**
Kwiecinski, M.,	**Plastic Design of Reinforced Slab-beam Structures**
Lencastre, A.	**Handbook of Hydraulic Engineering**
Lerouil, S., Magnan, J.P. & Tavenas, F.	**Embankments on Soft Clays**
May, J.O.	**Roofs and Roofing**
McLeod, I.	**Structural Engineering with Finite Elements**
Megaw, T.M. & Bartlett, J.	**Tunnels: Planning, Design, Construction**
Melchers, R.E.	**Structural Reliability Analysis and Prediction**
Mrazik, A., Skaloud, M. & Tochacek, M.	**Plastic Design of Steel Structures**
Pavlovic, M.	**Applied Structural Continuum Mechanics: Elasticity**
Pavlovic, M.	**Applied Structural Continuum Mechanics: Plates**
Pavlovic, M.	**Applied Structural Continuum Mechanics: Shells**
Shaw, E.M.	**Engineering Hydrology Techniques in Practice**
Spillers, W.R.	**Introduction to Structures**
White, R.G. & Walker, J.G.	**Noise and Vibration**
Zlokovic, G.	**Group Theory and G-Vector Spaces in Structures**

ENGINEERING HYDROLOGY TECHNIQUES IN PRACTICE

ELIZABETH M. SHAW, B.Sc., M.Sc.
formerly Department of Civil Engineering
Imperial College of Science, Technology and Medicine, University of London

ELLIS HORWOOD LIMITED
Publishers · Chichester

Halsted Press: a division of
JOHN WILEY & SONS
New York · Chichester · Brisbane · Toronto

First published in 1989 by
ELLIS HORWOOD LIMITED
Market Cross House, Cooper Street,
Chichester, West Sussex, PO19 1EB, England
The publisher's colophon is reproduced from James Gillison's drawing of the ancient Market Cross, Chichester.

Distributors:
Australia and New Zealand:
JACARANDA WILEY LIMITED
GPO Box 859, Brisbane, Queensland 4001, Australia
Canada:
JOHN WILEY & SONS CANADA LIMITED
22 Worcester Road, Rexdale, Ontario, Canada
Europe and Africa:
JOHN WILEY & SONS LIMITED
Baffins Lane, Chichester, West Sussex, England
North and South America and the rest of the world:
Halsted Press: a division of
JOHN WILEY & SONS
605 Third Avenue, New York, NY 10158, USA
South-East Asia
JOHN WILEY & SONS (SEA) PTE LIMITED
37 Jalan Pemimpin # 05–04
Block B, Union Industrial Building, Singapore 2057
Indian Subcontinent
WILEY EASTERN LIMITED
4835/24 Ansari Road
Daryaganj, New Delhi 110002, India

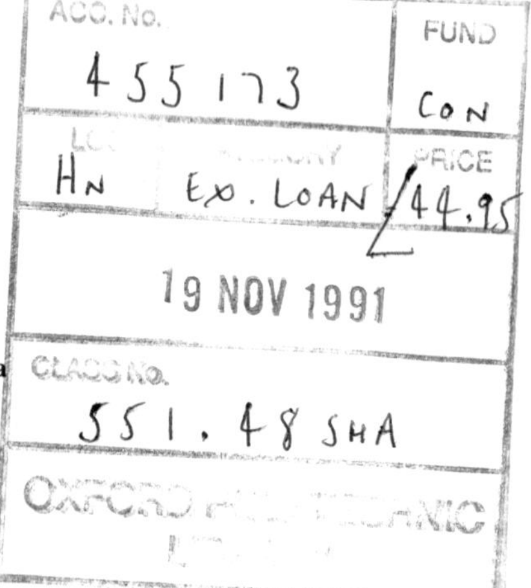

British Library Cataloguing in Publication Data
Shaw, Elizabeth M. (Elizabeth Mary)
Engineering hydrology techniques in practice.
1. Hydrology — For hydraulic engineering
I. Title
551.48′024627

Library of Congress Card No. 88–8403

ISBN 0–85312–943–6 (Ellis Horwood Limited)
ISBN 0–470–21309–4 (Halsted Press)

Typeset in Times by Ellis Horwood Limited
Printed in Great Britain by Hartnolls, Bodmin

Table of contents

Foreword

The greatly enhanced awareness today of environmental problems has led to a resurgence of interest in all aspects of water engineering. The need to reduce the pollution of inland waters and to reduce the risk of damage from flooding has led to a large increase in the number of engineers and scientists actively working in this field and in the complexity of the problems that they are facing. The resulting major industry has become increasingly an area of interdependence between specialists in a wide variety of disciplines, both scientific and engineering at root. The need for these two groups to communicate frequently and successfully at all levels has led to strains in the past and requires more attention in the future.

Engineering Hydrology is a subject in which engineers and scientists have traditionally worked in close collaboration and have jointly developed practical techniques drawing on their diverse experience and backgrounds. There is therefore a special need for excellent textbooks in this area which combine these various elements in an integrated whole. Elizabeth Shaw has already established a reputation as an Engineering Hydrologist that few can match. Her long-established textbook *Hydrology in Practice*, first published in 1983, now fills the need for a basic text that can be used both by advanced level students and also by practising hydraulic engineers. This new volume, *Engineering Hydrology Techniques in Practice,* goes on to build on the principles and methods so well set out in her first book. By adopting a wide-ranging yet systematic review of case studies she is following a format which has so often proved successful in other areas in the past. This approach enables the reader to relate his or her own immediate problem to the wealth of experience that Miss Shaw has acquired over a working lifetime in this subject while teaching in the Department of Civil Engineering at Imperial College, London. It very quickly becomes apparent to the reader that the author has also obtained her experience by working very closely with Civil Engineering consultants during most of this time. It is this, together with her own personal qualities that will ultimately ensure the success of Elizabeth Shaw's latest book.

The complexity, importance and interdisciplinary nature of Engineering Hydrology are together probably responsible for the popularity of MSc courses in this subject. The implication is that the serious student will benefit from having already completed a first degree course in one of the basic subjects and, again, may be expected to come from one of the cognate sciences or from a branch of engineering. The present volume will prove invaluable for such courses where the need is to give the student both a sound knowledge of the basic aspects of the subject and also a training in applied techniques, all in a short time.

For the consultant this collection of 65 case studies, arranged in 11 subject groupings, will have an immediate appeal as the range of topics covered is so broad. Each group is preceded by an introduction, and the volume is completed with a full list of references and a number of indexes which will add considerably to the ease with which it can be used.

This book has within it, in effect, the makings of an 'expert system' in engineering hydrology, but that goal must remain for the present in its early stages of development.

University of Bristol
9 January 1989

Robert Sellin

Preface

The inspiration for a reference book of case studies was generated by the constructive criticisms of my textbook *Hydrology in Practice* published in 1983 by Van Nostrand Reinhold (UK). Many academics wanted more examples of the application of hydrological techniques. When it comes to demonstrating the use of complex mathematical models or data-simulating methods, the presentation of model details related to a particular problem and the multiplication of possible results, become beyond the scope of a textbook.

In the original plans for this collection of case studies, the broad distinction between hydrological problems was recognized: those in which there is a surplus of water requiring drainage schemes and those in which the need for water resources is paramount. When a few studies had been assembled, it became obvious that 'real' applications cut across the boundaries of carefully chosen categories within the two broad subdivisions. The first problem to arise was the place of the hydrological studies for spillway design where the reservoir is part of a water resources scheme. These have finally been given a chapter to themselves, bridging the two contrasting sections. Within the surplus water section, the schemes fall easily into categories defined by the chapter headings. However, the classification of water resource schemes has been much more difficult and the chapter headings remain somewhat arbitrarily related to the assembled studies.

At first, the consideration of hydrometric schemes seemed outdated since in the UK these had been dealt with in the 1960s. However, after visiting the offices of several consulting engineers, it became apparent that the establishment of hydrometric networks formed a significant proportion of their overseas contract work. Hence the book begins properly with a chapter of hydrometric schemes. The final chapter cuts across all boundaries and provides examples of recent operational and management studies. These incorporate the use of the latest technological aids, the handling of large quantities of all kinds of data and the application of sophisticated mathematical techniques by computers. They also touch on other disciplines which are necessarily considered in the decision-making process.

The case studies have been assembled from academic research contracts, river and water authority in-house schemes and the worldwide commissions of consulting engineers. Many of the projects have been taken from consultants' reports to clients and permission to publish has been sought from the clients concerned. Each example has been presented in a uniform format and written from the engineering viewpoint. Following the title of the project, location and source of information, the engineering problem is posed, the clients are identified and the charge given to the hydrologist is outlined. After describing the studies made, techniques applied and results obtained, recommendations are made to the engineers and clients with regard to measures considered necessary to solve the problem. Naturally, the clients make the final decisions on any works. Again it has sometimes been difficult with the great variety of material studied to maintain the original plan of presentation and there evolved a certain flexibility in the section headings of each project following the posing of the problem.

One of the attractions of the civil engineering profession, and of the contributions of hydrologists, is that every problem is a new challenge and most projects are one-off jobs. Solutions to problems are related to the pertinent information available, to the diligence of the on-site research team and to the professional competence and experience of the project engineers. Thus this book contains examples of unique problems tackled in a singular fashion, to stimulate students and junior design engineers in considering means of solving their own problems. They must not be slavishly followed and transposed to other situations.

Most of the case studies pertain to schemes in the UK but 33 other countries are represented and the problems come from most climatic regions of the world. There are 11 chapters and under each heading there is a brief explanatory introduction. The case studies are then presented in chronological order. The resulting collection of studies represents a first attempt at providing illustrative material for teachers and practitioners alike. The applications of hydrology in engineering practice included in the book are not fully comprehensive and they have continued to expand in the 2 years during which the book has been compiled. The author looks forward to noting omissions and perhaps filling the gaps in future work. The book concludes with a general list of references, an author and geographical index and an index of hydrological subjects and techniques cited in the case studies.

Acknowledgements

This compilation of such a varied set of case studies demonstrating the application of hydrology in civil engineering practice would not have been possible without the helpful cooperation of many active practitioners. There are three groups of organizations to be acknowledged; the consulting engineers who carried out the studies on behalf of clients, the public authorities who may have had the double role of hydrologist and client and the major body of clients who commissioned the studies. In each case study, the source material is given and where this had already been published, no direct approaches to client or engineer were made for permission to publish. With much of the information obtained from unpublished confidential reports, requests for permission to publish were made to the clients, six of whom have not replied.

The author is most grateful for the help and cooperation received from the following organizations:

Consultants

Binnie & Partners
Binnie Dan Rakan
John S. Bonnington Partnership
Coode Blizard (formerly Lemon & Blizard)
Diyan Consultants
Gauff Ingenieure
Sir William Halcrow & Partners Ltd
Howard Humphreys & Partners
Hydraulics Research Ltd.
Sir M. MacDonald & Partners
Rofe, Kennard & Lapworth
WRc Swindon

Public Authorities

Institute of Hydrology, UK
Northumbrian Water

North West Water
Severn–Trent Water
South West Water
Wessex Water
Thames Water
Yorkshire Water

Clients

Department of Water Affairs, SWA/Namibia
Director General, Water Resources Bureau, Ministry of Construction, Seoul, Korea
Director of Water Supplies, PWD, Malaysia
Ministry of Agriculture & Natural Resources, Cyprus
Ministry of Planning, Republic of Honduras
Nairobi City Commission
National Electric Power Authority, Nigeria
Strathclyde Regional Council
West Wiltshire District Council

On a more personal basis, the author wishes to thank all those individuals who produced material, helped in the selection of case studies and read the first drafts critically and constructively. It is hoped that the following list includes all those who gave their valuable time and expertise to the appraisal of the case studies.

D. R. Archer
V. R. Baghirathan
K. T. Bass
D. C. Beale
J. Buckmaster
S. Chapman
T. E. Evans
Gong Shiyang
R. C. Goodhew
I. M. Goodwill
R. C. Gosling
M. J. Green
K. Guganesharajah
M. J. Hall
G. H. Hargreaves
R. W. Hatton
D. M. Helliwell
N. A. Hill
D. I. Hyde
P. Johnson
J. Johnston
P. M. Johnston
M. J. Lowing
M. Mansell-Moullin
J. C. Miles
P. Netchaef
P. E. O'Connell
N. C. Oxley
J. C. Packman
V. M. Ponce
R. K. Price
D. W. Reed
D. H. Rees
P. D. Walsh
P. Webster
H. S. Wheater
N. E. Whiter
J. Wild
P. R. Vaughan

The Librarians of the Department of Civil Engineering, Imperial College, K. Crooks and J. Underhill were ever helpful with advice on references and Hazel Guile kindly undertook the drawing of new and complex illustrations. Permission was obtained from Academic Press, American Society of Civil Engineers, Thomas Telford Ltd and World Meteorological Organization to reproduce drawings.

List of tables

List of figures

Chapter 4

Chapter 5

Chapter 6

1

Hydrometric schemes

INTRODUCTION

The successful application of engineering hydrology in civil engineering practice is dependent on the availability of reliable hydrometric data. Many of the sophisticated analytical techniques, often viewed with suspicion by practising engineers, have been developed as a consequence of the shortage of numerical information. Thus considerable efforts have been made in the design of hydrometric schemes and in the establishment of networks of measuring stations.

The population explosion of this midcentury coupled with widespread increases in living standards resulted in greater demands for water and consequently in the need for the evaluation of water resources. Existing sources and storages were insufficient to serve the new requirements and a great impetus was given to the expansion of water engineering. To establish a sound basis for this expansion there was an urgent requirement for the assemblage of measurements of rainfall, evaporation, river flow and groundwater. This was a worldwide phenomenon recognized in different degrees by the various countries according to their state of development. The advanced nations established or improved upon their own hydrometric networks and through the advice of their experts enabled organizations of the United Nations to lay down design criteria for the developing world (WMO, 1958; UN–WMO, 1960). Advances in the appraisal of existing networks and the planning of new national hydrometric systems in the 1960s were published in the *Casebook on Hydrological Network Design Practice* (WMO, 1972). Examples in the *Casebook* demonstrate the varying national needs over a wide range of topographical and climatic conditions in the inhabited world.

In the UK, a parliamentary statute for England and Wales, the Water Resources Act 1963, required the river authorities to prepare schemes and then to establish networks to measure the principal hydrological variables. The schemes were grant aided through the central authority, the Water Resources Board, which had additional duties such as national planning and sponsoring research. The designing of these hydrometric schemes had very little formal scientific basis but was carried out following the results of earlier experience and with detailed knowledge of the river catchments. The need for tools in the design of networks stimulated widespread research (Rodda, 1969) and some of the findings were presented at an international symposium (WMO, 1976).

The examples presented here begin with the initial scheme for a UK river authority area followed by three early overseas examples. A hydrometric scheme review based on a reappraisal of data requirements and on the cost of the network operation represents a second stage in UK hydrometry. With the emphasis changing recently from water resources to water quality, a sewage-monitoring scheme is included. In the UK, while the urgencies of network planning have given way to concern for network maintenance and the economical production of good reliable information, overseas in the developing world, hydrometric schemes are still providing the first phase in water resources development.

1.1 DEVON RIVER AUTHORITY HYDROMETRIC SCHEME

LOCATION The Devon River Authority area contained 10 main river catchments in southwest England.

SOURCES Gosling, R. C. (Chief Engineer) (1966) *Devon River Authority hydrometric scheme*, Report of the Engineer's Department.

Shaw, E. M. (1965) *A raingauge network prepared for the Devon River Board Area of the United Kingdom*, Publication No. 67. International Association of Scientific Hydrology, Quebec, pp. 63–70.

PROBLEM The Devon River Authority, which superseded the Devon River Board on 1 April 1965, was required by the Water Resources Act 1963 to provide under Section 15 schemes to measure rainfall, evaporation, surface water and water quality and under Section 18 a scheme for the measurement of groundwater.

River authority area

The total area of 6252 km^2 consisted of the major part of the county of Devon with smaller areas of Cornwall, Somerset and Dorset. The main catchment areas are shown in Fig. 1.1. The two outstanding topographical features are the granite mass of Dartmoor rising to 620 m, the source of the Rivers Torridge, Taw, Teign, Dart, Avon and Erme and Exmoor of Old Red Sandstone rocks rising to 520 m, the source of the Exe, tributaries of the Taw and the Lyn–Heddon streams. In the eastern projection of the area, the Chalk and Greensand of the Blackdown Hills, 180–240 m, are the source of the Axe, Otter and Culm, an Exe tributary. The average annual rainfall ranges from 2300 mm on Dartmoor to about 760 mm round the Exe estuary.

The variety in geological structure and composition and rainfall results in considerable differences in hydrological response and thus it was found necessary to consider each river basin separately. For all the variables to be measured, existing installations and their records had to be catalogued. There was considerable research

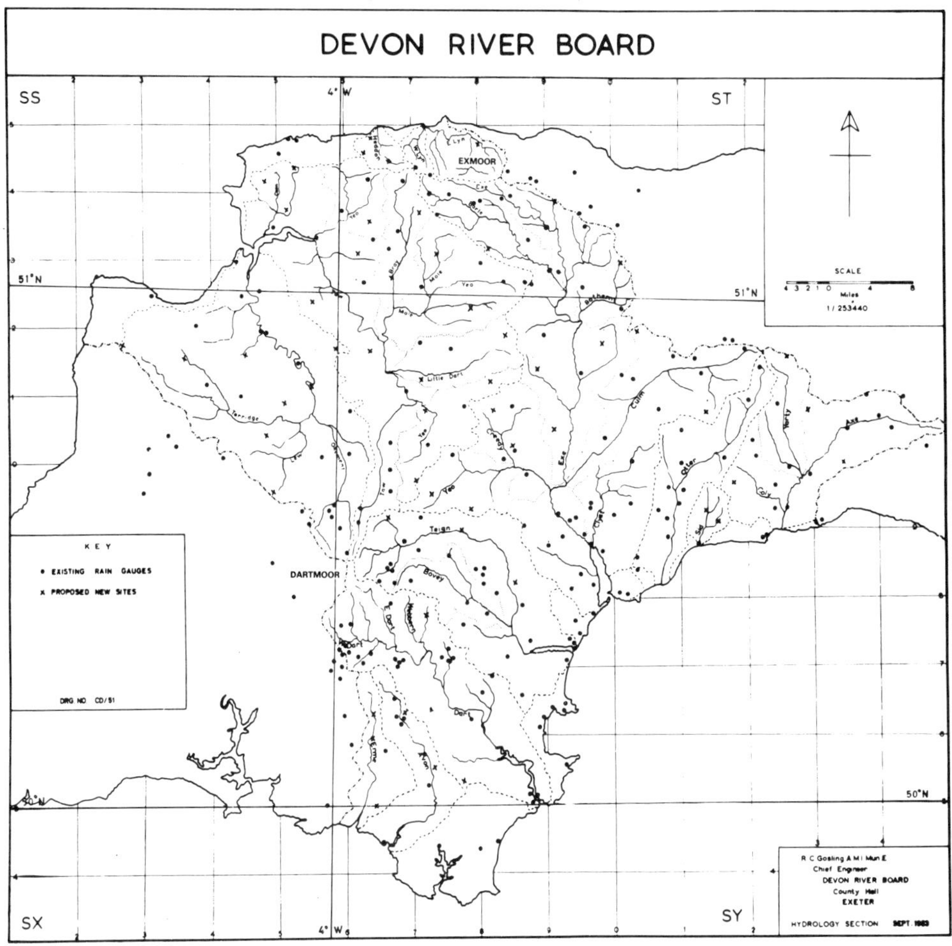

Fig. 1.1 — Devon River Board rain gauge network: ●, existing rain gauges; ×, proposed new sites.

and liaison with other interested organizations and private observers to assemble the current hydrological measurements before the catchment networks could be designed.

Rainfall

The existing rain gauge stations were plotted on 10 basic catchment maps (scale, 1:63 360) and their distribution examined. Since the primary aim was the evaluation of water resources, a regular coverage of each catchment was required for general water balance studies. New gauge sites were chosen to sample sparse areas and different altitudes, taking into account the methods to be used for determining areal rainfall and the availability of new observers. In assessing the number of new gauges required, the types of rainfall experienced in the region and the irregular topography were considered and a gauge density of 1 for every 26 km^2 seemed likely to provide a reasonable measure of the rainfall in the Devon area. The overall results are seen in Fig. 1.1 with the gauge numbers given in Table 1.1. The resultant gauge densities

Table 1.1 — The rain gauge network

Catchment	Area (km^2)	Existing guages	New gauges	Total	Gauge density (km^2 gauge^{-1})
Axe	404	12	4	16	25.3
Otter and Sid	286	13	2	15	19.0
Exe	1520	47	12	59	25.8
Teign	526	26	3	29	18.1
Dart	489	22	3	25	19.6
Avon	145	7	2	9	16.1
Erme	106	1	3	4	26.5
Torridge	847	17	7	24	35.3
Taw	1230	23	16	39	31.5
Lyn and Heddon	133	3	4	7	19.0
Total catchments	5686	171	56	227	(26.0)
Coastal areas	566	23	0	23	
Total	6252	194	56	250	25.0

differed in the individual catchments; the greater numbers in the Teign and Dart are associated with reservoirs and relatively fewer new gauges were required to serve the low relief areas of the Taw and Torridge middle reaches. For the new network of rainfall stations, the scheme provided for daily and monthly storage gauges but additional requirements included the installation of seven transmitting raingauges to measure intense rainfalls for flood warning purposes.

Evaporation

Measurement of evaporation included natural losses from open water surfaces and catchment losses from transpiration of vegetation. The national policy was to encourage the evaluation of potential evapotranspiration (evaporation plus transpiration with unlimited water) using the Penman formula. This required the measurement of air temperature, humidity, sunshine hours or radiation and wind speed at

2 m above a short grass surface and using monthly values, the calculation of an average daily evaporation rate.

There were eight existing stations in the Devon area measuring the necessary meteorological variables and another four private stations needing supplementary instrumentation. Evaporation is dependent on air temperature which in turn is related to altitude. Hence a hypsometric curve of the whole area (height above Ordnance Datum (AOD) versus cumulative area) was used to obtain the proportionate area between selected altitudes as a guide to the number of sampling evaporation stations required. The results of this exercise are shown in Table 1.2. For each

Table 1.2 — Evaporation stations

Height range (m AOD)	Percentage of total area	Existing stations	New stations	Total
Over 420	2	0	1	1
360–420	3	1	0	1
300–360	4	1	1	2
240–300	7	0	3	3
180–240	14	1	3	4
120–180	26	1	7	8
60–120	29	3	6	9
0– 60	15	5	0	5
Total	100	12	21	33

of the new 21 stations, site maps (scale, 1:25 000) were provided with indications of the site exposure and it was proposed to equip each station with an automatic recording weather station. A data logger would record measurements every 15 min from transducers sensing air temperature, humidity, net radiation and wind speed. Provision would also be made to record rainfall from a tipping bucket gauge. All the new evaporation stations would be at existing rainfall stations. It was also proposed that future developments would include the provision of instrumentation for the calculation of actual evapotranspiration.

Surface water

There were 12 river gauging stations in operation by the Devon River Authority prior to the hydrometric scheme. The major rivers, Rivers Axe, Otter, Exe, Teign, Dart, Torridge and Taw, all had established installations for water level chart recorders and were rated or calibrated by current meter. The Axe control was a compound Crump weir; there were two broad-crested weirs on tributaries of the Exe and Taw and the rest were river sections. In addition there were 11 stations operated by other authorities but the discharges were measured over a limited range and were only useful for low flow records.

In planning the new network of stations, each major catchment was studied individually. The natural characteristics were noted, together with the locations of abstractions by water boards and industrial concerns which included the diversion of flow through mill leats and irrigation channels. The type of abstraction was also identified be it from a spring, well, borehole or river intake and regular large daily quantities were recorded. Taking account of the survey information obtained, sites for new river stations were chosen, the numbers are summarized in Table 1.3. A

Table 1.3 — River gauging stations

Catchment	Area (km^2)	Existing stations	New stations	Total	Density (km^2 $station^{-1}$)
Axe		1	3	4	
Otter		1	1	2	
Exe		4	11	15	
Subtotal	2255			21	107
Teign		1	3	4	
Dart		2	1	3	
Avon			1	1	
Erme			1	1	
Subtotal	1509			9	168
Torridge		1	3	4	
Taw		2	10	12	
Subtotal	2146			16	134

standard specification was presented for the stations. Since a full weir river flow measurement would have been prohibitively costly, a low-level Crump section weir was proposed to act as a bed check and stable control section and as a measuring weir for minimum flows. Each station would be equipped with a current meter suspension cableway for calibration and a level recorder house. Both punched tape and chart recorders were specified. Varying site characteristics would determine the final installations. Water-quality-monitoring equipment was proposed for 11 stations.

Provision was also made for the continuous measurement of water levels in reservoirs in order to keep a record of their changes in storage.

Groundwater

The hydrogeology of the Devon River Authority area divides into two distinctive parts. The pre-Permian rocks of north, west and south Devon have low mass permeability and low yields. Areas of fissured rocks and pockets of alluvial deposits may provide local supplies. The Permo-Triassic and Cretaceous formations of east Devon contain the largest groundwater reserves and sustain the highest yields.

A systematic measurement of groundwater level variations was proposed for the area east of the base of the Middle Devonian rocks in south Devon and the Permian unconformity across central Devon. Considerable research was required to assemble existing information. Local and national authorities were visited to obtain locations of wells and boreholes and any available records. Approximately 200 wells and boreholes were visited and yields, rest levels and dimensions measured. A farm questionnaire revealed a further 1300 wells and boreholes. A total of 2000 sites with groundwater information was identified. In planning a network of new observation stations, two types were recommended, 110 providing manual level measurements twice a month and 35 with continuous water levels, recorded in deep boreholes. The groundwater stations were distributed as indicated in Table 1.4. Plans were included

Table 1.4 — Groundwater stations

Catchment area	Manual	Automatic
Axe	14	5
Otter and Sid	23	5
Exe–Clyst	10	4
Lower Exe	10	4
Middle Exe	9	4
Exe–Culm	10	4
Crediton trough	14	5
Teign	11	2
Dart	9	2
Total	110	35

in the scheme for occasional pump tests at selected boreholes to determine unknown aquifer characteristics such as transmissibility, storage coefficients and safe yields. A network of spring gauging stations at the headwaters of the east Devon Rivers Axe, Otter and Exe–Culm was also planned.

Data processing

The hydrometric scheme included full details of methods and instrumentation required for the processing and basic analysis of the vast quantities of data to be generated by the station networks. These were based on the technology of the day (1966) and since then have been superseded several times by instrumental developments and the microchip revolution in computers.

1.2 A RIVER GAUGING NETWORK FOR THE NICHOLSON BASIN IN QUEENSLAND, AUSTRALIA

LOCATION The Nicholson River drains the extreme northwestern corner of Queensland and flows into the Gulf of Carpentaria.

SOURCE Snowy Mountains Engineering Corporation (1972) A hydrometric network for a sparsely populated basin in north-western Queensland. *Casebook on hydrological network design practice.* World Meteorological Organization, V-4.1.

PROBLEM The Irrigation and Water Supply Commission of Queensland in 1964 required the Snowy Mountains Hydro-electric Authority to investigate and design hydrometric networks for the collection of water resources data in 15 river basins covering a total of 730 000 km^2 in the remote sparsely populated northwestern regions of the state. The planning of a network of river gauging stations in such remote terrain is demonstrated for the sample basin of the Nicholson River, 60 000 km^2, where the population density is barely one person per 100 km^2 (Fig. 1.2).

Nicholson Basin

The Nicholson River rises on the Barkly Tableland of the Northern Territory and is joined by the Gregory River and many tributaries draining the highly dissected centre of the basin. The wide coastal plain bordering the Gulf of Carpentaria is of low relief and the rivers tend to distribute their high flows in numerous channels. The flooding tendency results from heavy tropical cyclonic rainfalls capable of producing half the annual total. Fig. 1.2 shows the mean annual rainfall distribution with the wettest area over 800 mm along the coast and under 500 mm in the source regions. Situated in the tropics and influenced by monsoon winds, the main rains are in the wet summer season, and with high annual potential evaporation of about 2000 mm, the catchment runoff is limited to about 10 mm. Experience of gauged catchments in Queensland has shown that the seasonal variation in the monthly river flows is very high and the marked seasonal rainfall pattern results in many of the streams being intermittent in the dry season.

Network design

The basis of the design of the river gauging station network for the Nicholson River basin was that of a preliminary survey of water resources. The standard recommended types of station were adopted: primary stations intended to be permanent and to sample the range of catchment characteristics; secondary stations to sample significant tributaries on a short time basis until their flow pattern could be mathematically modelled satisfactorily; and special stations, in the area of the distributaries, partial record stations at which river levels were to be recorded continuously but with discharges only available for low to moderate flows.

The initial studies of the area were made on maps of scales 1:250 000 and 1:1 000 000 to assess the variations in topography and vegetation and, bearing in mind the limitation on costs, a minimum size of catchment to be considered was set arbitrarily at 250 km^2. Possible catchments were identified and, from aerial photographs on a 1:80 000 scale, examination of the channel geometry enabled suitable gauging sites to be chosen. The existence of natural bed controls, the containment of flood flows and the accessibility of the sites were among the criteria considered. Thirty-two possible sites were selected for closer inspection. This was carried out by low-flying high-wing aircraft before sending in a time consuming ground survey party

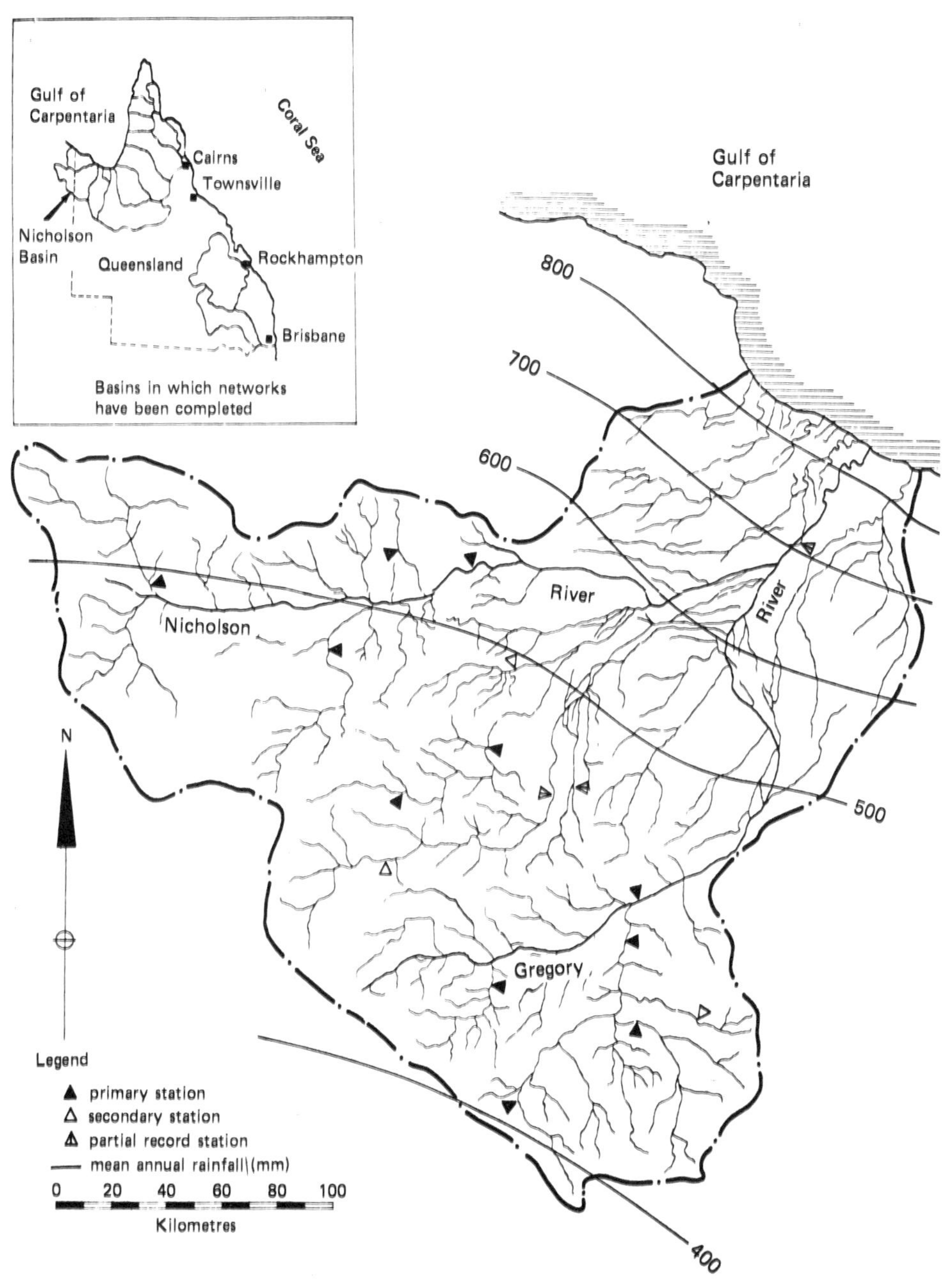

Fig. 1.2 — River gauging network in the Nicholson Basin.

to finalize details. For many sites the aerial reconnaissance was able to decide upon the location and the required instrumentation and was invaluable in identifying access routes for the ground surveys and construction teams. The final network design was prepared following the aerial reconnaissance and only minor alterations were needed following ground inspections.

The completed network for the Nicholson Basin, shown in Fig. 1.2 contains 17 river gauging stations on catchments ranging in size from 550 to 50 000 km^2. There are 11 primary stations, three secondary and three partial record stations and these will provide the basic river flow information on which to evaluate the water resources of the basin.

Conclusion

The network of river gauging stations was designed essentially from practical considerations. In such a pioneering environment, no stream flow records existed in or near the basin on which to base any analytical design methods. The expansion of the network in the future would depend on the economic development of the region and the accompanying management of the water resources. Then the requirement for further river flow data at prospective dam sites could result in the establishment of more stream gauging stations.

1.3 A RIVER GAUGING NETWORK FOR THE RHINE-MEUSE DELTA

LOCATION The delta of the Rivers Rhine and Meuse occupies the southwestern coastal stretch of The Netherlands bordering the North Sea.

SOURCES Van der Made, J. W. (1972) The hydrometric network in the Rhine–Meuse delta. *Casebook on hydrological network design practice.* World Meteorological Organization, V-6.1.

Van der Made, J. W. (1972) Streamflow and water levels — the Rivers Rhine and Meuse. *Casebook on hydrological network design practice.* World Meteorological Organization, I-5.1.

PROBLEM Following the disastrous storm surge of 1953 when a high tidal level nearly 3 m above normal inundated most of the delta islands, the extensive delta project was launched to close off all the tidal inlets except the Rotterdam Waterway in the north and the Western Scheldt estuary in the south, leading to Antwerp. The resulting fundamental change in the hydrological regime of the waters of the delta region called for a reappraisal of the existing water level and discharge measurement stations which were part of the main national network of The Netherlands.

The Delta Project

The completed delta project scheme is illustrated in Fig. 1.3. Dams have been built across the tidal inlets. In the southern part three freshwater lakes have been formed, Veere Lake, Zealand Lake and Grevelingen Basin. Water levels in these lakes will have relatively small variations since they will no longer be subject to tidal effects or river discharges. In the northern part the Haringvliet Dam incorporates controlling

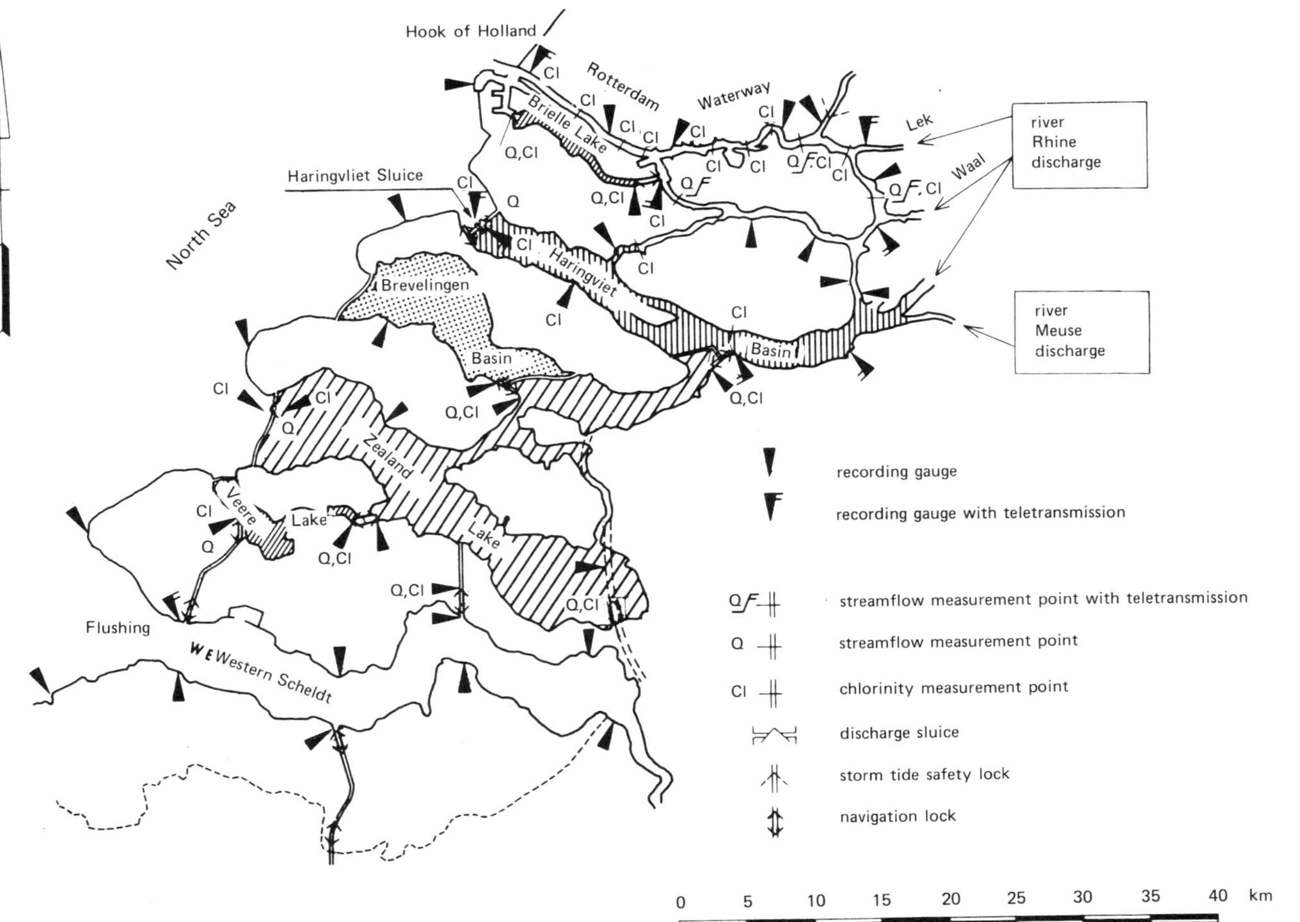

Fig. 1.3 — Proposed network for the Rhine–Meuse delta.

sluices and with the Rotterdam Waterway is still subject to tidal changes. These two branches of the delta receive the flows of the Rivers Rhine and Meuse which amount to 2200 $m^3 s^{-1}$ and 250 $m^3 s^{-1}$ on average at the respective Dutch borders. Interlinking waterways or canals between the lakes and basins have controlling navigational locks so that commerce is not impeded and flow measurements can be made at each of these controls.

Gauging network

The river flows of the two main branches of the Rhine, the Lek and Waal, together with the flow of the Meuse are known from upstream gauging stations and their distribution in the northern delta is monitored at three telerecording stations. One of these is on the main Lek channel and the other two are on cross-channels from the Waal distributary to the Rotterdam Waterway so that discharge into the Haringvliet Basin can be obtained from the remaining Waal and Meuse flows.

The water stage variations in Haringvliet Basin are less than previously but they are still affected by the tides via the two transverse channels and therefore existing water level stations will be continued. The other major influence on these water levels is the operation of the Haringvliet sluices for discharging river flows to the sea as follows.

(1) At low Rhine discharges, the gates remain fully closed and outflow is via the transverse channels to the Rotterdam Waterway.
(2) At medium river discharges and low flood waves, the gates are partly opened during low tide.
(3) For medium and high flood waves, the gates are fully opened during low tide. However, these control rules will be set aside during storm tides at sea in order to protect the low-lying land areas.

In addition to the stream flow transmissions, stage gauges at strategic points are fitted with teletransmission instruments and these data are analysed by an analogue computer programmed to provide control data for the whole of the northern delta area.

In the southern delta basin, the smaller level variations in the freshwater lakes require fewer gauging stations and the flow measurements at the navigation locks and drainage sluices provide most of the required information for the management of the lakes. Two teletransmission tide recording gauges at Flushing and the Hook of Holland provide instant warnings of probable high-tide levels to the Storm Tide Warning Service in The Hague.

The network map also shows a large number of chlorinity measurement points. These are particularly needed in the south to monitor the salinity levels in the freshwater lakes at the navigation locks. In the northern delta areas where the tides are free flowing in the Rotterdam Waterway, salt water intrusion is detrimental to groundwater supplies of freshwater and a close network of stations for continuous measurements of the conductivity has been installed.

Conclusion

The reappraisal of the network of surface water measurement stations in the delta has resulted in fewer manually operated stage gauges and a redistribution of recording stage gauges at strategic locations. The network with its new teletransmitting stations has been designed to provide real-time data for the control of the surface waters of the delta.

1.4 HYDROLOGICAL SERVICES IN KOREA

LOCATION The Republic of Korea, South Korea.

SOURCE Binnie & Partners and Hyundai Engineering Co. Ltd. (1978) *Hydrologic services Rural Infrastructure Project*, Final Technical Report to the Ministry of Construction, Seoul.

PROBLEM As part of a nationwide Rural Infrastructure Project, the consultants were asked to study the improvement in the collection and analysis of hydrological information for the whole country with particular attention to mountainous areas and small catchments. The project supported by the World Bank included a review of the existing networks for measuring hydrometric variables, evaluating their suitability and recommending any necessary modifications in organization and instrumentation. Importance was also attached to the processing of data by computer and their preparation for publication. Certain analytical studies were also requested and advice sought on the necessary qualifications and training for hydrometric staff.

Data requirements

The current state and anticipated development of the water resources of a country need to be assessed in reviewing hydrometric data needs. The consultants considered the problem in three phases.

(1) *Early phase* Total demand very small in relation to resources and mainly associated with local agriculture, domestic supplies and industry.
(2) *Middle phase* Industrial development accelerates, towns grow rapidly and agriculture is modernized. Demand for reliable water supplies increases. Large multipurpose schemes become necessary together with data needs for water management. There is rising concern for pollution and flood damage.
(3) *Late phase* Overall demand reaches a high proportion of available resources so that individual needs have to be compared and priorities assigned. Effluent treatment and pollution control become important and flood protection requires consideration.

The data requirements in the three phases range from simple evaluations of local resources from rainfall observations alone, through the middle phase of basin development with additional river gauging and sediment sampling with greater accuracy in water control for the paddy fields and with the beginning of pollution surveys, to the late phase when the complex systems using more of the resources require comprehensive information for reliable operation.

The Republic was considered to be developing rapidly through the middle phase with a growing population and an increase in water consumption per head. Expanding industry and increased urbanization were adding to the demands and providing the attendant pollution problems. Lack of natural fossil fuels was leading to plans for hydropower development.

Hydrometric network

In reviewing the existing measurement station networks, the consultants adopted the standard practice of considering the stations in three classes, the primary long-term stations with records of over 30 years for statistical analysis, the secondary medium-term stations operating for 5–10 years for amplifying regional variations or for calibrating rainfall-runoff models and the special short-term stations to provide data for specific water engineering projects.

Precipitation stations

About 349 rainfall stations give a good areal representative cover of the 99 715 km^2 of the Republic but only two of these above 600 m in altitude provide a poor sample of the extensive mountainous regions where another 30 gauges were recommended. In addition to the altitude variations, there are great contrasts between the cold snowy winters and the hot wet summers with over 80% of the annual precipitation and when an annual typhoon may bring very heavy rains any time between June and September. The network should also be enhanced by rain gauges at the small reservoirs to be instrumented.

River gauging stations

About 183 water level measurement stations operate within the Republic but owing to the wartime disruptions there are no continuous long-term river flow records. From only 23 sites (Fig. 1.4), have daily or monthly river flow data been derived but only 10 of these could be computed with confidence. Other stations have had a few discharge measurements made but inadequate to define rating curves. A major recommendation was the establishment of 41 permanent river discharge measurement stations. Their catchments should range from the three small representative basins to those of the major river basins where close to the tidal limits the residual basin runoffs would be measured. By siting these on the large rivers and main tributaries at existing water level stations the ratings should provide some long-term flow records. In conjunction with the 41 conventional stations, 18 major dams were to be regarded as part of the primary network.

For the small catchments, there was very little information and therefore a number of medium-term river gauging stations were recommended by instrumenting existing small irrigation reservoirs. A network total of 70 river gauging stations was recommended, 59 classed as primary and 11 as secondary.

River sediment stations

The changes in land use and construction of major storage reservoirs are expected to affect changes in the suspended sediment loads of the rivers. Therefore to monitor these changes a network of 20 sediment sampling stations each located at a discharge station and systematic reservoir sediment surveys were recommended.

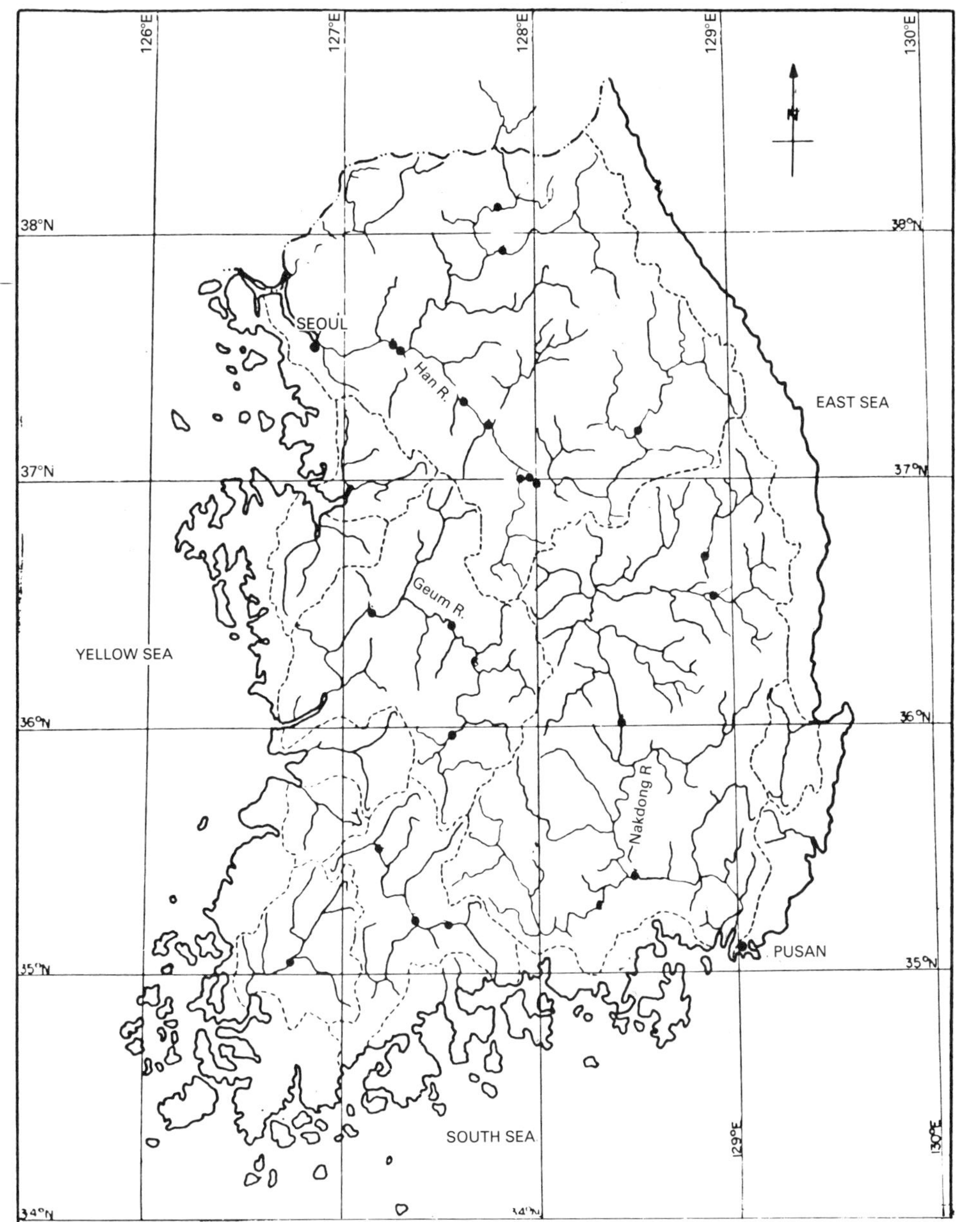

Fig. 1.4 — River flow stations of the Republic of Korea.

Instrumentation

The detailed reports of the consultants outlined the basic techniques of measurement to be applied at the precipitation, river flow and sediment load stations. Much of the existing equipment was unsuitable or no longer usable but the rain gauges were manufactured locally to standard designs. It was recommended that, for the river flow and suspended sediment measurements, standardized well-tried simple equipment should be used. Details of these were provided and it was considered that these might be manufactured locally using imported patterns. Newly created central stores should be responsible for spares and maintenance.

Data archive

The consultants were requested to review the Ministry of Construction's proposals to computerize the publication of daily rainfall and runoff records. Since the project entailed the assemblage from various sources of a large amount of data, a set of manual station files were established which would form a basis for a proposed new hydrometric service. Rating curves for many of the gauging stations were derived and from these mean daily flows (MDFs) were obtained on a desktop computer. All these daily flow data were then stored on magnetic tape. Thus important items of information suitable for computer storage were supported by an efficient manual data-processing and filing system. The files were identified by station numbers based on the major river basins. The computer files facilitated quality control of the data and an existing mainframe computer (a CDC-3600) had sufficient capacity and available computer time to handle the proposed archive. The staff to be responsible for the manual and computer operations should be trained to program the procedures so they can develop the system to serve future needs.

Analytical studies

The Ministry of Construction also included some analytical studies in the project specification. These were carried out with the help of specialists attached to the consultants' team. The following studies were based on countrywide data gleaned from the existing limited network of gauging stations.

(1) Average annual precipitation.
(2) Sedimentation of small reservoirs.
(3) Average annual runoff and river regulation.
(4) Flood flow frequencies.

The techniques used provided results suitable for preliminary designs but detailed calculations with local data would be recommended for particular major schemes. Two subsidiary studies were particularly pertinent to Korean conditions.

(5) The effects of record length.
(6) Maximum observed rainfalls.

The defining of optimum record lengths helped in the planning of the hydrometric network and the maximum rainfalls formed a basis for the evaluation of design storms.

Conclusion
The five volumes of the report represented detailed work carried out over 20 months by the varying efforts of some 14 engineers and hydrologists from the UK and Korea with the assistance of numerous organizations in both countries.

1.5 NORTHUMBRIAN WATER RIVER GAUGING NETWORK

LOCATION Northumbrian Water is the most northerly water authority area in England.

SOURCE Northumbrian Water Authority (1980) *Review of NWA gauging station network*, Internal Memorandum.

PROBLEM The rapid financial inflation in the late 1970s resulted in a large increase in the costs of the services provided by the water authorities. An appraisal of the cost-effectiveness of the river gauging network was requested. For any engineering project, the civil engineer is responsible for assessing the cost of the works, the benefits likely to accrue from the scheme and for assuring an acceptable cost–benefit relationship. In this problem, the hydrologists were asked to justify the operation and maintenance of the network of river gauging stations. They responded by identifying the needs for the hydrometric data at all the established stations, taking into account the reliability of regional flows and correlated flows between catchments.

Station network
The extent of the Northumbrian area, 9210 km^2, with its principal rivers is shown in Fig. 1.5. The 46 stations marked on the map which provide river flow records fall clearly into four groups: the northern rivers and the three catchments of the Rivers Tyne, Wear and Tees. The natural flows at three of the stations, namely Kielder, Derwent and Burnhope, are now obtained from the respective reservoir records. The remaining 'primary' stations consist of an instrumented river level recorder house and a fixed means of calibration by a weir or current meter cableway or both. In addition, four 'secondary' stations with temporary recorder installations have been established for particular investigations in connection with urban and water quality problems.

Data needs
In reviewing the future data needs from the gauging stations, the value of the data being collected was divided into three categories.

Category A Data essential for a given authority function.
Category B Data having an identifiable benefit but not essential for a given authority function.
Category C All other data.

For the wide variety of authority functions, the river flow data needs were

Fig. 1.5 — River flow stations of Northumbrian Water. ●, river flow stations; K, Kielder; D, Derwent; B, Burnhope.

considered in several groupings and within each group the data could be current operational information often passed to a control by telemetry and/or historic records to be archived.

Water resources and supply

(1) Prescribed flow conditions need to be monitored and recorded at five abstraction points which are subject to residual flow conditions.
(2) The river regulation and transfer scheme associated with Kielder Reservoir requires that flows at abstraction and transfer points for the tunnel connecting the Tyne, Wear and Tees be monitored for the control of reservoir releases, for the maintenance of flows to meet demands and residual flow conditions with the minimum of transfer pumping costs.
(3) The protection of river abstractions against pollution requires detailed knowledge of the travel times of upstream spillages to an intake at varying states of the rivers. This is based on flow monitoring at the gauging stations and field investigations using injected tracers.
(4) Yield assessments of reservoirs need natural flow data for their continuing appraisal since the development of conjunctive use schemes changes the conditions of their evaluation. Gauging stations downstream of dams and at significant points in regulated catchments provide important historic information for assessment of reservoir yields.

Design floods

Historic flow records are required for determining the probability of occurrence of floods of specified magnitude applied in the design of flood alleviation schemes with their cost–benefit assessments. Possible flood magnitudes are also needed for the design of river bridges and culverts and as information for development plans and insurance claims. Reliance on real flow records has been reduced by the application of regional modifications of The Flood Studies Report (NERC, 1975) and peak levels could be recorded by simpler instrumentation. In urbanized catchments, flood analysis is more complex and secondary gauging stations need to be installed in problem catchments.

Flood warning

An effective flood warning scheme is of great value in public relations and is essential for the protection of life and property in areas prone to inundation but it needs to be based on accurate timely flow forecasts. These are dependent on 12 of the primary gauging stations which form a flood warning telemetry network for most of the area; in the Tees catchment all the stations are linked in a telemetry network.

Water quality

(1) The close relationship between river discharge and water quality emphasizes the need for knowledge of flows in order to determine the effluent standards for discharge into the rivers especially under low-flow conditions.
(2) Water quality sampling undertaken at 11 sites near the tidal limits of the major rivers to assess total pollutant loads requires continuous-flow data.
(3) Special studies are needed with regard to sewage treatment in particular areas and to the proportion of baseflow composed of industrial effluent.

Hydrological studies

Most of the recent studies have related to Kielder resources or flood frequency but there are a variety of other special studies requiring good-quality river flow data. These include investigation of applications for abstraction and impounding licences, concern for the fish life of the rivers, preparation of drought orders, assessment of afforestation effects in reservoired catchments and the investigation of groundwater contamination from industrially polluted streams.

A sample of the data categories allocated to the river gauging stations is given in Table 1.5 showing selected stations from each group.

Table 1.5 — Station data category ratings

Gauging station		Water resources				Design floods	Flood warning	Water quality			Hydro-logical studies
Station	River	(1)	(2)	(3)	(4)			(1)	(2)	(3)	
Northern rivers											
Etal	Till	C	C	C	C	C	C	C	C	C	C
Mitford	Wansbeck	A	B	A	B	A	A	B	A	C	C
Tyne catchment											
Rede Bridge	Rede	B	B	A	B	B	A	B	C	C	C
Wide Eels	East Allen	C	C	C	C	C	C	C	C	C	C
Bywell	Tyne	A	A	A	B	B	A	C	A	C	C
Wear catchment											
Witton Park	Wear	B	B	B	B	B	B	B	C	C	C
Lanchester	Browney	C	C	C	C	B	C	B	C	C	C
Sunderland Bridge	Wear	B	A	A	B	B	A	B	C	C	C
Tees catchment											
Middleton	Tees	B	A	B	B	B	A	C	C	C	C
Broken Scar	Tees	A	A	A	B	B	A	C	A	C	C
Leven Bridge	Leven	C	C	C	C	C	C	B	C	C	C

Conclusions

The recommendations resulting from the review of the data needs of the individual river gauging stations were based on the following premises.

(1) Stations with at least one A — to be retained.
(2) Stations with no A but at least one B — operation to be reviewed.
(3) Stations with no A or B — to be abandoned.

Thus, in the selected examples in Table 1.5, the Till and East Allen stations would be removed, the Witton Park, Lanchester and Leven Bridge stations would be reviewed and the rest would be retained as primary network stations.

The final proposals were summarized as follows.

Continue station operation for continuous data	28 primary stations 4 secondary stations
Review requirements further	10 stations
Abandon station, remove recorders and cableways, leave weirs	8 stations

The abandoned stations were to be visited to collect flood levels after significant storm events.

As a result of this review, it was expected to make substantial savings in maintenance and running costs without undue loss of hydrometric information required.

1.6 GLASGOW COMBINED SEWER FLOW SURVEY

LOCATION Shettleston–Tollcross urban area in Glasgow, Scotland.

SOURCES Green, M. J. (1982) The role of flow surveys in the hydraulic analysis of sewerage systems. In: *Restoration of sewerage systems.* Thomas Telford, Paper 19, 193–199.

Green, M. J. (1982) Hydraulic analysis of sewerage systems. In: *Proceedings of the 1st International Seminar on Urban Drainage Systems.* Featherstone, R. E., & James, A. (eds.), Pitman.

PROBLEM The UK has many urban sewerage systems that are now over 100 years old and receiving loads above the original design capacity from expanding urban areas. Many sewer collapses annually have led to the need for a planned policy of renovation, renewal and/or replacement. Before a rehabilitation method can be selected, the strategic sewers must be identified and the performance of the sewerage system determined. Following several years of research, the Wallingford Procedure (NWC–DOE, 1981), a suite of computer programs, provides a powerful modern method of analysis. One particular mathematical computer program (WASSP-SIM) allows the user to simulate flows under free flow and surcharge conditions. Any model is only as good as its input data and sewer record input data are often inadequate. If rehabilitation projects are to be based on the output of the model, then the model must be verified by comparing it with an actual field event. The Shettleston–Tollcross urban area of Glasgow has demonstrated how the modelled catchment can produce misleading results.

Sewer flows

The measurement of sewer flows, required for model verification in the older combined sewerage systems taking sewage and stormwater, is a difficult problem. The corrosive and widely ranging flow conditions can now be monitored by instrumentation developed to measure depth or depth and velocity. If surcharging occurs, it is essential to derive the flows from depth and velocity measurements since flows calculated from depth only can result in an underestimate of the peak. One survey meter that is now available consists of a Doppler velocity meter and pressure transducer (for depths) mounted on a stainless steel ring which is expanded to grip the inside of the sewer pipe. The connecting circuitry, solid state logger and batteries are contained in a waterproof box suspended beneath the manhole cover. The measurements are logged at a pre-set time interval between 1 and 60 min according to the response time of the sewerage system. To calculate the flow, the measured Doppler velocity has to be related to average velocity by laboratory calibration and *in situ* calibration using a portable velocity meter. The manual check measurements enable any sensor drift to be identified when the recorded data are retrieved. The check measurements are also used to produce a rating-type curve for the instruments measuring depth only.

Shettleston–Tollcross survey

This urban catchment of 3 km^2 containing residential properties, derelict land, parkland and some industry is drained by a combined sewerage system of egg-shaped and circular sewers (Fig. 1.6). There are four known overflows and a cross

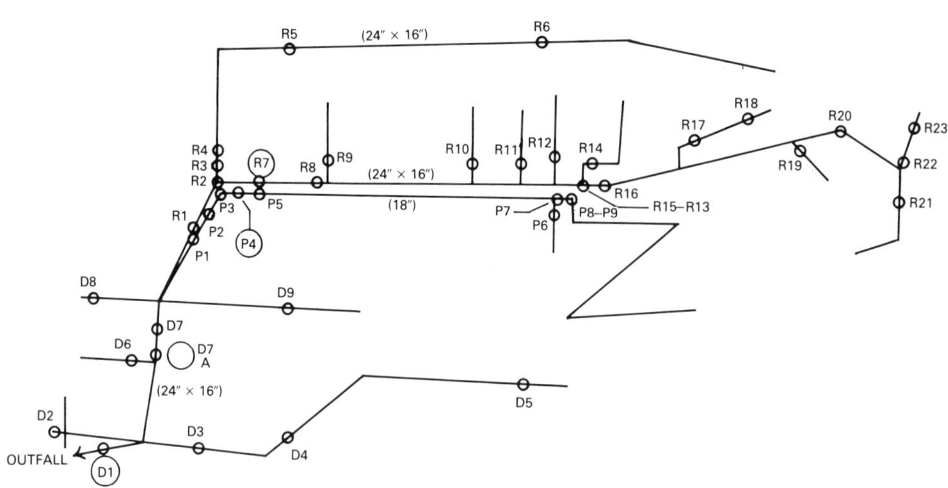

Fig. 1.6 — The Shettleston–Tollcross catchment: D5, depth monitor only; (D1), velocity–depth monitor; the dimensions in parentheses are the sewer dimensions.

connection between the two main arms of the system. Attempts to analyse the hydraulic performance of the system by modelling have resulted in unrealistic flow predictions.

The flow survey lasted for 5 weeks and covered two useful storm events with rainfalls of 2 and 5 mm. The system was monitored by 42 pressure transducers of which four had Doppler velocity meters (Fig. 1.6). Flows from depth only measurements were calculated using the Manning equation and checked with flows by manual velocity and depth measurements. Surcharge conditions on the day of the 5 mm storm showed that the depth–flow relationship could not be used at a number of sites. The analysis was therefore confined to the velocity–depth meter sites and sites where surcharge did not occur and good depth–flow relationships existed.

The catchment was modelled to simulate flows resulting from the 5 mm storm. The catchment data were taken from the sewer plans with spot checks on the ground for road widths, pipe sizes, lengths and gradients. The computer program was modified to take egg-shaped sewers and the required 2 min. rainfall values were averaged from three rain recorders.

Good agreement was obtained between measured and modelled flows in the well defined subcatchments. On the main sewer at R7 when surcharge occurred both velocity and depth monitors were required to give comparable flow measurements. The hydrograph at the catchment outfall D1 for the larger storm showed a wide discrepancy between measured and simulated flows (Fig. 1.7). A subsequent ground

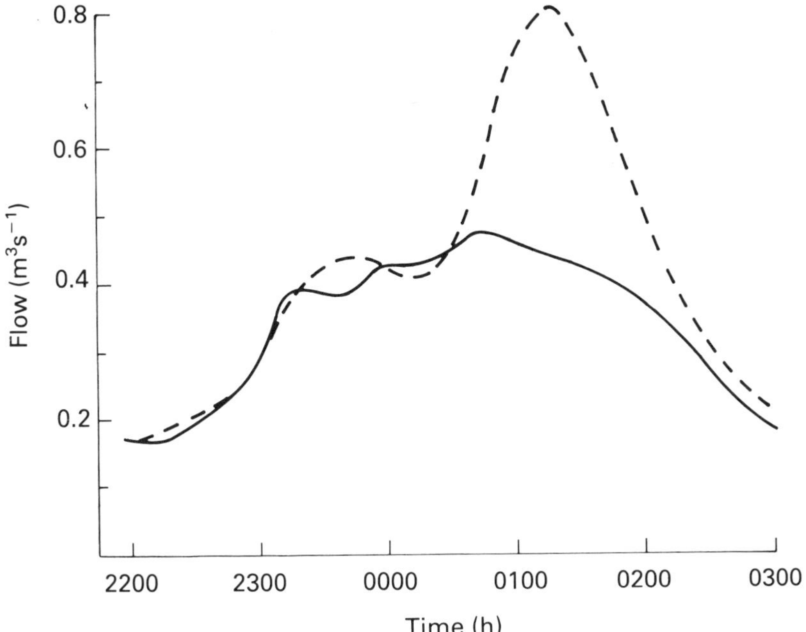

Fig. 1.7 — Hydrographs for site D1: ———, measured, ---, modelled.

survey located three overflows in the sewers below R2 and P3 which had not been noted for input into the computer model. A similar discrepancy was observed in a tributary sewer from an industrial area where the modelled flow volume was 2.5 times the recorded volume. Following a ground survey the discrepancy was identified to be the attributing of incorrect runoff areas to the model.

Conclusion

The need for flow surveys in sewerage systems was well demonstrated by the Shettleston–Tollcross survey. Wherever complex and hydraulically overloaded systems are to be investigated, *in situ* flow surveys should be used to confirm results from computer models. Although the velocity and depth monitoring is required for only a few weeks to sample significant storms, the establishment of the network of monitors is costly. It has been suggested that such surveys will be economic for large rehabilitation and sewerage schemes when the cost is less than 10% of the estimated capital works.

1.7 PALAWAN HYDROMETRIC NETWORK

LOCATION Palawan is an island in the Philippines.

SOURCE Sir M. MacDonald & Partners (1985) *Palawan hydrometric network operations manual*, Vol. 1, *Overview*, Report for the Republic of the Philippines National Council on Integrated Area Development.

PROBLEM The Palawan Integrated Area Development Project established a hydrology unit in 1980 to take over existing hydrometric stations and the EEC agreed to fund an expansion of the hydrometric network. The availability of a reliable water resources data bank is essential to the economic development of the island province. The planning of the hydrometric network of stations to monitor rainfall, climate, river flow, groundwater and water quality together with a computerized data-processing system located in Palawan constituted a programme of work to be undertaken by Sir M. MacDonald & Partners in association with Hunting Technical Services Limited. The main objective of the network is to provide high quality data in sufficient quantity to meet the needs of the development planners for such schemes as reservoir construction, domestic water and irrigation supplies and hydroelectric power generation. Operational data for flood and drought forecasting and other environmental activities are also to be provided. The data will be made available in published books for annual hydrological data and on demand for detailed immediate information.

Network design

The island of Palawan (12 000 km^2), long and narrow with short steep rivers draining from a backbone mountain range provides many difficulties for land communications (Fig. 1.8). Some of the western areas are difficult of access even though only a few kilometres distant. The planning headquarters in Puerto Princesa City is centrally located on the eastern side. Palawan has an equatorial climate with significant rainfall around the year and seasonal maxima dependent on location and occurrence of typhoons.

The first stage of the programme was the establishment of the station networks to monitor rainfall, climate and river flow. The network criteria were considered under two headings: the strategic and the physical criteria. The first took into account the developer's needs in irrigation, domestic water supply and hydropower generation, existing hydrometric stations and the necessity to define regional and seasonal patterns of hydrological variables and catchment water balances. The physical criteria pertained to the actual siting of the stations with due regard paid to standard

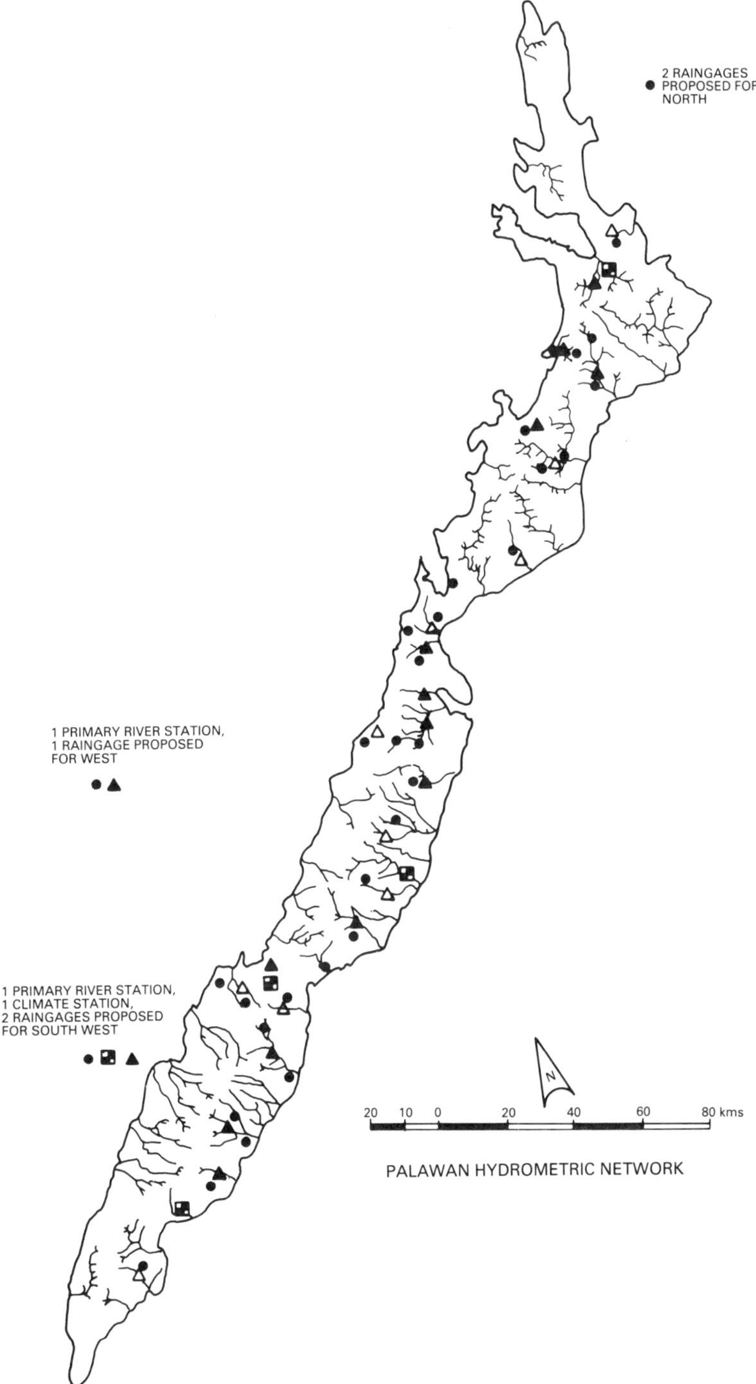

Fig. 1.8 — Palawan hydrometric network: ▲, 15 primary automatic river stations; △, 10 secondary manual river stations; ●, 35 automatic rain gauges; ◪, five automatic climate stations.

requirements for the installation of rain gauges, of climatological instruments and of equipment for river level and discharge measurements. All station sites were to have reasonable all-weather access and to be safe from interference or damage from extreme natural phenomena. The locations of the hydrometric stations comprising 35 automatic rain gauges, five automatic climate stations and 25 river level stations (15 automatic) are shown in Fig. 1.8. The rain gauges sample a range of altitudes from 3 to 470 m above sea level. The climate stations sample the range between irrigated lowlands at 7 m and an upland zone at 95 m above sea level. The river flow stations will provide data from a wide variety of catchments; the smallest is only 1.5 km^2 and the largest 126 km^2.

The specifications for the construction and instrumentation of each station, detailed by the consultants, endeavoured to provide for minimum capital outlay and recurrent maintenance costs consistent with the correct exposure of the instruments. A standard design for each station type was recommended. Since the funding source was the EEC, the instrumentation was mainly of European origin but a good supply of spares was included in the budget and the authorities were advised to have repair technicians trained in Europe by the equipment suppliers. The network is designed to be operated by 10 established staff, of whom four are to be part time. To keep costs to a minimum, the island is to be divided into three subnetworks and, on this basis, the important field operations are prescribed fully with detailed instructions on maintenance of sites and servicing of instruments to be carried out each month. Particular attention to the supervision and payment of the 25 river gauge keepers is necessary to ensure the continuous running of the flow stations.

Data processing

All the hydrometric stations are identified by a locational code number and the automatic measurements are logged onto solid state memory modules. The rainfall data are recorded on cartridges by loggers in the tipping-bucket rain gauges. The climate data are recorded on to mass storage units by the automatic weather stations and the water level data are recorded on cachettes by electronic recorders. The three different types of record are fed into a microcomputer at headquarters, with provision of back-up devices for alternative processing if the computer fails (Fig. 1.9). Presentation of the data is effected by two dot matrix printers and a graph plotter. Two main categories of computer software are a set of commercial packages containing system software such as a Fortran compiler and word processing system and a suite of hydrological data-processing and analysis programs. The latter consists of 37 programs and subroutines from Sir M. MacDonald & Partners computer library specifically tailored to the requirements of the Palawan hydrometric network.

Conclusion

The first stage of the hydrometric network program was completed in November 1985. In addition, an *Operations manual for the Palawan hydrometric network* of 10 volumes was produced.

Vol. 1, *Overview.*
Vol. 2, *Field Team's Handbook.*
Vol. 3, *Rainfall database user's manual.*

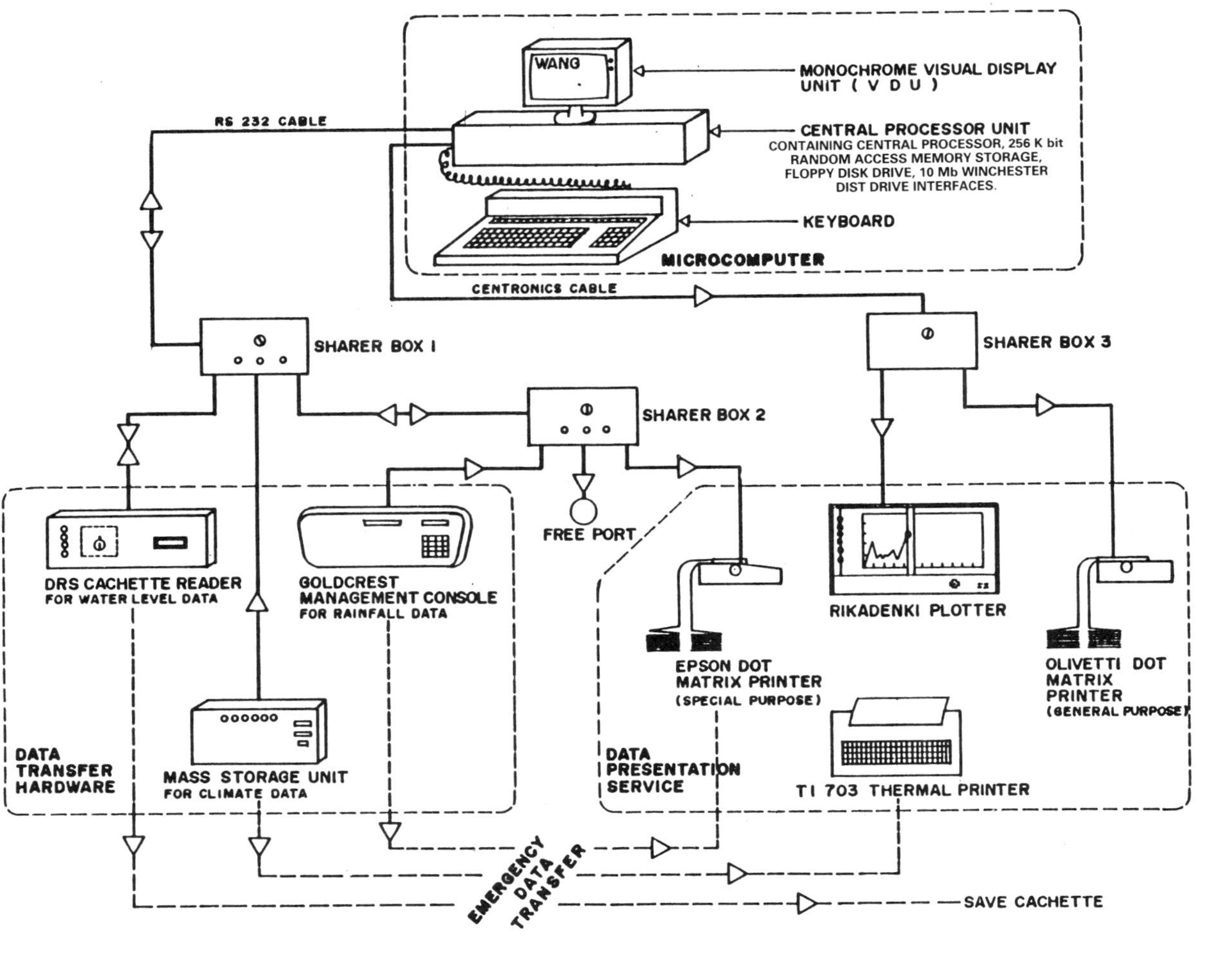

Fig. 1.9 — Data-processing system.

Vol. 4, *River stage and discharge database user's manual.*
Vol. 5, *Climate database user's manual.*
Vol. 6, *Miscellaneous computer program descriptions.*
Vol. 7, *Computer programmer's guide.*
Vol. 8, *Annexes — technical and management information.*
Vol. 9, *Maps (1:50 000 scale).*
Vol. 10, *Photographs.*

The proposed second stage will be concerned with groundwater- and soil-moisture monitoring stations, fluvial sediment and water quality monitoring, a sediment analysis laboratory and expanded computer software for the new data.

1.8 HYDROMETRIC SCHEME FOR THE MANICALAND PROVINCE OF ZIMBABWE

LOCATION The Manicaland Province of Zimbabwe lies along the eastern border with Mozambique, south of latitude 18°S.

SOURCE Johnston, J. C. (1986) An example of the utilisation of hydrological data in the Manicaland Province of Zimbabwe. *Engineer (Zimbabwe)* May 1986.

PROBLEM The Zimbabwe Water Act 1976 permits cost-free abstraction of river water by a farmer living on riparian land, but he must present details of his proposed irrigation scheme with his application for water to the Administrative Court. The Hydrological Engineer has to give evidence to the Court that there will be enough water to satisfy the abstraction Right for 7 out of every 10 years and that no Rights already granted will be adversely affected. The expansion in the number and size of irrigation schemes in the farming areas (Fig. 1.10) from 1965 to 1975 led to a rapid increase in the abstraction applications and the Hydrological Engineer had to prove with definite evidence when a stream was already fully committed and no further Rights could be granted on it. This need for reliable stream flow data led to a hydrometric network of minor stations to extend greatly the knowledge of baseflow runoff in the province.

Hydrometric network expansion

The developing farming districts along the eastern border region of Zimbabwe have favourable rainfall, topography and soil conditions combining to produce many perennial streams. The Eastern District Mountains provide the main watershed between streams flowing directly into Mozambique and those forming the headwaters of the Save River in Zimbabwe. The mountains experience an average annual rainfall of 1500–2500 mm and while most of this falls in the hot wet season (November–April), significant amounts are received in the cooler dry season from May to October. Thus the riparian farmers in these favoured areas are able to make the maximum use of their arable land and to produce crops in the normally dry season with the aid of irrigation water. With the expansion in water demand,

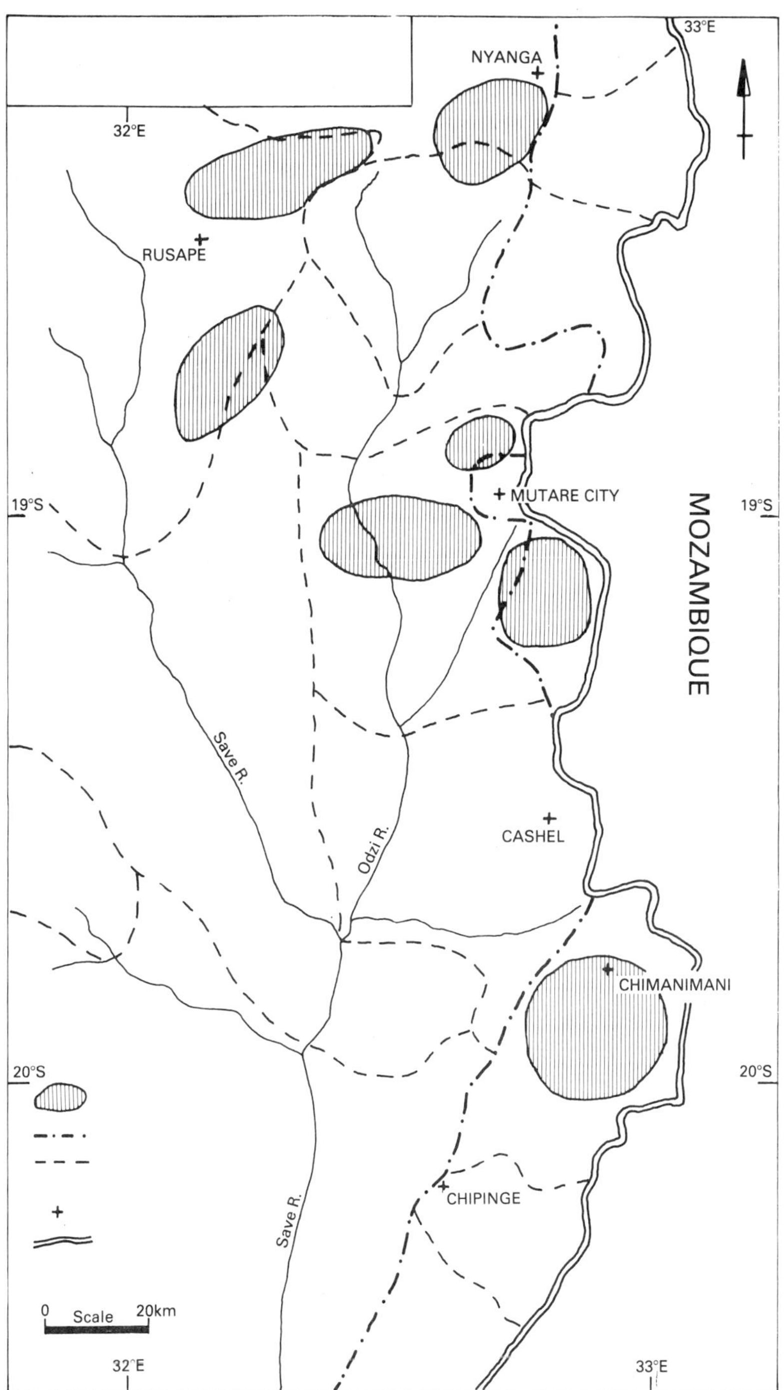

Fig. 1.10 — Manicaland Province, Zimbabwe: ▨, baseflow gauging districts; —·—, major watershed; ---, main catchment boundaries; +, towns; ═, national border.

especially from the smaller streams, there came the danger towards the end of the winter season of flows being insufficient to meet all the Rights allocated.

Prior to 1965, manual low-flow gaugings had been carried out towards the end of each dry season from September onwards until the main rains came. With such a large area to cover, although over a hundred gaugings were made, the actual lowest flows were measured in only a few districts each year. To obtain wider and more detailed information on stream flows, the network of measuring sites was expanded and changed from current meter gauging stations to manually read stage post and weir notch installations. In the years 1965–1971, about 130 measuring weirs varying in discharge capacity from 70 to 5000 $l\,s^{-1}$ (from 0.07 to 5.0 $m^3\,s^{-1}$) were constructed throughout the Province of Manicaland. To record the low flows, the weir stage gauges were read regularly by travelling observers once each week. The hundred hydrometric measurements previously limited by the end of the dry season rose to 8000 observations taken throughout the year. This consistent regular series of observations ensured that the low flow levels on all the gauged streams were recorded every year and the resultant excellent hydrographs (Fig. 1.11) formed a

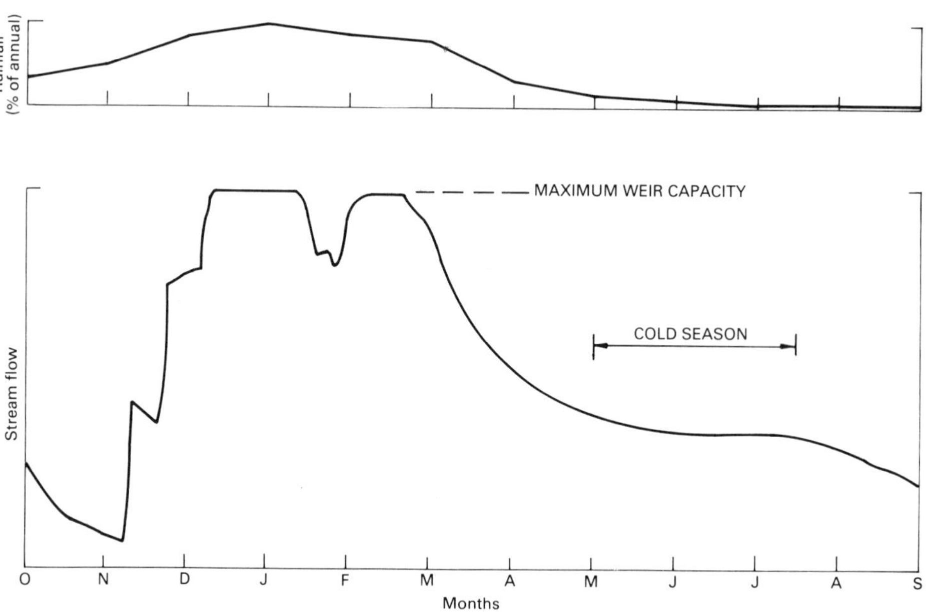

Fig. 1.11 — Typical rainfall and stream flow patterns for an eastern district catchment.

basis for the determination of available water. An automatic recorder station was built in the northeast Rusape group of gauge post stations and another among the stations south of Mutare. Single weekly readings abstracted from the automatic recorder charts gave computed flow values within ±3% of flows fully derived from the chart traces.

Conclusions

When a few years of records had been assembled from the expanded network of gauging stations, it was possible to establish typical dry-season hydrographs for the several farming districts. From these it was feasible to give an acceptable estimate of the lowest-flow conditions on any stream from a single gauging during the dry season and thereby the Administrative Court could judge applications for abstraction Rights.

The low-flow hydrometric network led to more efficient use of the available water while protecting existing holders of water Rights from shortages due to over-allocation to new applicants. The prevention of excessive Right awards also mitigated against over-investment in irrigation schemes likely to fail economically through water shortage and thereby also prevented friction between farmers and landowners.

Further benefits from the continuous flow records throughout the year are likely to be realized in the future. Given reliable knowledge of the wet season flows, provision for partial storage of excess water could be made, particularly for those farmers without riparian Rights. The financial gains from irrigated crops help with the development of more complex utilization schemes to the greater benefit of the whole community.

2

Agricultural drainage

INTRODUCTION

In most countries where stabilized agriculture and horticulture are practised, the land must be adequately drained. This lesson has been learnt at great cost in some semiarid regions with only seasonal rainfall where extensive irrigation schemes have not paid sufficient attention to the removal of surplus water. Adequate soil drainage to prevent the accumulation of damaging salts or the formation of iron pan horizons is essential for most farming land from old established permanent grassland to the intensively cultivated arable land growing cash crops.

Civil engineers are usually consulted for the development or expansion of agricultural land to advise on the design of suitable drainage schemes. Historically, the most renowned schemes are those in The Netherlands where over many centuries land has been reclaimed from the sea. It was the expertise of Dutch engineers that played a large part in the drainage of the lowlying fen country in the UK. For large projects involving the layout of new drainage channels, the modification of existing watercourses or perhaps the installation of pumps if the land is below natural drainage levels, the civil engineer is charged with the investigations, plans and costings of the schemes. At local levels in the consideration of the drainage required for individual fields and crops, it is the agricultural engineer who provides the specialized knowledge. There must be close liaison between the two with overall direction by the project engineer.

Hydrological information is required at all stages in the investigations and in the designing of a scheme. Net areal rainfall, from design rainfalls and measured or calculated evaporation, must be established for the area to be drained. The transpiration from the natural vegetation or the relevant crop water use needs to be assessed and deducted from the net rainfall to evaluate the runoff from the area. From this information, the rates of runoff for designated areas can be calculated in order to specify the field drainage pipe network and the geometry of the open drainage channels. Particular attention to channel water levels in relation to the field drain outfalls is important when calculating the discharge capacities of the main drainage courses. Controlling sluice gates and pumping stations are the principal civil engineering construction tools of land drainage schemes. A comprehensive account of lowland drainage practice in UK is given in the new *Water practice manual, River engineering* (Brandon, 1987, Chapter 16).

The sample case studies range from detailed considerations of the improvement of small areas in the UK to extensive major plans in developing countries. The hydrometric techniques applied in the studies depend on the size of area involved and the information available. The drainage of small areas in the UK relied on the evaluation of water balances requiring rainfall and evaporation measurements, and for probabilities of occurrence of design discharges the methods of The Flood Studies Report were used (NERC, 1975). The extensive schemes of the developing countries required the application of regional hydrometric data and catchment models to derive runoff and river flows.

2.1 CROSSENS DRAINAGE SCHEME

LOCATION The Crossens catchment area, part of the Lancashire Plain, is drained into the Irish Sea north of Southport.

SOURCES Hall, M. J., & Prus-Chacinski, T. M. (1975) Forecasting runoff volumes for the drainage of peat lands. *J. Agric. Eng. Res.* **20**, 267–278.

Prus-Chacinski, T. M., & Harris, W. B.(1963) Standards for lowland drainage and flood alleviation and drainage of peat lands with special reference to the Crossens Scheme. *Proc. Inst. Civ. Eng.* **24**, 177–206.

PROBLEM The original swampland behind coastal dunes has been drained since 1692 but the subsequent drainage operations have barely kept pace with the needs of agricultural developments. A drainage improvement scheme was implemented between 1958 and 1961 designed to provide protection against floods with a recurrence interval up to 20-25 years. Since the pumping system design criteria were based mainly on rainfall data, a hydrological study was made to investigate the dependence of runoff volume on both the magnitude of a storm event and the catchment wetness with the aim of forecasting flood volumes for improving pumping operations.

Crossens area

The Crossens catchment area of 131 km^2 may be divided into three separate drainage systems (Fig. 2.1). The Western System (Three Pools) receives stormwater overflows from 9 km^2 of Southport. The downstream reaches of the Eastern System (The Sluice) and of the Western System (jointly called the Wings) are embanked where they cross the low-lying Middle System (Back Drain) and prior to 1961 discharged by gravity outlets into the Crossens tidal channel. The Middle System has been pumped since 1852. The most significant hydrological feature is the large proportion, 40% of

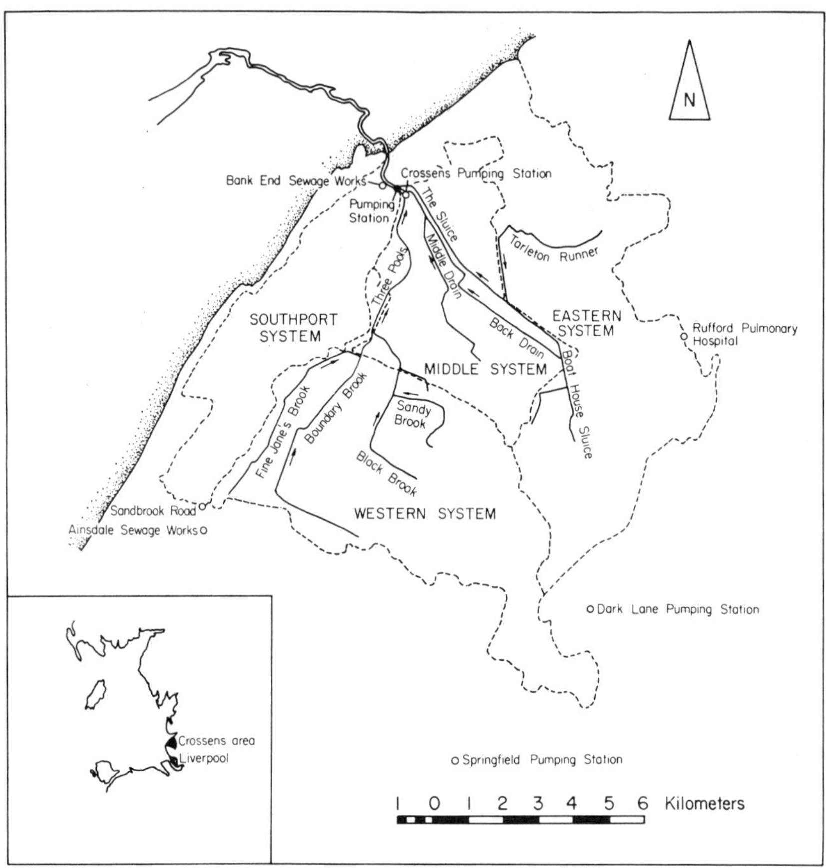

Fig. 2.1 — Crossens catchment areas: – – –, catchment boundaries; ———, main channels; ○, rain gauges.

the area, covered by peat ranging greatly in type and depth. The peat covers most of the Middle System, the Wings being composed mainly of sands and clays with clay soils predominating in the Western System.

Hydrometric data

Hourly measurements of the water levels in the three drains and the Tidal Pool together with logs of the number of pumps running were available for 7 years from the commissioning of the new pumping station in 1961. By subtracting the drain levels from the Tidal Pool level to give the head differences and using the relevant pump characteristics, the rates of discharge from each catchment area were estimated. Since the intake pools for the Eastern and Western Systems were connected, the joint Wings areas were treated as one catchment.

The daily areal rainfalls over the catchment were evaluated using Thiessen polygons and observations at four rain gauge stations. Autographic gauges at

Ainsdale Sewage Works and Bank End Sewage Works provided concurrent hourly cumulative rainfall totals. These were averaged and scaled according to the daily areal rainfall to give mean hourly rainfall over the catchments to compare with the pumped discharges. Significant storms producing more than 17.5 mm of rain in 24 h or 25 mm in 48 h were identified and abstracted for detailed analysis. An example of a wet spell with 102 mm in 9 days when all 13 pumps were operating is shown in Fig. 2.2.

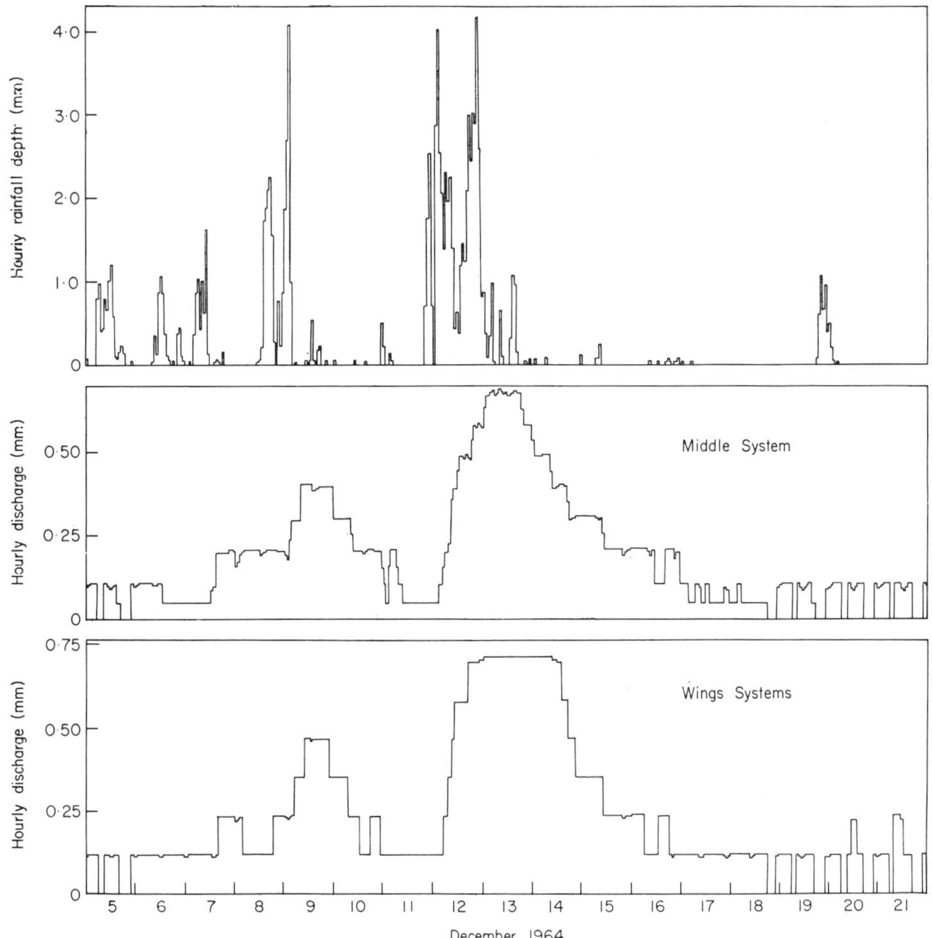

Fig. 2.2 — Concurrent hourly totals of rainfall and pumped volumes from the Wings and the Middle System.

From closer inspection of the records, it was noted that the subjective operation of the pumps, usually between 08.00 and 17.00 hours when the staff were on duty, prevented the establishment of reliable relationships between the rainfall incidence and the time variations of the discharge. The problem was to determine the times at which storm runoff ceased. After considerable experiment and close study of the pumping logs, the method adopted was based on the rate of change of water levels in the intake pools. When pumping had ceased after a storm and the rate of rise of water level did not exceed 100 mm h^{-1} over several hours, storm runoff was deemed to have ended. Following this tedious preliminary data identification, 32 storm sequences on the Wings and 28 on the Middle System were available for hydrological analysis.

Forecasting pumping volumes

The hydrological factors which influence the volume of storm runoff were defined in two categories.

(1) Storm rainfall.
   (a) Total depth of areal rainfall.
   (b) Total storm duration.
   (c) Total duration of rainfall given by total storm durations minus intermediate dry periods.
   (d) Average rainfall intensity taken to be total rainfall divided by total rainfall duration.
(2) Catchment wetness.
   (a) In the absence of soil moisture measurements, the antecedent precipitation index (API) was based on the daily rainfall measured by the Crossens Pumping Station rain gauge. The daily API values are given by

$$\mathrm{API}_i = K(\mathrm{API}_{i-1} + P_{i-1})$$

   where P is the daily rainfall and K is a recession constant related to evaporation loss. Hence K varies with the season. A mean value for K of 0.96 was chosen and monthly values of K were obtained by seasonal adjustment using monthly mean potential evaporation values.
   (b) Mean monthly potential evaporation calculated from 8 years of data at Squires Gate Airport, Blackpool.

Multiple-linear-regression analysis was applied to the problem of relating causative factors to pumping volumes. With the dependent variable the pumping volume and the independent variables being selected from the four storm rainfall and the two catchment wetness variables, a standard computer package was used which incorporated test statistics to assess the significance of the regression. A 5% level of significance was applied for all the trial runs and an additive relationship between the independent variables was found to yield the best results.

For the Wings catchment, the following relationship was obtained:

$$R = 0.500P + 0.140\mathrm{API} - 13.20$$

where the total runoff R, the total rainfall P and API are in millimetre depth over the catchment area.

For the Middle System, the comparable equation was

$R=0.275P+0.128\text{API}-9.60.$

The related statistics are given in Table 2.1. The equations were tested on 10 storms

Table 2.1 — Multiple regression statistics

	Multiple correlation coefficient	Runoff explained variance (%)	Standard error (mm)
Wings catchment	0.939	88.3	4.5
Middle System	0.975	95	2.4

not used in the analysis and all but one storm on the Middle System produced runoff volumes within two standard errors of the observed values.

Conclusion

Since the Middle System is nearly all covered with peat, the runoff predicting equation could be used for peatlands with similar climate and land use. However, the varied sands and clays of the Wings catchments makes the first equation of little use beyond the Crossens catchments. The prediction equations were derived for total rainfalls up to 60 mm and API values ranging from 4 to 65 mm and, although total storm duration was not statistically significant, it should not exceed 48 h. For operational purposes, the API values should be updated continuously and both equations can be used to forecast runoff volumes from immediate storm rainfall measurement. Thus the Station Superintendent can plan during the initial phases of runoff the timing and number of pumps needed to discharge the predicted runoff volume.

2.2 MAIN STELL LAND DRAINAGE

LOCATION The Main Stell stream and its tributaries form the headwaters of the River Tame, a tributary of the River Leven which flows into the River Tees upstream of Stockton.

SOURCES Archer, D. R. (1981) A catchment approach to flood estimation. *J. Inst. Water Eng. Sci.* **35**, No. 3, 275–289.

Archer, D. R., & Kelway, P. S. (1987) A computer system for flood estimation and its use in evaluating the flood studies rainfall-runoff method. *Proc. Inst. Civ. Eng.* Part 2, No. 83, 601–12.

Northumbrian Water Authority (1986) *Main Stell land drainage scheme,* Internal Report.

PROBLEM The full potential of the gently undulating agricultural land in the catchment area of the River Tame headwaters cannot be fully realized at present by the 15 farmers and landowners. The effectiveness of the existing underdrainage (field drains) is impaired by high groundwater levels extending under 175 ha of land bordering the water courses. Considerable remedial works to the natural streams are required to rectify the present situation and to provide for extensions to the field drainage system. The water courses have been statutarily designated Main River and thus the Northumbrian Water Authority is empowered to undertake a land drainage improvement scheme. The drainage standard for grade 3 category agricultural land, permanent grass for sheep and cattle and some cereal production prescribes that it should not be flooded more often than once in 10 years. A further provision is that areas of clayey soils over 40 ha draining to Main River should not have their pipe drain outfalls submerged for longer than 24 h once in 5 years. The evaluation of the required design flows was made by the hydrologists of the Headquarters Operations Investigations Department.

Design flows

The catchment area to Tree Bridge (Fig. 2.3) is 19.52 km^2. The highest point of the catchment boundary, 322 m at Roseberry Topping, is a noted vantage point of the Cleveland Hills; the lowest parts of the catchment near the outfall are around 80 m above sea level. There are no continuous flow records on the Main Stell stream but the River Leven has two river gauging stations, at Leven Bridge (196 km^2) from 1959 and at Easby (14.8 km^2) 4.5 km south of the Main Stell from 1971. Estimates of the annual flood flow at Tree Bridge and the flood flow with a 10 year return period were made with the Northumbrian Water Authority's flow prediction program FLOWPRED with due regard paid to the measured flows in the area.

Of the three basic calculation modules of the computer package FLOWPRED, the estimation of floods of specified return period was the module used in this problem. The analytical methods of FLOWPRED are based on those of The Flood Studies Report (NERC, 1975) and for design floods both the statistical method and the rainfall-runoff method provide estimates. The catchment characteristics of the Main Stell catchment were derived by the methods described by Sutcliffe (1978) from requisite measured parameters (Table 2.2). Climatic and soil variables for 10 km^2 areas defined by the Ordnance Survey National Grid have been determined from the data maps of The Flood Studies Report, Vol. V (NERC, 1975) and these are abstracted from file storage by the computer system for the catchment area under consideration. For this study, the following particulars from the gridded matrices of information were required.

SAAR	Standard average annual rainfall 1941–1970 (mm).
PERC-M5	The M5–2 day rainfall as a percentage of SAAR (%).
R	The ratio of M5-1 h to M5-2 day rainfalls (%).
SMDBAR	Effective soil moisture deficit (mm).
SOIL	Winter rainfall acceptance potential (factor).

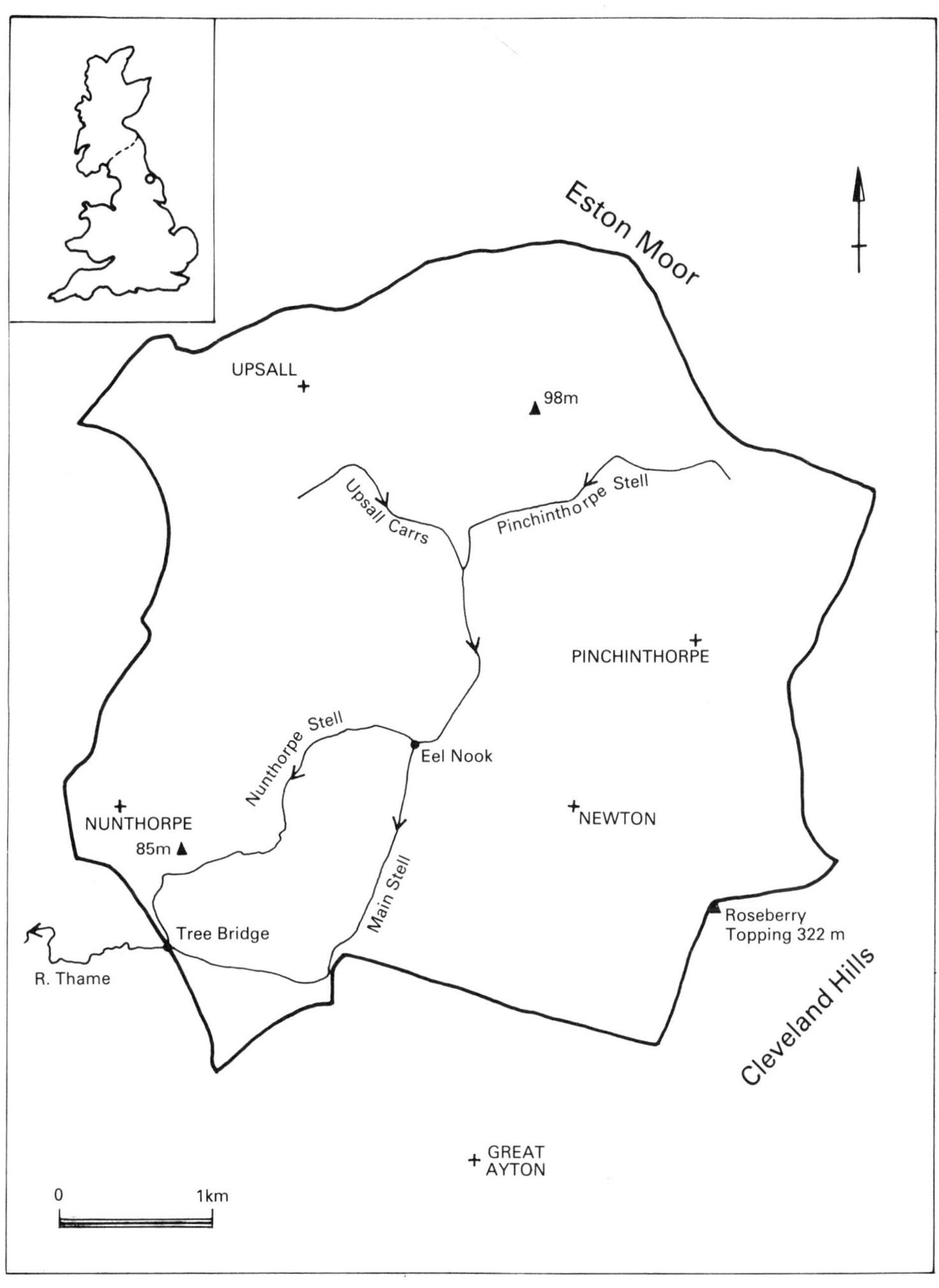

Fig. 2.3 — Main Stell catchments

Table 2.2 — Main Stell catchment parameters

AREA (km^2)	19.52
S1085 (main stream slope) ($m\ km^{-1}$)	5.38
MSL (main stream length) (km)	7.70 km
STMFRQ (stream frequency) (stream junctions km^{-2})	0.61
URBAN (catchment urbanized) (%)	0.0
LAKE (area draining through lake (%)	0.0

From these catchment and climatic characteristics, a rainfall profile and catchment unit hydrograph and hence a design flood hydrograph were derived. The computer package allows for changes to be made to the information on file where specialized local knowledge is obtainable. For example, a real unit hydrograph or specific storm features may be substituted for the synthetic values. The results of the computations may be presented as hydrograph tabulations and/or in graphical form. The FLOWPRED system was used to provide flood estimates using The Flood Studies Report methods on 47 gauged catchments as a basis for comparing the results from catchment characteristics and those from measured data. Regional correction factors for the mean annual flood (MAF) $\overline{Q}$ and flood growth rates were thus obtained. In this instance, application of these factors reduced $\overline{Q}$ but increased the growth rate. The mean of the adjusted flows from statistical and rainfall-runoff methods was adopted for the 10 year return period (Table 2.3). From the two methods, 7.90 m^3 s^{-1} was recommended for the 10 year return period flood flow.

Table 2.3 — Design floods for the Main Stell at Tree Bridge

	Statistical method			Rainfall-runoff method			Adjusted mean
	FSR[a]	Correction factor	Stat.[b]	FSR[a]	Correction factor	R–R[b]	
$\overline{Q}$ ($m^3 s^{-1}$)	5.39	45	3.71	6.36	40	4.54	4.12
Growth factor			1.91			1.91	
Q_{10} ($m^3 s^{-1}$)			7.10			8.70	7.90

[a]The Flood Studies Report value.
[b]Adjusted value.

Remedial works

In designing the remedial works the following conditions were laid down.

(1) The water level in the improved channels for average daily flows should be 150 mm below underdrainage outfalls.

(2) For outfall invert levels, 1.15 m was to be subtracted from lowest adjacent field levels, thus allowing 1 m cover above the pipe soffit and a diameter of 150 mm for the pipe.
(3) The 10 year flood flow to be contained within the channel without freeboard.

The hydraulic simulation of the flows was complicated by the stream bifurcation between Eel Nook and Tree Bridge and the Authority's backwater program was adapted to model the divided flow conditions. To achieve the stated objectives, the traditional methods in land drainage were employed, either enlarging the channel section and/or improving the bed slope to provide a uniform flow profile. Each channel section had special problems to be overcome such as concern for a railway bridge foundations, existing footbridge widths and a necessary culvert under a disused railway with a gas main buried in the embankment. A new channel was recommended along one stretch to avoid remedial works in a forestry plantation. The final designs following detailed survey work on the ground and continuous collaboration with the landowners were finally costed and must be approved by the landowners before the works are carried out.

2.3 AGUAN RIVER BASIN DRAINAGE

LOCATION The Aguan River flows to the north coast of Honduras, Central America.

SOURCES Sir William Halcrow & Partners (1983) *Hydraulic master plan for the Aguan River Basin,* Vols 1 and 2, Report to the UN and Republic of Honduras.
SOGREAH (1982) *Modelo Matematico de Simulacion,* Informe Final ONU.

PROBLEM The Aguan River Basin is an area of Honduras of great potential economic development. At present the fertile alluvial valley soils are underdeveloped for commercial agriculture since they require improved drainage. The valley downstream of Saba (Fig. 2.4) is subject to frequent and extensive flooding. Four overall alternative master plan strategies have been defined with regard to national and regional development aims but attention here has been confined to the hydrological assessments of the drainage problems.

Aguan catchment

The drainage basin of the Aguan River totalling about 10 300 km^2 is subdivided at Saba into the Upper Aguan with two major right-bank tributaries draining an area of about 7550 km^2 and the Lower Aguan with numerous small tributaries contributing from both banks. Lying between 15°N and 16°N, the area is dominated by the tropical easterly trade winds. The monthly mean temperatures range from 23°C in April–May to 30°C in December–January. Monthly minima and maxima can range from 16 to 36°C.

A network of 27 rain gauge stations provided data to assess the rainfall

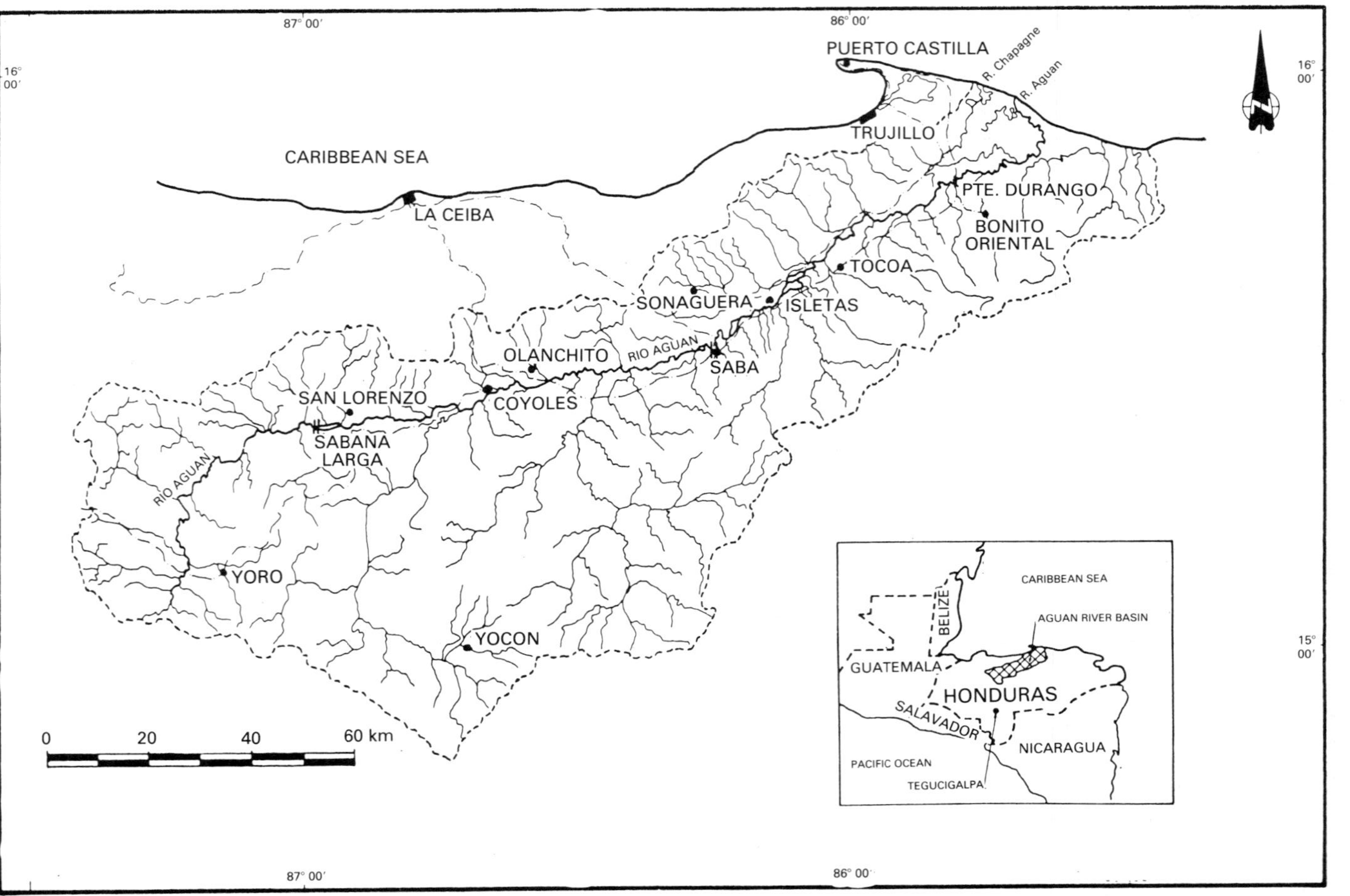

Fig. 2.4 — Aguan River Basin.

characteristics. The salient features are demonstrated in Table 2.4. The stations, located on Fig. 2.4, indicate a distinctive areal rainfall distribution. Yocon with an annual average of 2070 mm is in the upper reaches of the river and Isletas with 1616 mm is in the flood plain region. Coyoles with only 940 mm in the centre of the catchment lies in the rain shadow of the coastal mountains which receive the full benefit of tropical storms in the prevailing winds. Owing to its exposed position, the Lower Aguan has no marked dry season. The hurricane season extends from June to November with Honduras being affected about once every 5 years.

River flow monitoring only began in 1980; thus, with the shortage of records, surface water resources have to be evaluated from the longer rainfall records. Mean discharge values for the period 1973–1981 are given in Table 2.5 for two gauging stations shown on Fig.2.4. The monthly flows reflect the seasonal rainfall pattern with minima in April and maxima in October the stormy season. The mean annual flow at Sabana Larga is 16 $m^3 s^{-1}$ and at Saba 74 $m^3 s^{-1}$. The annual yield at Saba is estimated at 2327 Mm^3 giving a unit yield of 0.30 $Mm^3 km^{-2}$.

Flood hydrology

In studying the drainage problems of the Aguan River valley, particular attention was paid to the causes and frequency of occurrence of major floods. Records of historic storms indicated that storms or hurricanes with 3–5 days duration were the main cause. Depending on the storm track the coverage of the catchment varied between 20–60%. Rainfall depth–duration–frequency values were calculated using the EV1 distribution fitted by the method of moments for 27 rainfall stations taking durations of 1–5 days for return periods of 2, 5, 10, 25 and 50 years. A sample of the results is given in Table 2.6; Trujillo has been added to show the full hurricane effects. In addition, by studying the areal extent of key storms and the available autographic records, depth–duration–frequency estimates over the subcatchments were evaluated.

The SOGREAH conceptual model which defines the catchment rainfall-runoff process by a set of three equations with seven parameters was used to estimate flows from the Upper Aguan at Saba. The model, calibrated from recorded rainfalls and the river flow data at Saba, divides the Upper Aguan into 17 subcatchments. Flood hydrographs were computed for each subcatchment and these were then routed downstream using the Muskingum method. Resultant flood hydrograph particulars, peak discharge Q_p, ($m^3 s^{-1}$) and time to peak T_p (h) for 3 and 4 day storms for three return periods at two stations on the main river are shown in Table 2.7. The model is able to produce hydrographs at other selected river points in the catchment enabling the effect of seven possible flood control reservoirs in the Upper Aguan to be studied.

For the Lower Aguan, floods from 18 tributary streams were synthesized using the unit hydrograph method. Dimensionless unit hydrographs were derived from catchment characteristics and by the US Soil Conservation Service synthetic hydrograph method (USBR, 1973).

Flood mitigation works

To investigate the extent of the flood flows in the Lower Aguan, the flows at Saba and from the downstream tributaries were made the input to the hydraulic model

Table 2.4 — Monthly mean rainfall at selected stations

Station	Mean rainfall (mm)											
(height)	Jan	Feb	Mar	Apr	May	June	July	Aug	Sept	Oct	Nov	Dec
Yocon (700 m)	100	17	15	43	230	315	255	257	329	300	144	66
Coyoles (305 m)	69	46	27	23	55	127	80	79	120	115	123	76
Isletas (50 m)	91	99	57	52	64	157	151	136	209	221	196	176

Table 2.5 — Monthly mean flows of Aguan River

Station	Mean flow ($m^3 s^{-1}$)											
(C/A)[a]	Jan	Feb	Mar	Apr	May	June	July	Aug	Sept	Oct	Nov	Dec
Sabana Larga ($1844 km^2$)	11.4	6.8	4.2	3.0	2.7	12.9	22.2	19.9	27.9	33.3	26.6	18.6
Saba ($7558 km^2$)	66.7	44.2	28.7	22.3	23.6	52.7	67.9	71.7	117.3	157.6	135.4	97.2

[a]Catchment area.

Table 2.6 — Rainfall depth–duration–frequency estimates

Duration (days)	3			4			5		
Return period (years)	5	10	25	5	10	25	5	10	25
Rainfall Yocon (mm)	151	183	222	161	195	227	168	204	249
Coyoles (mm)	114	130	152	122	141	164	135	143	144
Isletas (mm)	171	195	226	181	208	243	199	230	270
Trujillo (mm)	369	423	492	409	471	549	454	524	612

Table 2.7 — Design flood hydrograph particulars

Station	C/A[a] (km^2)	Storm (days)	Return period					
			5 years		10 years		25 years	
			Q_p	T_p	Q_p	T_p	Q_p	T_p
Sabana Laga	1844	3	491	52	592	50	727	50
Sabana Laga	1844	4	865	70	1020	68	1205	68
Saba	7558	3	1962	62	2378	60	2931	60
Saba	7558	4	2943	82	3518	80	4231	80

[a]Catchment area.

ONDA. The features of ONDA, a general-purpose computer program developed by Sir William Halcrow & Partners, used in the Aguan study were as follows.

(1) Steady and unsteady flow in open channels.
(2) Flow over weirs and localized hydraulic losses.
(3) Flow into and out of reservoirs.
(4) Channel junctions with up to six branches.
(5) Flow–time and stage–flow boundary conditions.
(6) Over-bank spills into flood compartments.

There was considerable data preparation required for the model application. River and flood plain cross sections were assembled from different surveys, flood hydrographs came from the SOGREAH model, and stage–discharge relationships at key locations and hydraulic particulars at bridges were measured. The computational size of the mathematical model is related to the number of channel subdivisions and nodes, the time step required and the number of iterations needed to attain mathematical stability. For the Lower Aguan, a reach length of 2 km requiring 65 cross-sections and a time step of 15 min, later reduced to 10 min, were adopted. There were 12 tributary inputs between Saba and the sea.

The model was calibrated on two major storm events using observed flood levels and considerable difficulties were encountered at Durango where flood water left the main river for a neighbouring channel to the sea, the River Chapagua. The natural river model representing the present state of the valley was applied to determine peak water levels and discharges for a range of return periods. Stage and flow hydrographs were produced for required locations.

The main aim of the flood study was to determine a suitable scheme to contain the floods and thereby to profit agriculture in the valley. Four separate designs were modelled.

(1) The Aguan dyked on both banks from Saba to the sea.
(2) The dyked river with a floodway on the right bank at Durango.
(3) The dyked river with a floodway on the left bank to the River Chapagua.

(4) The dyked river with both left- and right-bank floodways.

The dykes would be constructed to contain the 1 in 10 year return period floods as it had been shown by economic analysis that residual flooded areas for higher discharges would be acceptable. The results from numerous model runs incorporating the four designs were presented in tabular form giving the maximum stages and flows at selected points on the river for the range of return periods.

2.4 PEMALI–COMAL DRAINAGE

LOCATION The Pemali–Comal irrigation area is in Central Java Province, Indonesia.

SOURCE Baghirathan, V. R., & Hall, M. J. (1984) Drainage studies in Central Java. *ICID Bull.* **33** 1.

PROBLEM The Pemali–Comal Project (1970–1980) was a rehabilitation irrigation scheme on the coastal plain of north central Java to increase food production (Fig. 2.5). Of the 127 000 ha, about 21 000 ha are subject to

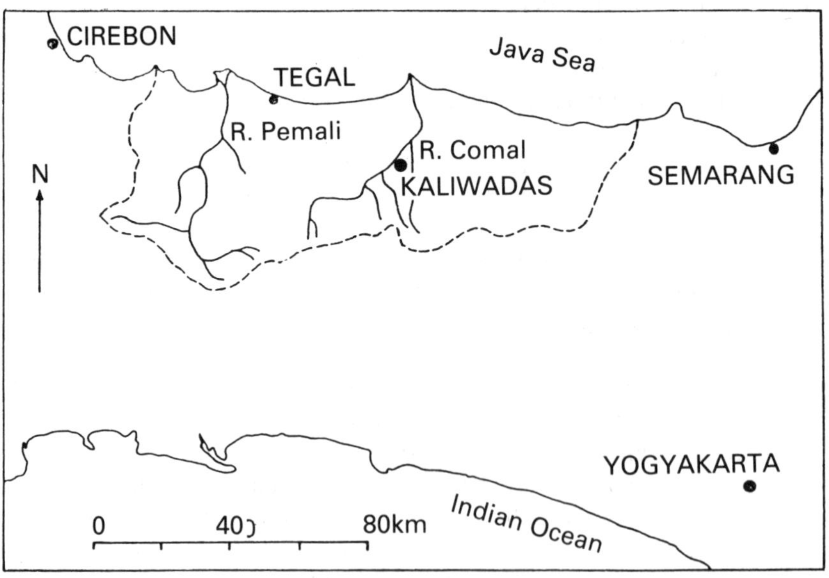

Fig. 2.5 — Pemali–Comal catchments.

chronic flooding owing to poor river flood protection works and to the inadequacy of the internal drainage systems which results in the loss of the rice crop. Uncontrolled frequent flooding in some areas makes about 10 000 ha unproductive in the wet season. The main objectives of the hydrological studies were to define design discharges for the rivers and to provide design capacities for the drains prior to the design of extensive improvement works which by their nature would involve considerable capital investment.

Pemali–Comal

The area of the irrigation scheme named from two of the major rivers extends 120 km along the coast and for about 40 km inland. The rivers rise in a mountainous region of high volcanic peaks (up to 3 000 m) and following a short precipitous course pass through a less steep foothill zone before crossing the nearly flat plain to build up deltas along the coastline with their heavy sediment load. Lying about 7°S of the Equator, the area is in the equatorial climatic region, average temperature 27°C and relative humidity 80%, but with a monsoonal rainfall pattern, 70% of the average annual rainfall coming between November and March. The topographic range produces marked rainfall gradients with annual totals ranging from 1 750 mm on the western lowlands to 4 000 mm in the southern mountains.

Design floods

The rainfall studies, derivation of depth–duration–frequency relationships and construction of a 3 day rainfall profile were based on a network of 33 daily rain gauges selected from 150 in the project area. The selected stations had good-quality records varying from 14 to 48 years in length. The depth–duration–frequency analysis was made by fitting EVI distributions to the annual maximum series at each station by the method of moments (NERC, 1975). A sample of the results is given in Table 2.8, the

Table 2.8 — Depth–duration–frequency values in the Pemali–Comal area

Return period (years)		2			10			25	
Duration (days)	1	2	3	1	2	3	1	2	3
Rainfall, Tegal (mm)	98	119	138	165	190	210	198	226	247
Kaliwadas (mm)	134	190	227	184	295	372	209	347	446

study being limited to composite daily durations owing to lack of autographic data. In addition an isohyetal map of the 2 year, 1 observer day rainfall depth was plotted. The 3 day rainfall profile was derived from selected larger events recorded at several sites on the coastal plain and this showed that 37.5% of the total fall occurred in the first 24 h and 70% in the first 48 h.

The river flow records from only seven sites of varying length for 4 to 10 years were poorly representative for determining flow regimes in the area and there were very many ungauged rivers needing to be assessed. It was decided to use regional flood frequency analysis by relating the magnitude of an index flood (the 2 year peak flow) to catchment characteristics and the more reliable rainfall measures using multiple linear regression. A regional 'growth curve' would then be constructed relating the quotient of the T year flood and the index flood to the return period T. The recorded peak flows at the individual stations were analysed taking the annual maximum series or the partial duration series depending on the length of record. The 2 year floods Q_2 ($m^3 s^{-1}$) were determined and regressed with the 2 year, 1 observer day rainfalls P_2 (mm) and the catchment areas A (km^2) as independent variables. The following model explaining 94% of the variance of the 2 year flood was obtained:

$$\ln Q_2 = 0.65 \ln A + 1.62 \ln P_2 - 5.45 \quad (2.1)$$

The growth factors Q_T/Q_2 obtained at each flow station were averaged and related to T to give the well-fitting growth curve

$$Q_T/Q_2 = 0.96 + 0.14Y \quad (2.2)$$

where

$$Y = -\ln\left[\ln\left(\frac{T}{T-1}\right)\right].$$

These relationships, applicable only to catchments with comparable characteristics in the area, were used to estimate the 25 year floods for the three major rivers (Table 2.9). In evaluating design discharges, the limitations of the basic data used to derive

Table 2.9 — 25 year design discharges

River	Location	Catchment area (km^2)	Flood Q_{25} ($m^3 s^{-1}$)
Pemali	Rengaspendawa	1111	1300
Comal	Ambo	730	1447
Cisanggarung	Ciledug	834	743

the model need to be considered; there could be wide confidence limits with some return periods.

Drainage studies

Although the rice plant grows in flooded fields, excess depths of water affect crop yields. In prescribing criteria for the drainage studies, yield reductions were assumed to be minimized if more than 100 mm depth of water covered the land for less than 3 days and if the average surplus was less than 200 mm above the soil surface. To meet these requirements, the internal drains needed to be designed to cater for the 1 in 5 year, 3 day maximum rainfall.

To accommodate the range of rainfall and topographic conditions within the irrigation area, a nomogram was constructed to assist in the computation of drainage coefficients to be used in the varying locations (Fig. 2.6). The nomogram was built up

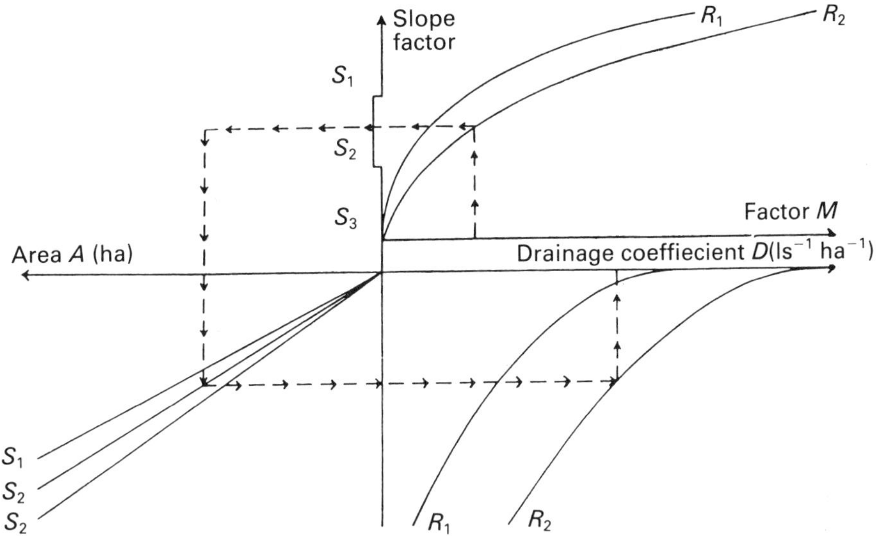

Fig. 2.6 — Drainage coefficient nomogram.

from runoff estimations by a form of rational method taking the 1 in 5 year, 3 day storm and allowing 10% of the rainfall as loss for evaporation and infiltration. The following drainage area characteristics were required: catchment area A (ha); catchment length L (km); average catchment slope S (non-dimensional). The rainfall R (mm) was determined from the nearest rainfall station and an areal reduction factor obtained from

$$F=1.39-0.15 \ln A.$$

The drainage factor $M=L^2/SF$ was then computed. To obtain a drainage coefficient D (l s^{-1} ha^{-1}) for a rice field, the diagram is entered with a value of M and following anticlockwise, the final axis gives the required drainage coefficient from which the design capacity is DA l s^{-1}. The drainage coefficients for the Pemali–Comal coastal plain varied from 6 to 10 l s^{-1} ha^{-1}, increasing from west to east.

Conclusion

The regional flood frequency analysis in the Pemali–Comal area provided design flows for the river engineering works. The regional approach to the derivation of the drainage coefficients by means of areal considerations of depth–duration–frequency, rainfall reduction factors and runoff coefficients and repeated discharge computations for the drainage blocks led to the nomogram for the convenience of the drainage design engineers.

Although the present results are acceptable, further recorded hydrometric data should be used to update the equations. The analytical technique only could be applied in other comparable areas using relevant local data.

2.5 DRAINAGE OF NEWBOROUGH FEN

LOCATION Newborough Fen northeast of Peterborough, England, is drained into the River Welland.

SOURCE Reed, D. W. (1985) Calculating inflows to Newborough Fen. *Conference of River Engineers, Cranfield.*

PROBLEM The rich agricultural fenland of eastern England is below sea level and the natural rivers are contained in embankments. The low-lying land is drained by man made ditches from which the surplus water is pumped up into the embanked rivers. In order to minimize energy costs it is necessary to devise pump operating procedures to avoid floods and yet maintain high water levels for the farm crops in the growing season. This regulation of the drain levels is dependent on knowledge of the input of water into the system which can be evaluated from the rainfall over the area and the catchment response. A problem for the hydrologist in a catchment with a controlled outflow is to determine the response pattern by evaluating the natural inflows.

Fen hydrology

The drainage system of Newborough Fen (32.5 km^2) is shown in Fig. 2.7. From experimental monitoring of water levels, it has been noted that following heavy rains, the catchment responds more or less uniformly, with levels at the pump intake rising almost as soon as those at remote sites. This may be attributed to the low gradient of the main drain and the consistently shallow water table over the nearly level catchment. Following these observations, the contents of the main drain can be treated as a reservoir whose capacity (stock) can be evaluated from the no-pumping rest water level and the drain geometry. By monitoring the stock variation ΔS and measuring the quantity O of water pumped, the natural inflow I to the main drain can be calculated by the simple water balance over a selected time period:

$$I=O+\Delta S \tag{2.3}$$

The drain geometry was modelled by relating the channel width W to the water level h and the distance x from the pumping station. Detailed surveys at the four water level stations on the main drain and surveys at a further 19 sections provided data for a regression analysis to give

$$W(h, x)=16.4+3.85h-1.18x \tag{2.4}$$

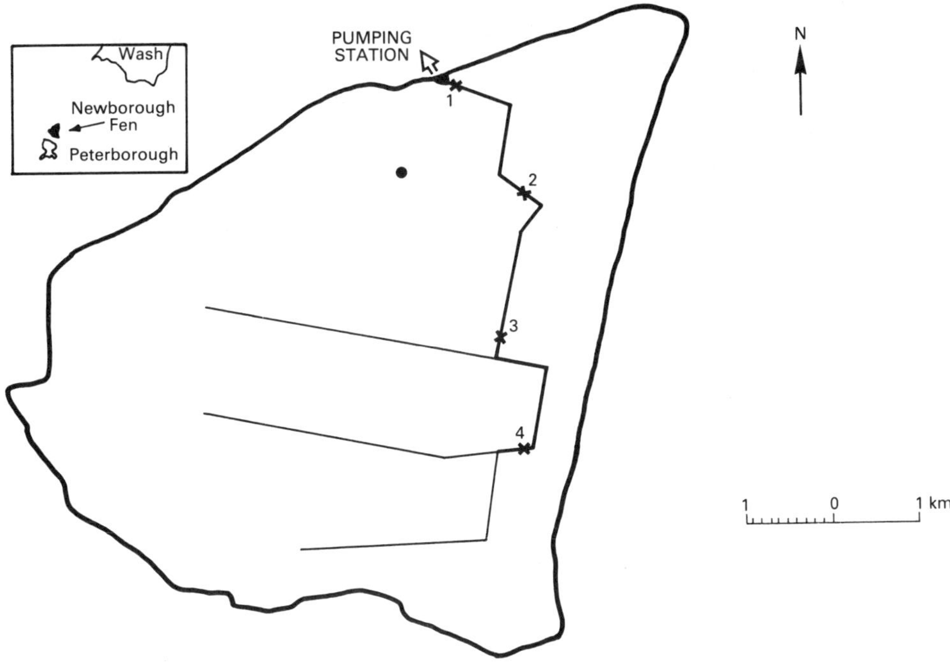

Fig. 2.7 — Newborough Fen catchment: ×, water level recorders; ●, tipping-bucket rain gauge.

with W and h in metres and x in kilometres. The shape of the drain defined by equation (2.4) has a tapering triangular section, but this is not important since the model is required only to give stock changes not absolute values.

The drain geometry model in general terms

$$W(h, x) = a_0 + a_1 h - a_2 x$$

was recalibrated with respect to the water discharged in 28 isolated pump runs using the water level data for sites 1–4. The stock depletion was taken as the difference between the common steady stock gradients, before and after pumping, fitted by least squares. To determine the model parameters, the best fit of the stock depletions to the pumped quantities was obtained by numerical optimization. The resultant model, adjusted to compare with equation (2.4), was

$$W(h, x) = 1.135(16.4 + 3.85h - 1.24x) \tag{2.5}$$

indicating a 13.5% increase in the widths. It was considered that this could represent the effect of side drains and bankside storage. The close fit of the drain geometry to the isolated pump runs was shown by a root-mean-square error of 0.6 Ml (6 min pumping) compared with a mean pump run of 9.8 Ml (105 min pumping).

Natural inflows

The recalibrated drain geometry model was used with water level data for sites 1–4 to calculate the stock as a sum over five reaches. Linear interpolation between the water level recorders was used for the first four reaches and the water level at site 4 was taken to represent the water level at the far end of the drain. The stock record was formed at 15 min intervals but, in applying the water balance equation to derive the natural inflows, a data interval of 3 h was preferred since the catchment response is slow. Fig. 2.8 shows an inflow sequence (in runoff (mm h^{-1}) terms) calculated over a

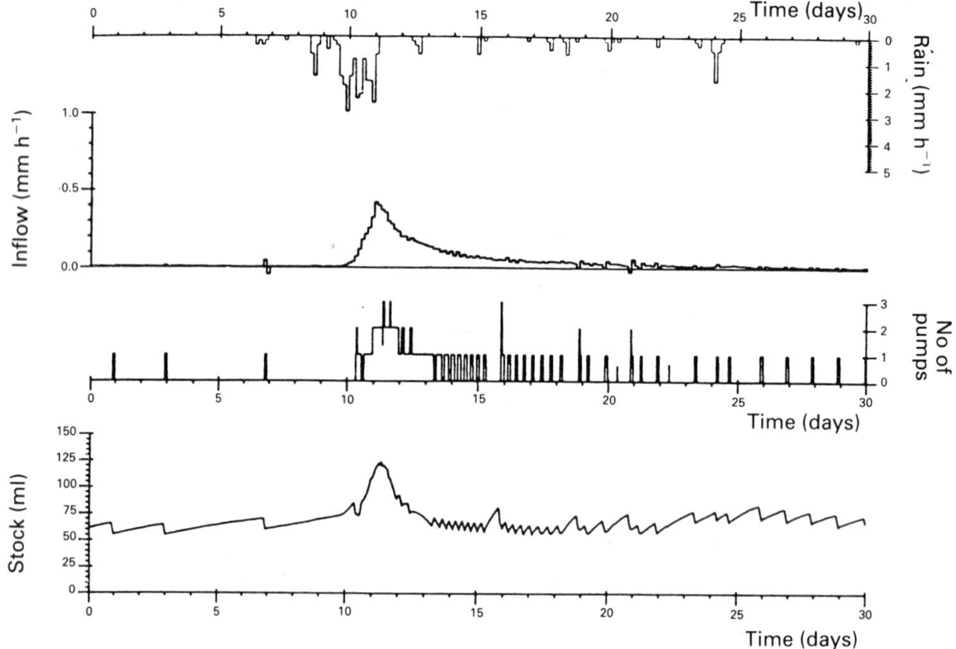

Fig. 2.8 — Rainfall–inflow–pump–stock data for 30 days commencing 16 April 1981.

30 day period, with the causal rainfall rates for comparison. The small perturbations in the inflows result from the back computations of the water balance being sensitive to minor timing discrepancies, given the relatively high pumping rates.

Application

From a series of rainfall rates and the inflows calculated from the water balance relationship, the catchment response can be represented by a unit hydrograph related to antecedent catchment wetness. It would then be possible to forecast, from recorded rainfall, inflows to the main drain 6–12 h ahead. This is of interest in the scheduling of pump runs to avoid peak electricity tariff periods.

In the longer-term management of the catchment on a seasonal basis, the inflow sequence could be used to develop more sophisticated pump-operating rules to take account of crop demands and varying surplus drainage through the farming year and so to optimize pumping requirements and to minimize costs.

3

Urban drainage

INTRODUCTION

The hydrological response of a natural catchment to precipitation is altered radically when areas of the land surface are made impervious by the building of towns and cities. The immediate effect is the production of runoff soon after the commencement of rainfall. There is a consequent decrease in the times to peak flow in the streams and an increase in the magnitude of the peaks and total discharge volumes compared with pre-urbanization conditions. A less obvious effect on the stream flow is the reduction in the baseflow component since there is less percolation and recharge to groundwater. This in turn leads to a lowering of the standards in water quality.

The consequences of urbanization feature widely in hydrological studies since they have been an increasing problem in recent years in both developed and underdeveloped countries (Packman, 1981). With the siting of a completely new town, the planning of the urban development, the proportion of impervious surfaces to gardens or natural land surfaces must be studied in order to assess the impact of rainfall events on drainage to the rivers. In particular, the locations of possible floods must be identified and their frequency of occurrence determined. The change in flood characteristics due to urbanization depends on the nature of the initial catchment conditions and also on the severity of the rainfall event (Hall, 1984). The shorter response time after development makes a catchment more vulnerable to intense rainfall from thunderstorms than to prolonged rainfall from depressions.

The design of civil engineering works for urban drainage requires hydrometric data for the catchment concerned before and after development. One of the basic methods of deriving measures of the changes in catchment response is to compare the two unit hydrographs: the first representing the response of the initial catchment state, the second giving the response after urbanization. The effect of urban development has been incorporated into the multiple-regression equations to obtain peak flows from catchment characteristics in regions of the UK (NERC, 1975). Catchment models are also used in the derivation of required rainfall-runoff relationships but the most important information required is the rainfall depth (or intensity)–duration–frequency data for the area. Analysis of recorded rainfall over several years is essential for the construction of the rainfall depth–duration–frequency curves but in some countries regional values may be considered adequate. For parts of the world where real data are lacking, the generalized rainfall–duration–frequency relationships of Bell (1969) provide an initial guide.

The case studies give examples of pre-urban development planning and the improvement of existing drainage in established urban areas. Of particular interest in the old urban areas of the Isle of Dogs and Brightlingsea, the first experiencing radical redevelopment and the second suffering from an outdated sewer system, is the use of the flexible Wallingford Procedure for identifying inadequate pipes and designing systems of drainage to prescribed standards (NWC–DOE, 1981). In this context where there is insufficient site information to make modelling practicable, the *Handbook of Hydraulic Engineering* (Lencastre, 1987) recently published in English provides invaluable design material.

3.1 RIVER CAM DRAINAGE

LOCATION The River Cam and Wicksters Brook flow from the Cotswolds into the River Severn estuary, UK.

SOURCES C. H. Dobbie & Partners (1978) *Flood alleviation on the River Cam and Wicksters Brook*, Feasibility Report to Severn–Trent Water Authority, South Gloucestershire Internal Drainage Board and Stroud District Council.

Severn–Trent Water Authority (1978) *River Cam and Wicksters Brook, Determination of design flood flows*.

PROBLEM Frequent flooding along the lower reaches of the River Cam and Wicksters Brook and the possible worsening of conditions from increased surface water runoff resulting from future urban developments in the catchments up to the year 2000 led the Severn–Trent Water Authority to instruct C. H. Dobbie & Partners, consulting engineers, to investigate the problem and to recommend alleviation measures. The hydrological studies were carried out by the Severn–Trent Water Authority.

The catchments

The two streams drain an area of about 45 km^2, approximately 70% to the River Cam. They rise on the Cotswold escarpment about 240 m AOD and flow in a general northwesterly direction to the lowlands bordering the Severn estuary at 7.6 m AOD. The Cam joined by Wicksters Brook discharges into a canal crossing the alluvial plain from which abstractions are made for water supply at a point south of the Cam confluence where there is also an outfall to the Severn. Both the Cam and Wicksters Brook pass under a motorway and trunk road before being embanked to ensure continuity of flow to the canal (Fig. 3.1).

A third of the catchment is composed of oolitic limestone overlying sandstone and the remaining two thirds is formed of Lias clays and marlstones. The land is

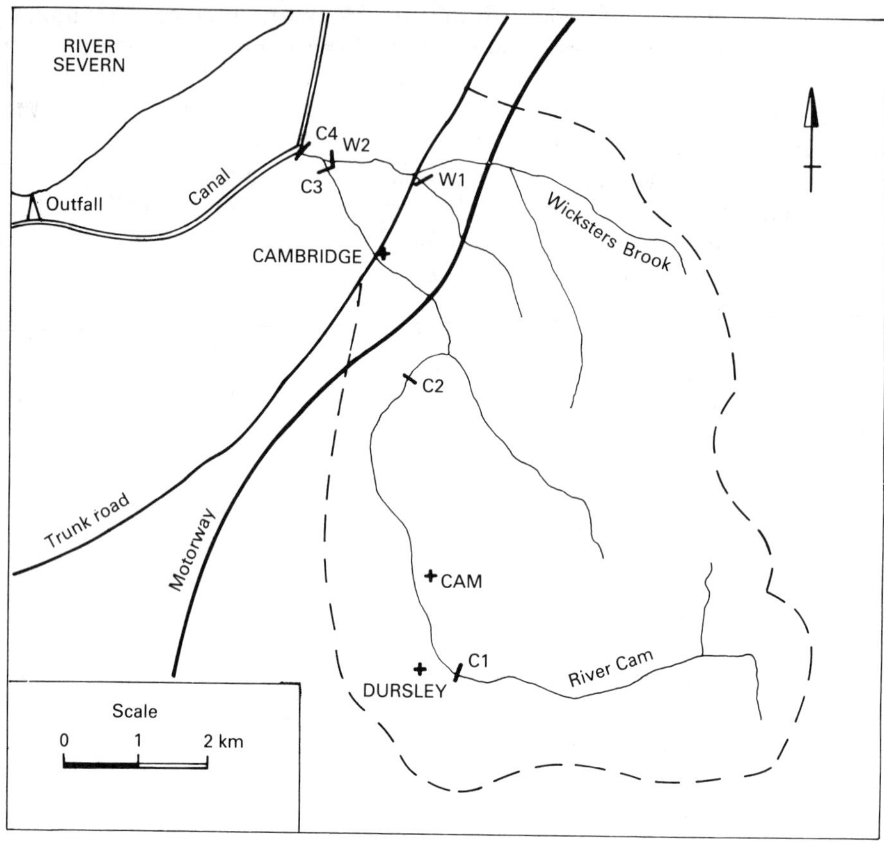

Fig. 3.1 — River Cam drainage.

predominantly kept as permanent grassland with woodland on the escarpment and upland valleys. The present urban areas are centred on Dursley and Cam. A slight decrease in population is projected in Dursley but an increase by a third at Cam.

Hydrological studies

The sites most vulnerable to flooding from heavy rains or snow melt are associated with inadequate culverts under the trunk road. At Cambridge the situation is aggravated by backing up of water from high levels in the canal. It was required to determine peak flows for a range of return periods at selected points, two on Wicksters Brook and four on the River Cam (Fig. 3.1) so that channel improvements and larger culverts could be designed. Flood hydrographs to give flood volumes were also requested at one or two points in case it was thought expedient to arrange for flood detention ponds.

There were very few hydrometric data available for the catchments. For serious flood events in December 1965 and July 1968 rainfall records from two gauges in the Cam catchment produced estimates of the rainfall recurrence frequencies of 1 in 10 years and 1 in 45 years respectively. Measurements of river flow only began in 1976 when a water level chart recorder was installed at Cambridge upstream of the trunk road. This was calibrated by current meter gaugings at a point downstream of the trunk road where the Cam enters its embanked course. All the peak flows in the following few months of record were contained in bank with the highest level equivalent to 9.7 m^3 $^{-1}$ being 1 m below local bankful level. From a partial duration analysis of 1.25 years of record, the six highest peaks gave a mean annual flood $\overline{Q}$ of 9.16 $m^3 s^{-1}$. Owing to the shortage of data on the Cam and the limited help from neighbouring catchments, it was decided to evaluate the design floods by the methods of The Flood Studies Report (NERC, 1975).

Design flood peaks

Estimates of the MAF, $\overline{Q}$ having a return period of 1 in 2.33 years were first calculated from the The Flood Studies Report regional regression equation. The necessary catchment characteristics for the key points on the streams (Fig. 3.1) are shown in Table 3.1. The equations used for $\overline{Q}$ are given by

Table 3.1 — Catchment particulars in 1976

	Wicksters Brook		River Cam			
	W1	W2	C1	C2	C3	C4
AREA (km^2)	2.4	14.8	12.8	22.4	30.2	45.0
Stream length (km)	2.7	5.9	5.7	10.2	12.8	14.6
S1085 ($m\ km^{-1}$)	7.53	4.48	12.83	9.96	8.89	7.93
STMFRQ ($j\ km^{-2}$)	0.42	0.88	0.86	0.71	0.83	0.84
SOIL	0.40	0.40	0.28	0.32	0.34	0.37
URBAN (%)	0	0	3.7	16.8	12.5	9.3
RSMD	32.1	32.4	34.4	33.7	33.6	32.8
SAAR (mm)	746	762	842	823	815	790
M5-2 day rainfall (mm)	53.0	54.2	57.5	56.9	56.7	55.8

$$\overline{Q} = \text{Constant} \times \text{AREA}^{0.94}\text{STMFRQ}^{0.27}\text{S1085}^{0.16}\text{SOIL}^{1.23}\text{RSMD}^{1.03}(1 + \text{LAKE})^{-0.85}$$

and by

$$\overline{Q} = 0.0945\text{AREA}^{1.00}\text{STMFRQ}^{0.62}\text{S1085}^{0.51}$$

following a regional analysis of rural catchments in the Severn–Trent region. In the light of further research, adjustments to the first estimates of $\overline{Q}$ were made for small catchments and for the urbanization:

$$\bar{Q}_s = 0.0014\text{AREA}^{0.92}\text{SAAR}^{1.13}\text{SOIL}^{2.10}$$

and

$$\frac{\bar{Q}_u}{\bar{Q}_r} = (1+\text{URBAN})^{2n}\left[1+\text{URBAN}\left(\frac{24}{\text{PR}_r}-0.3\right)\right]$$

where

$$\text{PR}_r = 102.4\text{SOIL} + 0.28(\text{CWI}-125) + 0.1(P-10) - 1.9$$

with parameters and variables defined following Table 3.4 in the case study in section 3.2.

Then applying the appropriate growth curve $Q_T/\bar{Q}$ from The Flood Studies Report (NERC, 1975) modified where appropriate for urban effects, values of Q_T were obtained for the required return periods $T = 1$, 5, 10, 20 and 100 years. By calculating the Q_T for the natural catchments and accounting for the urban proportions in 1976, the increased effect on the peak flows due to the urbanization could be assessed. The results are summarized in Table 3.2 and the increasing urban effects demonstrated in Fig. 3.2. With the expected urban proportion reaching 18.2% in the C2 catchment area by 1991, the new estimated peak flows would be raised by 45–47% of the natural peak flows.

Design flood volumes

For determining flood volumes, flood hydrographs were derived by The Flood Studies Report unit hydrograph no-data method. A triangular unit hydrograph was defined from the three parameters: time to peak T_p (h), peak flow Q_p ($m^3\,s^{-1}$ $100\,km^{-2}$) and the base time T_B (h). The design storm duration D was obtained from T_p and the mean annual rainfall SAAR and the rainfall depth over the duration D was derived from the M5-2 day rainfall and the relevant table of M5-1 day/M5-2 day rainfall ratios from The Flood Studies Report. From the required flood return periods of 1 and 10 years were obtained the related rainfall return periods for which the rainfall depths were then estimated from the MT/M5 growth factors. The point rainfalls were converted to catchment areal rainfalls by applying an areal reduction factor and the rainfall patterns over duration D were obtained from a standard profile. Assuming the worst initial conditions, the rainfall patterns were convoluted with the synthetic unit hydrograph.

The design flood hydrographs for Wicksters Brook, W2, and two points on the River Cam, C2 and C3, are shown in Fig. 3.3. Some adjustments to the computed hydrographs were made based on the short measured flow records. The related flood volumes are given in Table 3.3. It will be noted from Fig. 3.3 that the peak flows Q_T obtained by the unit hydrograph method are higher than those derived by the statistical method. The flood hydrograph shapes may vary with the choice of rainfall profile and, while the 10 year peak may be considered too high, the flood volume is considered reliable.

Table 3.2 — Design peak flows and yields

	Wicksters Brook		River Cam			
	W1	W2	C1	C2	C3	C4
$T = 2.33$ years						
$\overline{Q}$, natural ($m^3 s^{-1}$)	0.70	3.84	3.06	5.53	7.75	10.99
$\overline{Q}$, 1976 ($m^3 s^{-1}$)	0.70	3.84	3.29	7.74	9.69	12.97
1976 yield ($m^3 s^{-1} km$)			0.26	0.35	0.32	0.29
$T = 1$ year						
Q_T, natural ($m^3 s^{-1}$)	0.58	3.19	2.54	4.59	6.43	9.12
Q_T, 1976 ($m^3 s^{-1}$)	0.58	3.19	2.73	6.42	8.04	10.76
1976 yield ($m^3 s^{-1} km$)			0.21	0.29	0.27	0.24
$T = 5$ years						
Q_T, natural ($m^3 s^{-1}$)	0.89	4.88	3.89	7.02	9.84	13.96
Q_T, 1976 ($m^3 s^{-1}$)	0.89	4.88	3.98	9.52	11.82	15.82
1976 yield ($m^3 s^{-1} km$)			0.31	0.43	0.39	0.35
$T = 10$ years						
Q_T, natural ($m^3 s^{-1}$)	1.04	5.72	4.56	8.24	11.55	16.38
Q_T, 1976 ($m^3 s^{-1}$)	1.04	5.72	4.90	11.53	14.44	19.32
1976 yield ($m^3 s^{-1} km$)			0.38	0.51	0.48	0.43
$T = 20$ years						
Q_T, natural ($m^3 s^{-1}$)	1.25	6.84	5.45	9.84	13.80	19.56
Q_T, 1976 ($m^3 s^{-1}$)	1.25	6.84	5.72	13.62	16.96	22.70
1976 yield ($m^3 s^{-1} km$)			0.45	0.61	0.56	0.50
$T = 100$ years						
Q_T, natural ($m^3 s^{-1}$)	1.80	9.87	7.86	14.21	19.92	28.24
Q_T, 1976 ($m^3 s^{-1}$)	1.80	9.87	8.88	20.12	25.19	35.02
1976 yield ($m^3 s^{-1} km$)			0.69	0.90	0.83	0.79

Conclusions

The capacity of the channel downstream of Cambridge was less than the flow from a 1 in 5 years return period storm. Thus considerable channel improvements were required. It was also seen that the canal discharge system needed improvement and several methods of increasing the efficiency of outfalls were suggested. In all, 12 possible schemes were proposed involving increases in channel capacity and enlarged culverts under the trunk road, taking into account the calculated effects of increased urbanization. Three schemes producing benefit-to-cost ratios in the range 1.06–1.07 were particularly recommended to the clients.

Implementation

Having selected one of the schemes, site works commenced in June 1980 and substantial completion of the constructed scheme was achieved in December 1981.

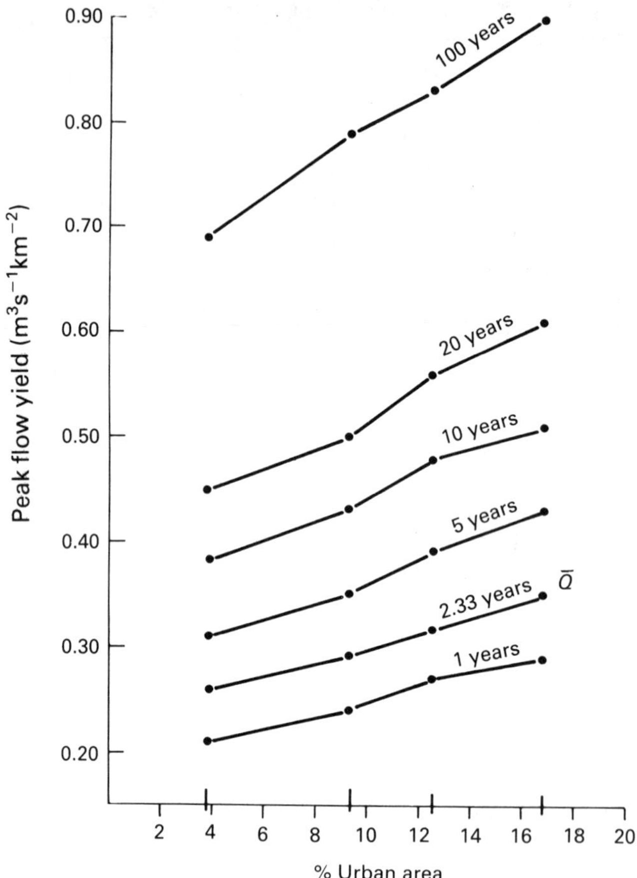

Fig. 3.2 — Design peak flow yields for Cam catchments.

The chosen scheme used the above design flows supplemented by the 1 in 20 year volumes for a flood storage area in the Wicksters–Cam–Trunk road triangle of grade 4 agricultural land.

3.2 WARMINSTER SURFACE WATER DRAINAGE

LOCATION Warminster, Wiltshire, is situated on the Were Stream, a tributary of the River Wylye which joins the River Avon at Salisbury.

SOURCES Lemon & Blizard (1983) *Warminster surface water drainage — feasibility study*, Report to Wessex Water Authority and West Wiltshire District Council.

Reed, D. W & Robinson, M. (1982) *Warminster flood study*. Report to Lemon and Blizard. Institute of Hydrology.

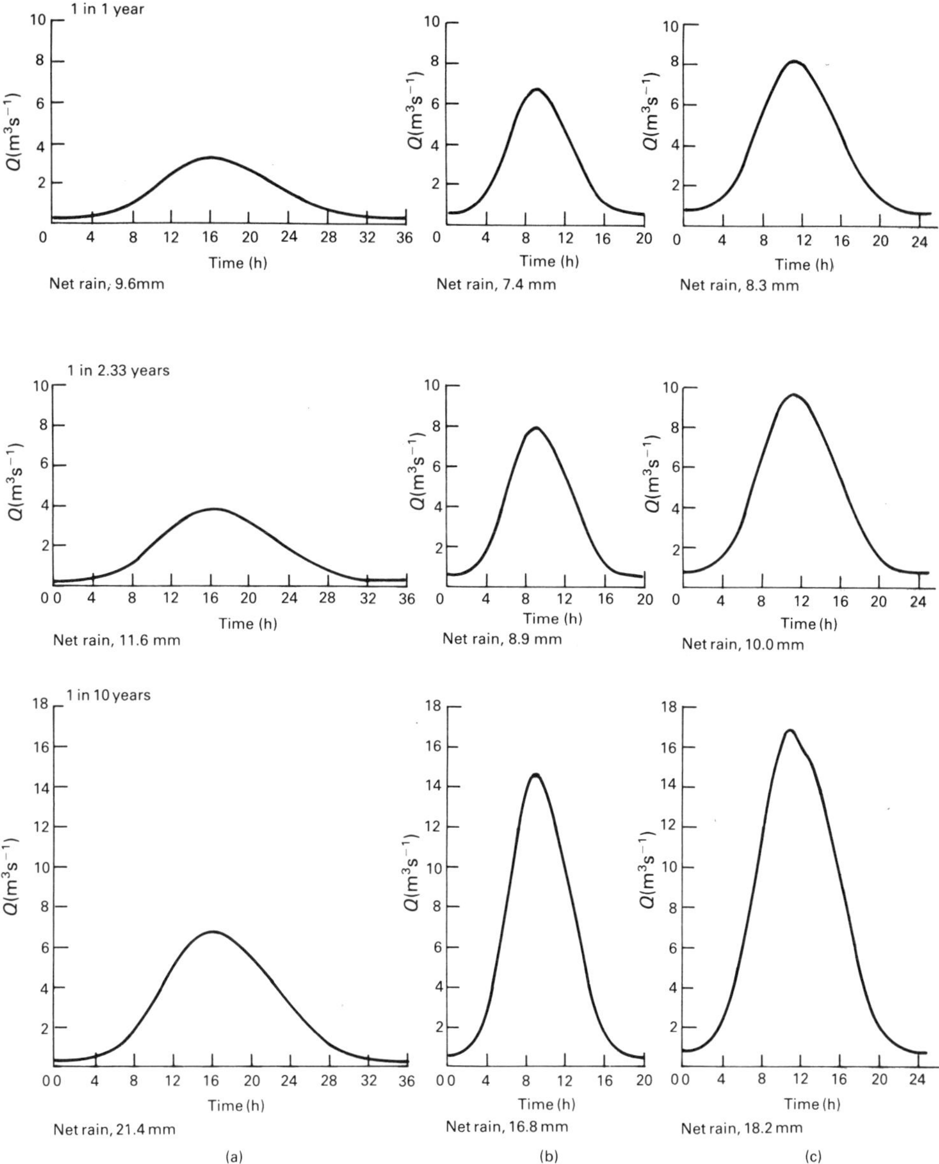

Fig. 3.3 — Design flood hydrographs: (a) Wicksters Brook, W2; (b) River Cam, C2; (c) River Cam, C3.

Wessex Water Authority, Avon and Dorset Division, 2 Nuffield Road, Poole, Dorset BH17 7RL.

PROBLEM Expansion of the impermeable area due to the steady development of Warminster has increased the discharge and volume of

Table 3.3 — Design flood volumes

	River Cam		Wicksters Brook	River Cam
	C2	C3	W2	C3 + W2
Design baseflow (m^3)	42 950	72 350	44 755	117 105
$T = 2.33$ years				
surface runoff (m^3)	199 410	300 570	171 285	471 855
flood hydrograph (m^3)	242 360	372 920	216 040	588 960
$T = 1$ year				
surface runoff (m^3)	165 540	249 530	142 695	392 225
flood hydrograph (m^3)	208 490	321 880	187 450	509 330
$T = 10$ years				
surface runoff (m^3)	375 130	547 070	316 465	863 535
flood hydrograph (m^3)	418 080	619 420	361 220	980 640

surface water runoff to the existing sewers and watercourses. Local plans for the development of further large blocks of land has led the responsible authorities to consider the improvement of the existing surface water drainage system to accept even larger discharges and to mitigate against the possibility of future flooding. To assess the frequencies of flood discharges under both present conditions and those likely to result from the future development, the Institute of Hydrology was asked by Lemon and Blizard to investigate and advise.

Investigation

The study area was defined by a critical culvert in Warminster (Fig. 3.4). The catchment area to this point on the Were Stream, 13.2 km^2, is composed of chalk across the northern three-quarters of the area while the remainder in the south is mainly upper greensand and chert beds with alluvium along the valleys. There were no hydrometric data available for the Were Stream catchment; so rainfall-runoff studies were made for comparable chalk catchments in the neighbourhood using historical flood events. It was noted that there was a rapid storm response from greensand areas but that it took a long-period rainstorm to produce a response from the chalk. With these qualitative guidelines, the hydrologists then applied the techniques of The Flood Studies Report (NERC, 1975) to the Warminster catchment.

Statistical approach using catchment characteristics

In order to apply the regression equations of The Flood Studies Report, Vol. I, catchment characteristics must be measured on the relevant Ordnance Survey (OS) 1:25 000 topographic maps and climatic characteristics obtained from the published

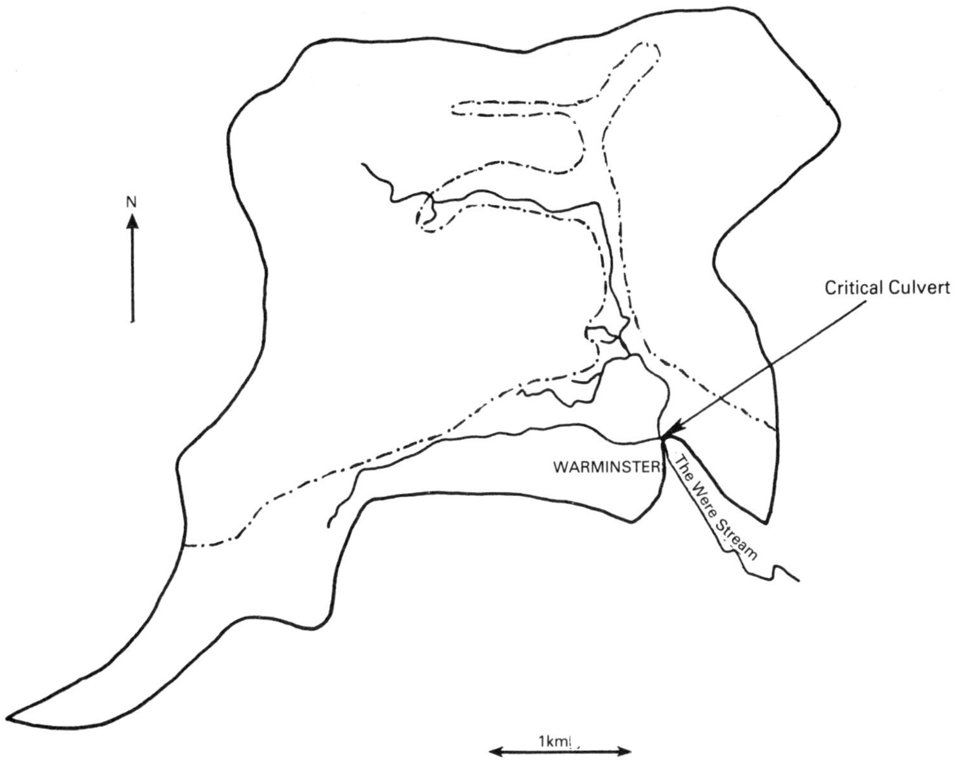

Fig. 3.4 — Warminster catchment area: —·—, limit of chalk.

maps and tables of The Flood Studies Report, Vols. II and V. The required data are given in Table 3.4. SOIL is an index derived from five soil classes, URBAN

Table 3.4 — Catchment and climatic characteristics

Catchment		Climatic	
AREA (km^2)	13.2	SAAR (mm)	870
MSL (km)	3.74	M5-2 day rain (mm)	51
STMFRQ (junctions km^{-2})	0.407	*r* (M5-1 h/M5-2 day)	0.35
S1085 (m km^{-1})	2.93	M5-24 h/M5-2 day	0.87
SOIL	0.15	M5-24 h (mm)	44.4
LAKE	0	M5-1 day (mm)	40.0
Urban area (km^2)	0.84	ARF	0.963
URBAN	0.064	SMDBAR (mm)	9.0
		RSMD (mm)	29.5

is the urban fraction of the catchment and LAKE is the catchment fraction draining through a lake or reservoir. RSMD is the net 1 day rainfall of 5 year return period, i.e. catchment M5-1 day rain minus the mean soil moisture deficit (SMDBAR). For further definitions see the original references (e.g. Sutcliffe, 1978).

The equation for the MAF $\overline{Q}$ is

$$\overline{Q} = 0.0234\text{AREA}^{0.94}\text{STMFRQ}^{0.27}\text{S1085}^{0.16}\text{SOIL}^{1.23}\text{RSMD}^{1.03}(1+\text{LAKE})^{-0.85}$$

where the multiplying factor 0.0234 is for region 7, the south coast area of England. This gives an estimate for $\overline{Q} = 0.78\ \text{m}^3\,\text{s}^{-1}$. Further studies by the Institute of Hydrology of the urban effects on catchment floods led to the following relationship (NERC, 1979). The mean annual flood Q_u with urban influences is given by

$$\overline{Q}_u = Q(1+\text{URBAN})^{1.5}\left[1+0.3\text{URBAN}\left(\frac{70}{\text{PR}_r^{-1}}\right)\right]$$

where PR_r, the percentage rural runoff, is obtained from

$$\text{PR}_r = 102.4\text{SOIL}+0.28(\text{CWI}-125).$$

The catchment wetness index (CWI) is related to average annual rainfall (SAAR) and for this catchment is 120 (The Flood Studies Report, Vol. I, Fig. 6.62). The catchment mean annual flood $\overline{Q}$ is thereby increased by a factor of 1.18 to give a value of $0.92\ \text{m}^3\,\text{s}^{-1}$ for $\overline{Q}_u$. The flood discharges for selected return periods were then obtained from the region 7 growth curves giving values of $Q_T/\overline{Q}_u$ for return periods T (The Flood Studies Report, Vol. I, Table 2.39). The full results are given in Table 3.5.

Rainfall-runoff study: a short duration storm on the non-chalk area

A 25 year design storm was applied to a synthetic unit hydrograph with the significant components derived from catchment characteristics: non-chalk area, $3.6\ \text{km}^2$; chalk area $9.6\ \text{km}^2$; non-chalk SOIL, 0.192; URBAN, 0.23; MSL, 2.6 km; S1085, 8.2 m km. The time to peak of the unit hydrograph is given by

$$T_p = 46.6\text{MSL}^{0.14}\text{S1085}^{-0.38}(1+\text{URBAN})^{-1.99}\text{RSMD}^{-0.4}$$

Using the non-chalk data, $T_p = 4.1$ h and an appropriate data interval $T = 1$ h. To determine the storm duration D,

$$D = \left(1+\frac{\text{SAAR}}{1000}\right)T_p = 7.67\ \text{h}$$

and this is rounded to the nearest odd multiple of T so that $D = 7$ h. To evaluate the rainfall depth in 7 h expected once in 25 years, data are taken from The Flood Studies Report and given in Table 3.4.

$$\text{M5-2 day rainfall} = 51 \text{ mm}$$
$$r = 0.35$$
$$\frac{\text{M5-7 h}}{\text{M5-2 day}} = 0.63$$
$$\text{M5-7 h} = 0.63 \times 51 = 32.1 \text{ mm.}$$

The rainfall growth factor of 1.44 is applied to obtain the M25-7 h rainfall, i.e. $1.44 \times 32.1 = 46.2$ mm. Then using the procedure outlined in The Flood Studies Supplementary Report No.5 (IH, 1979), the percentage runoff for rural conditions PR_r leads on to PR_u, the percentage runoff for urban conditions.

$$\begin{aligned} PR_r &= 102.4\text{SOIL} + 0.28(\text{CWI} - 125) - 1.9 + 0.1(P - 10) \\ &= 20.0 \\ PR_u &= (1 - 0.3\text{URBAN})PR_r + 70(1 + 0.3\text{URBAN}) \\ &= 23.4. \end{aligned}$$

Continuing the Flood Studies Supplementary Report No. 5 procedure, a short cut to unit hydrograph convolution is possible which is specially useful for determining peak flows alone. The ratio D/T_p i.e. $7.0/4.1 = 1.71$ and its relationship with RC, a routing constant,gives a value of 0.302 for RC.

Then an estimate of the required peak flow is given by

$$\begin{aligned} q &= \text{RC}\frac{PR_u}{100}\frac{P}{D}\text{AREA} \\ &= 1.68 \text{ m}^3\,\text{s}^{-1}. \end{aligned}$$

The estimated average baseflow for the whole catchment, $0.31 \text{ m}^3\,\text{s}^{-1}$, is added to give an estimate of the peak flow resulting from a 25 year rainstorm:

$$Q_p = 1.99 \text{ m}^3\,\text{s}^{-1}.$$

The rainfall-runoff method was applied to storms of different return periods and, following the specifications of Flood Studies Supplementary Report method, the flood peaks obtained were assumed to have the same return period. To test the response of the whole catchment, the rainfall-runoff method was reapplied with a long duration hypothetical design storm (11 days) applied separately to the chalk and non-chalk areas. This extension to the analysis used the response characteristics of the chalk area inferred from a neighbouring gauged catchment. The pattern of response is shown for the 50 year event in Fig. 3.5 and the full results are given in

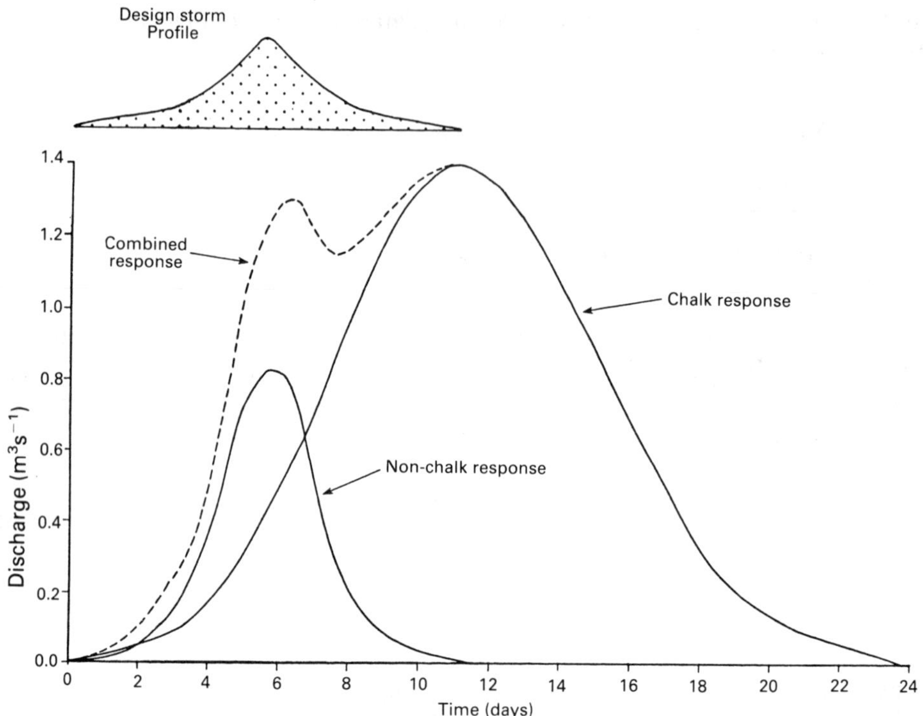

Fig. 3.5 — Chalk, non-chalk and combined response hydrographs for 50 year long-duration storm.

Table 3.5. In the proposed further development of Warminster, the URBAN fraction was assessed to increase from 0.23 to 0.31 of the non-chalk area. Peak flows were estimated using the changed condition and the short storm on the non-chalk area. The results are in the final column of Table 3.5.

Conclusion

From the range of peak flow estimates provided in Table 3.5, the design consultants were able to advise the clients on improvements to existing culverts and drainage channels and on further provisions that would be required for storm drainage following the future urban developments. Fig. 3.5 illustrates that, although producing a much lower flood peak than the short-duration storm, a long-duration rainfall event could generate an appreciably prolonged response from the chalk area, militating against a storage method for alleviating flooding from the Were Stream.

Table 3.5 — Peak discharge estimates

Flood peaks return period	Peak discharge ($m^3 s^{-1}$)				
	Statistical method	Rainfall-runoff method (baseflow added)			
		Short storm	Long storm		Short storm
		Non-chalk	Chalk	Non-chalk	Increase URBAN
2.33	0.92	1.12	0.81	0.46	1.30
5	1.18	1.41	1.07	0.62	1.65
10	1.49	1.67	1.20	0.71	1.97
25	1.97	1.99	1.39	0.83	2.36
50	2.41	2.35	1.50	0.90	2.77
100	2.93	2.75	1.64	1.00	3.24

3.3 SURFACE WATER DRAINAGE OF KUWAIT ZOO

LOCATION Southwest of Kuwait City at the head of the Persian Gulf.

SOURCE IH (1982) *Surface water drainage of Kuwait Zoo*, Report to J. S. Bonnington Partnership, Architects and Engineers. Institute of Hydrology.

PROBLEM For a new zoo in Kuwait, provision had to be made to control the surface water resulting from the heavy rainstorms, a common feature of the rainfall pattern in arid zone regions. The irregular quadrilateral-shaped plan of the zoo is shown in Fig. 3.6 with the natural drainage areas defined by six zones. Zones 1 and 4 drain to the north under the motorway and any increase in peak runoff rates from the zoo would not have to exceed the capacity of the culverts. Drainage from the other zones focusses on the natural low point and a natural lake (retention pond) to contain surface runoff was envisaged. A retaining embankment round the zoo would exclude drainage from zones 5 and 6. However, to contain surface water on the zoo site, the Institute of Hydrology were asked to advise on the runoff volumes, infiltration rates, evaporation losses and pondage required.

Investigations

From previous experience, it was suggested that infiltration ponds would become ineffective and to retain all surface runoff on the site was too ambitious. It was recommended that the runoff should be controlled so that the natural runoff rates would not be exceeded after the zoo development. Thus detention storage should be provided to absorb the peak flows. These may be onstream or offstream detention ponds according to requirements, each type having appropriate hydraulic outlet controls (Hall & Hockin, 1980). In the design and operation of the ponds, all the runoff is assumed to pass through the pond; the losses by infiltration or evaporation, difficult to quantify, nevertheless occur and form a safety factor.

The overall drainage design is shown in Fig. 3.7. The detention pond in zone 1

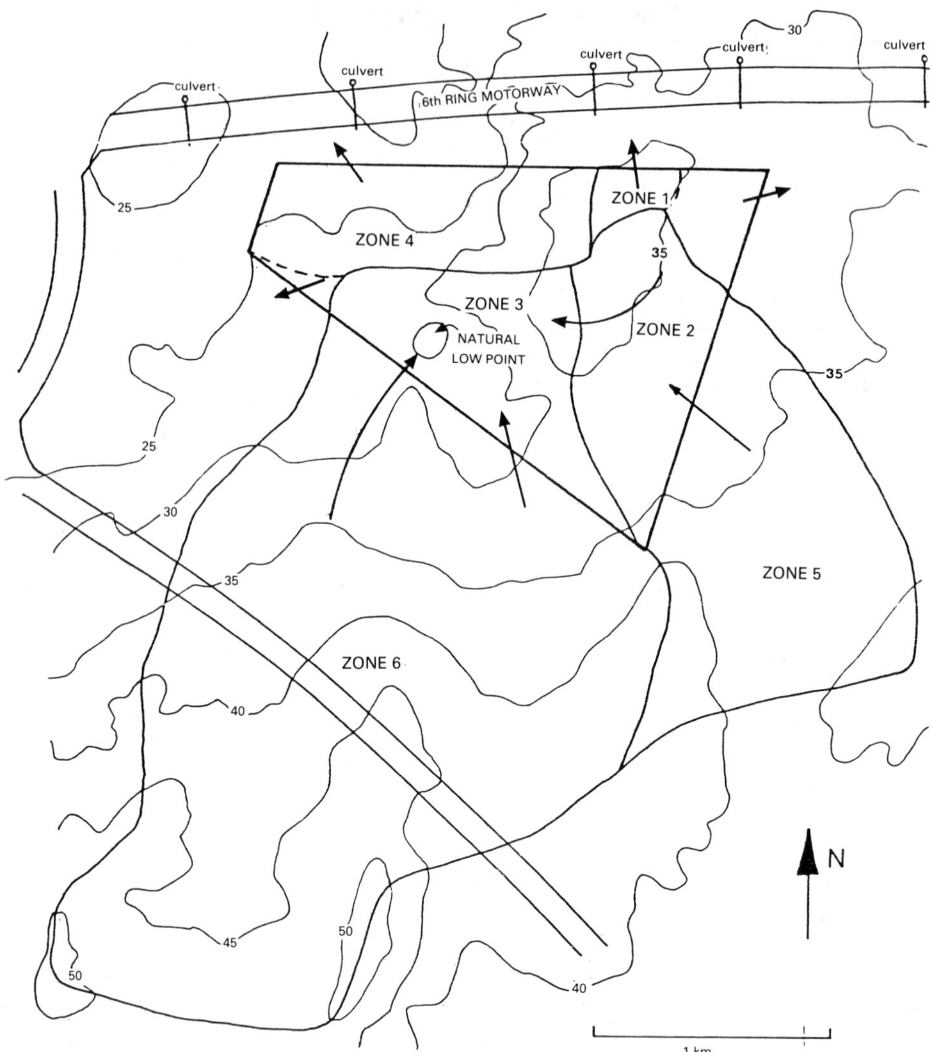

Fig. 3.6 — Natural drainage areas for Kuwait Zoo.

discharges to the north. Possible ponds in zones 2 and 3 would be connected by deep drains to the pond in zone 4 which also discharges to the north. Simple retention storages are indicated for zones 5 and 6 with an emergency overflow for zone 6 via the deep drain through zones 3 and 4.

The zoo drainage scheme was specified to accommodate a design storm with a return period of 5 years and the necessary information on rainfall intensity–duration–frequency was available from data published in Kuwait. Using the 5 year, 1 day rainfall of 32 mm and volumetric runoff coefficients of the US Soil Conservation service (USDA, 1972), an estimate of the development effects on the runoff volume

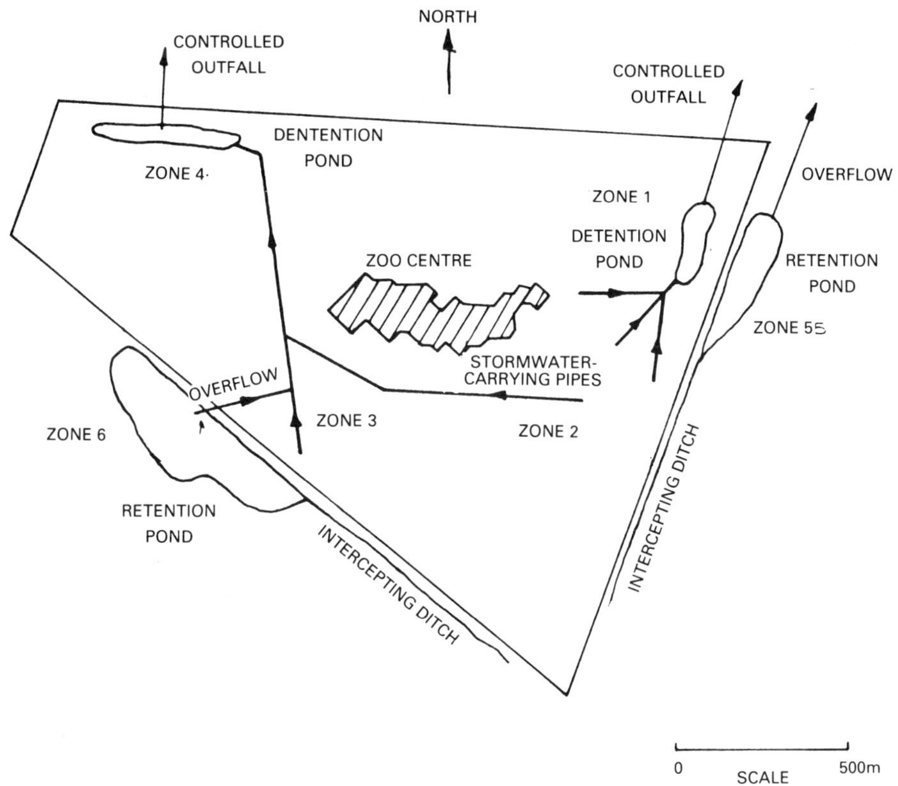

Fig. 3.7 — Final design stormwater drainage layout for Kuwait Zoo.

from the zoo area was made (Table 3.6). These figures of increased volume disguise the considerable effects of increased rate of runoff. To examine these the time sequence of the runoff, i.e. the hydrograph and the peak rate of runoff are needed.

Pre-development runoff

The Flood Studies Report (NERC, 1975) triangular unit hydrograph method was used to derive pre-development response for zones 1 and 4, yielding discharges at the two zoo outlets. The times T_p to peak were obtained from the drainage area characteristics, whence the peak flows Q_p were obtained from

$$Q_p = \frac{220}{T_p} \text{ per } 100\,\text{km}^2$$

with the base time T_B of the unit hydrographs given by $2.525T_p$. The unit hydrograph ordinates ($m^3\,s^{-1}$) corresponding to a rainfall of 10 mm in 5 min were obtained. The

Table 3.6 — Changes in runoff volume from 5 year, 1 day rainfall

Zone	Area (ha)	Paved area (ha)	Pre-development runoff (m^3)	Post-development runoff (m^3)	Increase in runoff (m^3)
1	29.0	7.9	3770	5020	1250
2	46.0	10.7	5980	7670	1690
3	63.0	14.3	8190	10450	2260
4	62.0	3.3	8060	8580	520

5 year storm rainfalls for 60, 120 and 180 min durations were taken as 21, 26.4 and 30 mm respectively and symmetrical storm profiles at 5 min intervals calculated from storm patterns experienced in the Kuwait area. Adopting runoff coefficients appropriate to a desert area of 0.23, 0.26 and 0.27 for the 60, 120 and 180 min storms, the convolution of the unit hydrographs and residual rainfalls gave the following peak runoff rates (Table 3.7). The 120 min storm may be taken as the critical storm for each zone and these flows 0.27 and 1.15 $m^3 s^{-1}$, for zone 1 and zone 4 respectively, form the target maximum outflows to be allowed from the zoo area after development.

Table 3.7 — Pre-development peak runoff rates

Storm duration (min)	Peak runoff rate ($m^3 s^{-1}$)	
	Zone 1	Zone 4
60	0.27	1.11
120	0.27	1.15
180	0.25	1.07

Post-development runoff

A simplified version of the Wallingford Procedure (NWC–DOE, 1981) was used for assessing the flows and the required pipe sizes for the drainage design outlined in Fig. 3.7. Certain modifications to the UK package were necessary; the local Kuwait rainfall curves were used and percentage runoff was taken as 90% from paved areas and 20% from unpaved areas. Pipe routing used the Muskingum–Cunge method and the detention ponds were modelled by the level pool method. Each zone was modelled as a single catchment using the adapted sewered subarea component of the Wallingford Procedure. The model was run for each of the 5 year storm durations 60, 120 and 180 min and experiments were made with ponds of different types, sizes and controls. The peak post-development runoffs were as shown in Table 3.8. These runoff rates are considerably greater than the pre-development values in Table 3.7.

Recommendations

To accommodate these discharges and to provide the detention storage required, various options were proposed and a choice of schemes presented to the clients. The drainage pipe data resulting from the model runs are shown in Table 3.9. Ponds were considered in zones 1, 2 and 4. Zone 3 was omitted owing to the limited available storage. A selection of the derived specifications are given.

Zone 1 Sloping base type, area of 4400 m^2 with a 165 mm flume (slot) control, outflow peak 0.27 $m^3 s^{-1}$ and 1932 m^3 storage.

Three options of ponds in zones 2 and 4 are examples of the several schemes evaluated.

Option I

Zone 2 Sloping base type, area of 2000 m^2 with three 375 mm orifice controls, outflow peak 1.12 $m^3 s^{-1}$ and 810 m^3 storage.

Zone 4 Sloping base type, area of 16 000 m^2 with two × 450 mm orifice controls, outflow peak 1.13 $m^3 s^{-1}$ and 8410 m^3 storage.

Option II

Zone 2 No pond.

Zone 4 Sloping base type, area of 18 000 $m^3 s^{-1}$ with two 450 mm orifice controls, outflow peak 1.11 $m^3 s^{-1}$ and 8620 m^3 storage.

Option III

Zone 2 No pond.

Zone 4 Sloping base type, area 20 000 m^2 with 650 mm flume (slot) control, outflow peak 1.09 $m^3 s^{-1}$ and 9840 m^3 storage.

These storages are much greater than the increased volumes of The Flood Studies Report, Table 3.6, but the extra storage is needed to control runoff rate. In each case the stored volume represents about 40% of the total storm runoff.

In all options, the deep interzone drains had to be considered. Option I was set aside since the holding pond could not be accommodated. For the last two options, pipes of 1200 mm diameter would be required but the deep carrier pipes from zones 2 and 3 should have a 1275 mm diameter. For the ponds in zones 1 and 4, emergency overflows should be provided. The capacity of the large carrier pipes would also take any overflow from the external zone 6 which would have a time lag on the zoo peak discharges.

Table 3.8 — Post-development peak runoff rates

Storm (min)	Peak runoff rate ($m^3 s^{-1}$)			
	Zone 1	Zone 2	Zone 3	Zone 4
60	2.16	2.79	3.97	2.74
120	1.39	1.96	3.73	1.80
180	1.18	1.67	2.33	1.52

Table 3.9 — Pipe data

	Zone 1	Zone 2	Zone 3	Zone 4
Pipe length (mm)	480	650	800	680
Slope	1/116	1/290	1/180	1/60
Diameter (mm)	900 (outfall)	1200	1275	900

3.4 DRAINAGE OF THE ISLE OF DOGS

LOCATION The Isle of Dogs is part of the docklands of east London, UK.

SOURCES Hall, M. J. & Hill, N. A. (1983) The analysis and design of sewerage systems using the Wallingford procedure. In: *Hydraulic aspects of floods and flood control*. British Hydromechanics Research Association, pp. 371–380.

Bennett, M. B., Harmond, I. G., Legg, R. A. & Lewin, J. (1987) The design and construction of the Isle of Dogs Pumping Station. *Institution of Water Environment Management, Metropolitan Branch Meeting*.

PROBLEM The 350 ha of the Isle of Dogs is contained within a large meander of the River Thames (Fig. 3.8). About 54 ha comprises the locked water bodies of the West India Docks and the Millwall Docks. With the closing of the docks in 1980, much of the associated property became derelict and a year later the London Dockland Development Corporation was given the power and resources to develop a first enterprise zone on the Isle of Dogs. As part of the planning and construction of mains services, the engineering consultants, Sir William Halcrow and Partners, were commissioned to investigate and advise on the drainage of the new development area.

Data

The existing sewer network was derived from maps from the responsible authorities, the London Borough of Tower Hamlets, the Thames Water Authority and the Port of London Authority. Any anomalies and deficiencies in information were resolved

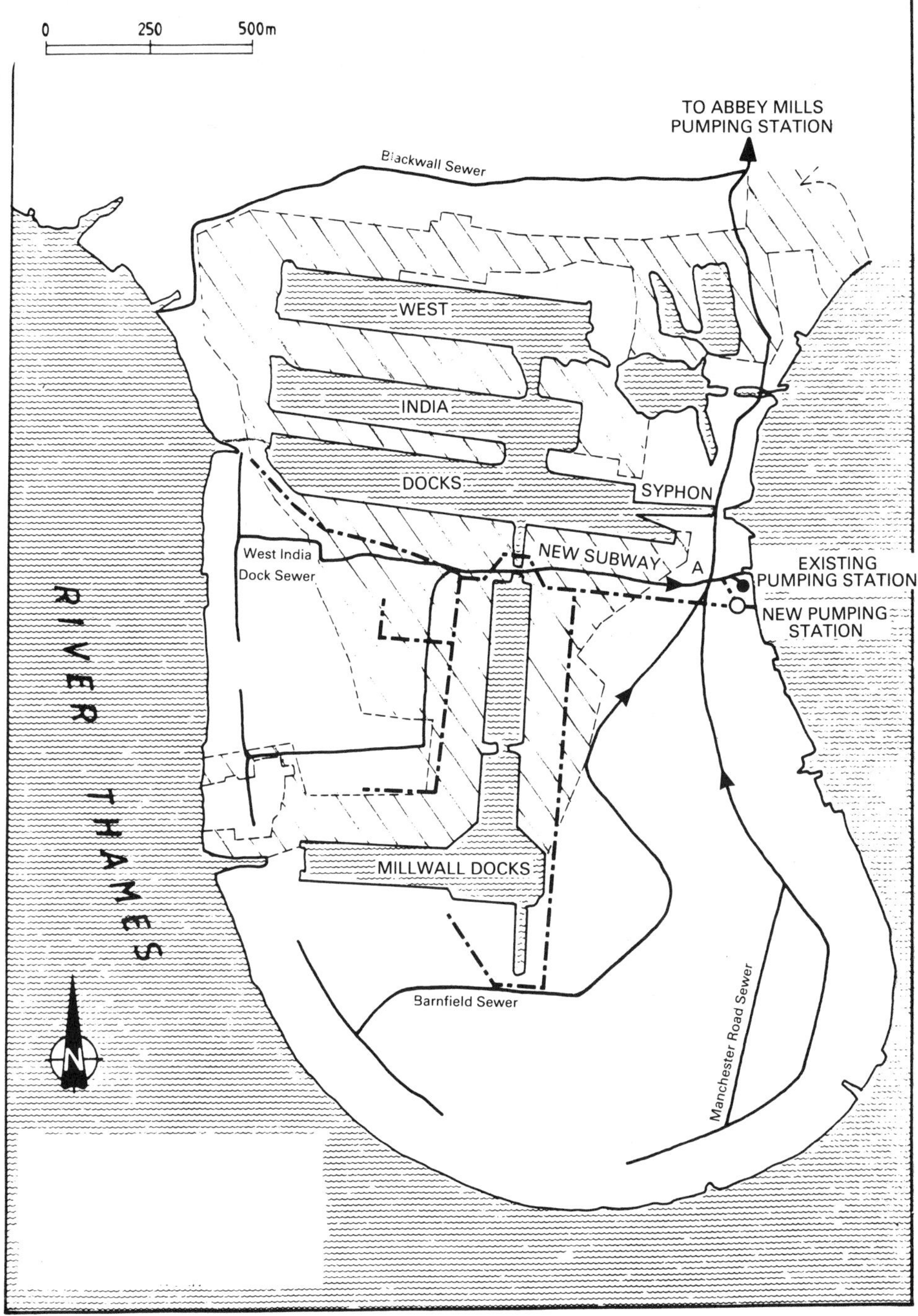

Fig. 3.8 — Main drainage of the Isle of Dogs: ———, existing combined sewer; —·—, new surface water sewer; \\\\\\ enterprise zone boundary.

by site inspections. The sewers range from the old egg-shaped sections of the early nineteenth century to modern tunnels laid in the 1960s. In this low-lying area, mostly at only 2 m AOD and below mean high tides, all sewers have very low gradients. Subcatchment areas to each sewer were estimated from the OS 1:1250 scale maps and the impervious areas contributing to each pipe were determined by inspection. By sampling different classes of land use, evaluating their percentage imperviousness and determining each class area within a subcatchment, an overall estimate of the impervious area for the subcatchment was obtained.

From a study of current and possible future populations of the area, a dry weather flow of 340 l $head^{-1}$ day^{-1} was adopted, increased by a factor of 2.3 for planning purposes to allow for commercial and industrial usage and infiltration.

The layout of the present main combined sewers is seen in Fig. 3.8. Most of the discharge drains to the Thames Water Authority's Abbey Mills Pumping Station and thence by trunk sewers to the Beckton Sewage Treatment Works. The sewers south of the West India Docks join together and pass under the main entrance to the docks by means of an inverted syphon. A connection to the existing pumping station allows excess sewage to be discharged directly to the River Thames when the syphon capacity is exceeded. North of the syphon, the sewer is joined by the Blackwall Sewer, which serves the area north of the West India Docks.

Network model

In view of the susceptibility of the combined sewers to surcharge during periods of heavy rainfall, the adoption of the methods of the Wallingford Procedure (WASSP) (NWC–DOE, 1981) was most appropriate. However, the relatively large area, high impermeability and large very-low-gradient sewers proved to be a severe test for one of the first commercial applications of WASSP. A number of modifications were made to the WASSP suite of programs to enable them to cope with the size and problems of the network and exhaustive tests were carried out on the existing sewer network model. No unexplained differences were found in results compared with those from the Rational and Transparent and Road Research Laboratory methods and from information such as flooding and surcharge records. The WASSP techniques were considered satisfactory without further calibration using site rainfall and flow measurements for which time was not available. Sensitivity tests using the simulation program were carried out to check on the required assumptions and over a range of storms up to a 1 in 20 year return period, a 5 min time of entry was found appropriate with the pipe friction and other variables to be used as recommended in the WASSP manual.

Of major importance was arrangement for the satisfactory drainage of the enterprise zone (Fig. 3.8) and effects on low-lying areas outside. There was considerable uncertainty about unmapped sewers from the old riverside warehousing and abandoned industrial sites at the heads of the main sewer runs. From exhaustive model runs under different conditions it was concluded that the development area could not be drained satisfactorily by the existing sewers. It was decided to provide a new surface water sewer network for the enterprise zone with pipes up to 1500 mm diameter draining to a new pumping station and thence to the river. Further foul sewerage connected to the existing trunk system was also recommended. By removing the dockland surface water contribution mainly from the West India Dock

Sewer System, the risk of flooding from the existing network would be considerably reduced. Fig. 3.9 shows that the reduction in peak flow is about 16% and the reduction in runoff volume about 10%.

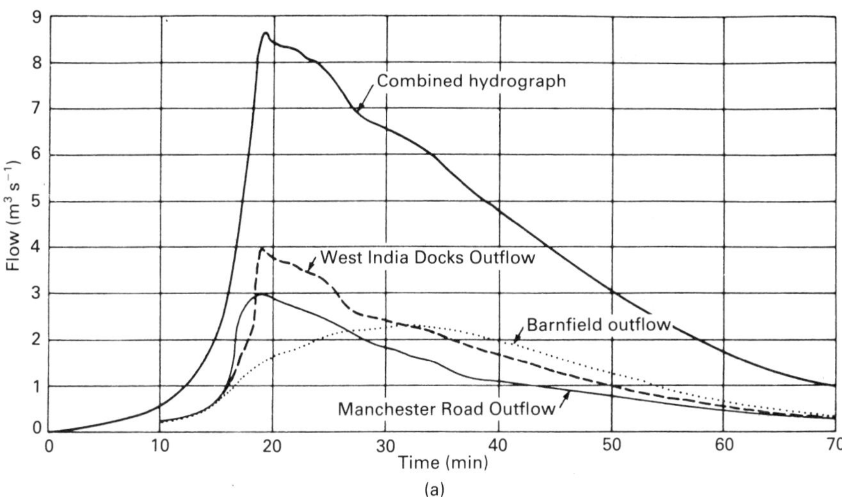

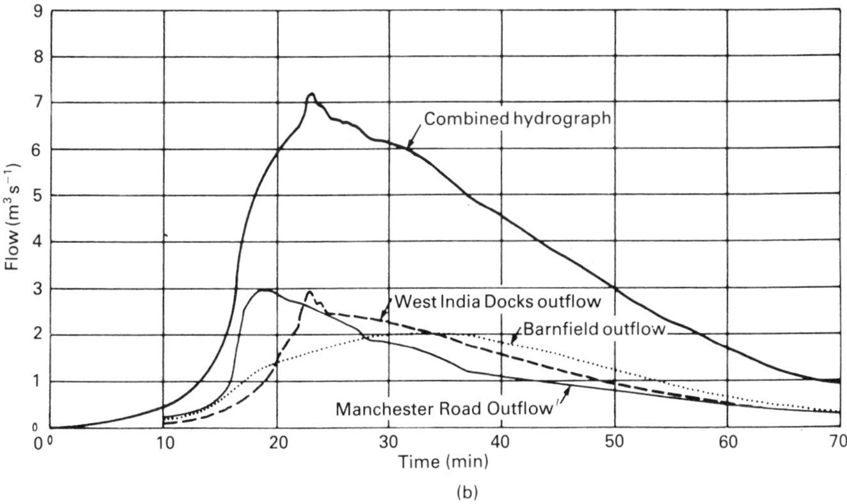

Fig. 3.9 — Runoff hydrograph at junction A for 5 year, 30 min storm: (a) with existing docklands; (b) without most of docklands surface water.

New surface water sewer design

A number of different schemes were prepared in conjunction with the development plans of the London Dockland Development Corporation. In designing the new sewers, low-flow behaviour was studied in addition to the provision for flood flow rates. A gravity system using a disused dock as a retention basin to be discharged into the river at low tides was considered. Composite schemes using two or more pumping stations to several outlets were more elaborate but each plan was assessed for suitability, flexibility and cost. The schemes were initially planned using the WASSP Rational method requiring minimal data and computer time, with the hydrograph and simulation methods needing more detailed data used to check and refine the final schemes.

For the final design shown in Fig. 3.8, the WASSP Optimising method was applied to assess the relative merits of large low-gradient pipes and smaller steeper pipes. The final scheme with pipes up to 1500 mm diameter included the construction of a subway under the waterway connection between the West India Docks and Millwall Docks and a new pumping station to replace the existing Isle of Dogs Pumping Station on a nearby site.

Conclusion

In the application of the Wallingford Procedure to the drainage problem of the Isle of Dogs, the simulation of surcharged pipes allowed the designers of the new network to keep surface flooding to acceptable levels. With the peakiness of the hydrographs, the duration of flooding would be short and the volume of surface flooding would be small since the overflowing pipes tended to be in the upper areas of the system.

The final plan to convey the surface water from the western part of the enterprise zone to the new pumping station incorporated the 1650 mm West India Dock Sewer, twin 300 mm water mains and a 450 mm gas main in a deep services subway under the dock crossing. The surface water sewer was then joined by new eastern draining surface water sewers draining the eastern areas and entering into tunnels to pass under the other deep combined sewers and other obstructions.

The design requirements of the new pumping station were determined as $5\ m^3\ s^{-1}$ from a 10 year storm for the enterprise zone surface water drainage and $4.7\ m^3\ s^{-1}$ from the combined sewer overflows, the design capacity of the existing pumping station. The nominal capacity of the new pumping station was fixed at $12\ m^3\ s^{-1}$ including standby provision, to be provided by 12 large submersible pumps.

3.5 STORMWATER DRAINAGE FOR JEDDAH, SAUDI ARABIA

LOCATION The port of Jeddah is centrally located along the Red Sea coastal strip of Saudi Arabia.

SOURCES Barnsley, R., Price, R. K. & Packman, J. C. (1984) Application of the Wallingford Procedure to the design of urban stormwater drainage for Jeddah, Saudi Arabia. *International Conference on Planning, Construction, Maintenance and Operation of Sewerage Systems*. British Hydromechanics Research Association, Paper F3, pp. 265–277.

Kattan-Gibb and Hydraulics Research Station (1982) *Study and design of stormwater drainage for Jeddah southern area*. Interim Study Report to Jeddah Municipality.

PROBLEM The project area consists of 42 km^2 of the southern sector of the City of Jeddah (Fig. 3.10). While about 80% is at present developed for a

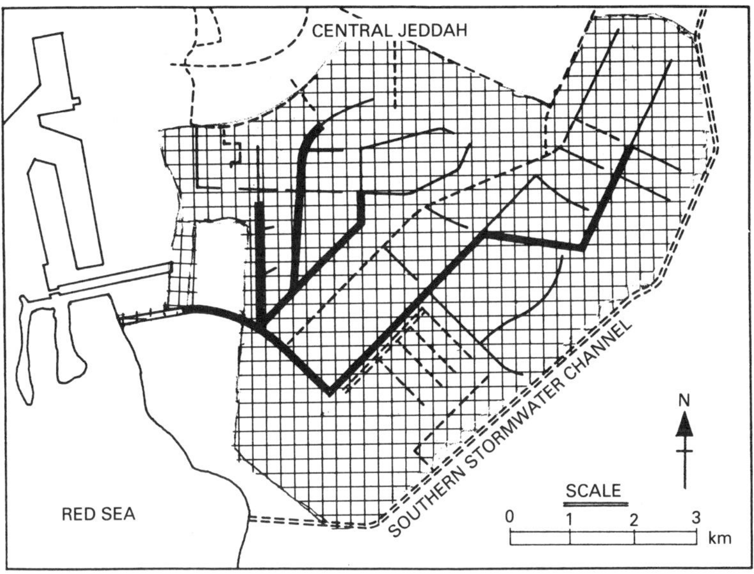

Fig. 3.10 — Jeddah project area and proposed stormwater drainage system; ▭, proposed open channel; ▬, proposed culvert sewer; ——, proposed pipe sewer; ----, existing stormwater sewers; ⊞, project area.

wide variety of uses, with rapid industrial growth, full development is expected in the near future. In the final plan, 46% will be devoted to residential and commercial usage.

Although the rainfall is low (average, 61 mm year^{-1}), it occurs during the winter in short thunderstorms. In some years, 70–80 mm may fall in a few hours. In order to design an acceptable stormwater drainage system, the Hydraulics Research Station and Institute of Hydrology were asked to advise on the applicability for the project of the Wallingford Procedure (NWC–DOE, 1981) known as WASSP.

Investigations

The extent of the project area is shown in Fig. 3.10. The southern boundary is a stormwater channel designed to intercept several natural wadis and to divert their flows from the development area. Initially a single stormwater sewer 12 km long drained a narrow strip of land across the centre of the area and two small areas had also been catered for.

The investigations were carried out in two phases. First, a study phase in which the Modified Rational method in WASSP was adjusted to deal with Jeddah conditions and operated to produce several possible drainage schemes. Second, a detailed design phase when the Simulation method in WASSP was used with a range of design storm profiles to derive the least cost scheme for the desired level of protection.

The first main problem was the derivation of rainfall intensity–duration–frequency data for the project area. The records of three autographic rain gauges within 130 km of Jeddah and representative of the same rainfall regime were used. After some consideration, it was decided that the records were independent and could be combined to give a 20 station year record. The annual exceedence series for selected durations were abstracted and intensity–duration–frequency curves compiled graphically. The results did not follow the trends observed elsewhere (Bell, 1969), probably owing to the arid climate of Jeddah. A scaling factor was derived to convert the daily estimates to those obtained from the Jeddah daily record and this was applied to shorter durations. Similarly adjustments to the frequency curves were made (Fig. 3.11). In experimenting with possible drainage schemes, the 2 year return

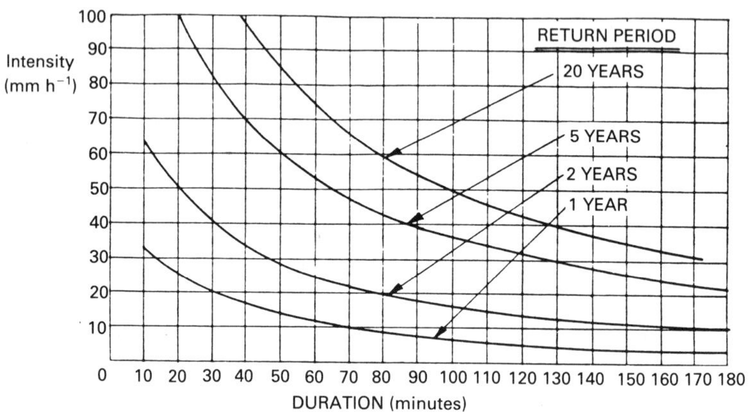

Fig. 3.11 — Jeddah rainfall intensity–duration–frequency.

period curve was used to provide a set of comparable feasible drainage nets. In the application of the modified Rational method to each drainage reach, the peak discharge

$$Q_p = 2.78 C_v C_r i A \qquad (1\,\mathrm{s}^{-1})$$

with $C_r = 1.3$, i (mm h^{-1}) the rainfall intensity and A (ha) the catchment area. Values for C_v, the volumetric runoff coefficient, needed careful consideration. For fast-response areas such as road surfaces, C_v was 0.9 but for slow-response areas values ranged from 0.25 to 0.55, i.e. permeable areas to high-density housing and commercial areas. A long time of entry of 20 min was allowed on account of the low gradients and large areas draining to each sewer. When all the necessary 15 items of information were fed into the computer, the program took only 15 s of central processing time to design a 29 branch drainage system with 130 reaches. The output contained the following information for each reach

(1) Longitudinal gradient of the drain.
(2) Effective catchment area AC_v.
(3) Time of concentration.
(4) Peak discharge.
(5) Pipe diameter or culvert dimensions.
(6) Pipe full capacity.
(7) Pipe or culvert capacity.
(8) Warnings of low velocity or surcharging.

When the cheapest scheme had been identified in relationship to other existing services, it was redesigned using the WASSP Simulation method with the 1 year return period rainfall intensity–duration curve and the 2 year curve truncated so that higher intensities were reduced to 20 mm h^{-1}.

In applying the Simulation method, 1 and 4 mm depression storage for fast- and slow-response areas were used for areas with gradients less that 2%. Surface runoff was modelled using a quasi linear reservoir, with delay dependent on rainfall intensity. For fast response areas, the delay time K was defined by

$$K = 0.393 S^{-0.264} a^{0.333} I^{-0.39} \qquad \text{(min)}$$

where S (%) is the slope, a (m^2) the area and I (mm min^{-1}) maximum average rainfall intensity over 10 min.

For slow-response areas, a 20 min delay time was adopted. Additional items of data required for the Simulation method included storm profiles, surface slope index, number of manholes and particulars of flooded areas. The output from the Simulation program contained the discharge in the drain, the incremental and cumulative discharges from the drain onto the surface and flood levels for each time increment. In addition, the maximum flood depth was calculated for each reach, assuming uniform flooding along the reach over a width of 100 m. This phase was much more expensive in computer time, the simulation of the same drainage system with a 1 min time increment took 10 min of central processing time.

Conclusions

In applying the Wallingford Procedure, the hydrological considerations are closely linked with the drainage system design process. The values of C_v are the most important factors in the procedure and the lack of rainfall-runoff data for the area meant that calculated estimates had to be made. In a region with irregular rainfall provided occasionally by intense storms, considerable judgement must be made in assessing the protection from floods to be afforded by drainage systems. In this example, the consultants recommended the scheme be designed to the 2 year truncated intensity–duration curve with the following resultant final overall catchment particulars: catchment area, 30.51 km^2; mean volume runoff coefficient, 0.42; time of concentration near outfall, 100 min; design discharge at outfall, 70 $m^3\,s^{-1}$.

3.6 SWINDON URBAN DEVELOPMENT

LOCATION Swindon is a prosperous town in north Wiltshire, England.

SOURCE Hydraulics Research (1986) *River Ray catchment study, Swindon*, Summary 143. Hydraulics Research Ltd, Wallingford.

PROBLEM Major urban extensions to Swindon by the year 2000 are planned within the catchment area of the River Ray, a tributary of the Thames (Fig. 3.12). Thames Water wished to determine the development

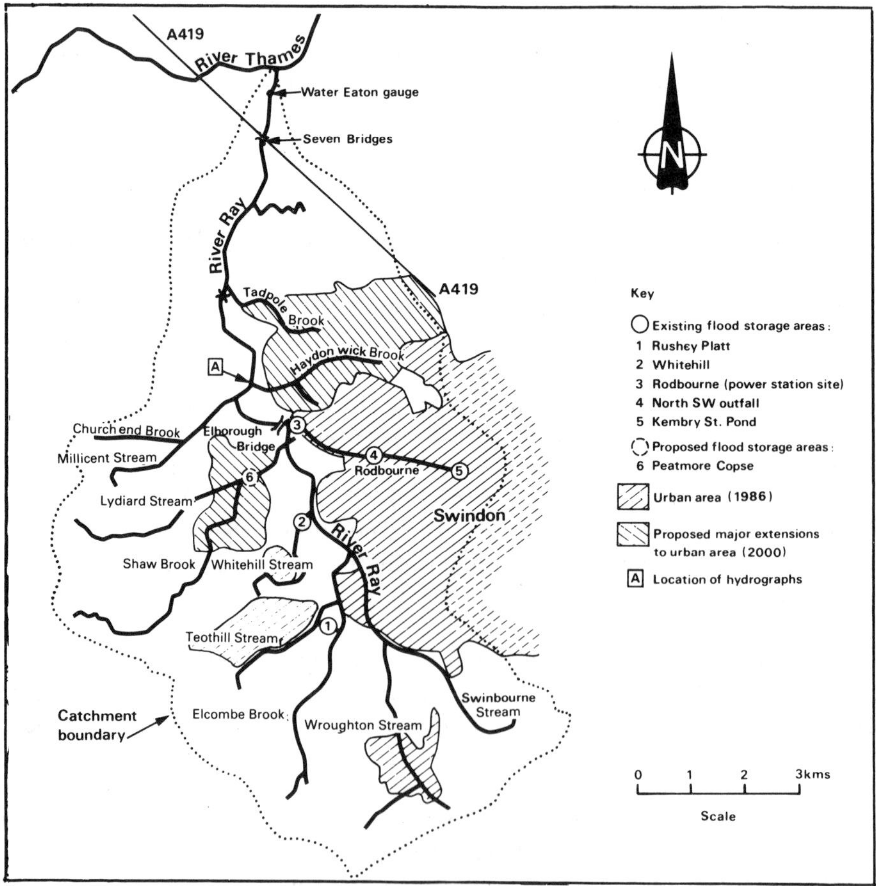

Fig. 3.12 — Swindon urban development.

effects on the catchment drainage and on the river flood levels. Hydraulics Research Ltd with Howard Humphreys, consulting engineers, and Silsoe Agricultural College were commissioned to develop a catchment model for the following.

(1) To assess existing drainage of the Ray catchment.
(2) To assess drainage and flood levels after the urban developments.
(3) To appraise agricultural land flooding and to propose appropriate alleviation measures.

A combination of four computational models was used to study the problem.

Ray catchment

The River Ray rises in the chalk of the Marlborough Downs and flows northwards over sandstones and clays to join the Upper Thames, forming a total drainage area of 80 km^2. Of the many small subcatchments, five already have flood storage ponds designed to a variety of protection standards according to location but, downstream of the present confines of Swindon, the agricultural land is subject to inundations across the flood plain. In devising the further control measures required after yet more urbanization, it was proposed to keep peaks low to inhibit flooding of arable land at the expense perhaps of prolonging inundations of less vulnerable riverside pastureland.

There are two main areas of development, to the west of Swindon draining into Lydiard and Shaw streams, which is already under construction and to the north, draining into Haydon Wick and Tadpole Brooks. These will add 5.5 km^2 to the urban area of the Ray catchment and increase the population by 20%.

Computer models

A preliminary design for the pipe network for the northern area based on the developer's plans was prepared and analysed using the Wallingford storm sewer package of WASSP models developed at Hydraulics Research Ltd. (NWC–DOE, 1981). The changes in the runoff characteristics of the subcatchments following urbanization were assessed. Hydrological parameters from the WASSP models were fed into the comprehensive catchment model RIBAMAN, for river basin management (Price, 1980). The routines of this model simulate runoff, unsteady river flood routing and flood storage in both on-line and off-line storage ponds. The model acts in turn on each of the subcatchments and main river reaches and produces flow hydrographs for selected points on the drainage system. A steady state model FLUCOMP constructed to predict maximum flood levels from the peak flows generated by RIBAMAN was applied to the lower reaches of the Ray taking into account the detailed channel and flood plain geometry and particulars of bridges, weirs and sluices. The output is in the form of flood levels at each cross-section which are plotted on plans to give outlines of flood limits. An agricultural flood damage model was used to estimate likely benefits arising from reduction in flood risk.

RIBAMAN was calibrated using autographic rainfall records and the Water Eaton river gauging station records for three out-of-bank floods and four further major in-bank flows.

Model applications

Design storms with return periods of 2, 5, 10 and 25 years following The Flood Studies Report methods (NERC, 1975) were run in RIBAMAN and FLUCOMP to establish the present flood risk and potential flood damage to the agricultural land. To establish the expected changes in catchment response due to the urban expansions, the models were modified to investigate the two stages.

(1) The western subcatchments.
(2) The western and northern developments together.

An additional 8% urban increase was assumed in other partly urbanized catchments and the 20% increase in population which will affect sewage works discharges. The resultant predicted flood peak discharges for the whole catchment are given in Table 3.10. The interesting feature of these results is the decrease in

Table 3.10 — Predicted flood peak discharges at Water Eaton

Return period years	2	5	10	25
Design storm (mm)	30.9	36.8	46.7	55.9
Flood peak discharge present catchment ($m^3 s^{-1}$)	13.74	16.07	20.06	23.79
Flood peak discharge after western development ($m^3 s^{-1}$)	15.55	18.22	22.62	26.00
Flood peak discharge after total development ($m^3 s^{-1}$)	14.39	16.82	20.70	23.53
Flood peak discharge after total development and increasing pond storage ($m^3 s^{-1}$)	13.81	16.10	19.23	23.17

flood peaks from the western development situation alone to the total development situation. The development of the Lydiard–Shaw reduces times to peak to match those of the main stream and so the peaks are enhanced. The rapid response of the Haydon Wick catchment after development produces an earlier peak and thus at the confluence there is a double peak in the Ray with the later peak from upstream being attenuated. In the lowland area, the effect of the total urban development is to increase the peaks slightly but more seriously the duration of high flows is markedly increased by about 3 h for the 5 year flood. By increasing the storage capacity of the upstream storage ponds, the peaks were similar but again the flood durations were prolonged.

The application of the agricultural flood damage model showed that benefits to be gained by further reducing flood risk were small and could not justify new storage ponds.

Conclusion

This computer study demonstrated that the regulation of surface water floods could be assessed for different catchment states and storm events to minimize flooding and damage to agricultural land. The package of models will provide the clients with a valuable tool for investigating alternative flood mitigation strategies in the event of further urban development.

3.7 BRIGHTLINGSEA SEWERAGE STUDY

LOCATION This small port is 12 km southwest of Colchester on the River Colne estuary in East Anglia, UK.

SOURCE Beale, D. C. & Leonard, O. J. (1987) Brightlingsea WASSP study. *National Hydrology Symposium*, British Hydrological Society, Hull, Paper 17.

PROBLEM The original combined sewerage system draining to the West Dyke was constructed in the last century and up to the 1960s the town expansions were connected into the existing system. Then a storm water sewer was laid through the centre of the town and subsequent new developments were sewered on a separate system. However, the two systems are linked by overflows and, during intense storms, flooding occurs at several locations from the overloaded old sewers in the lower part of the town (Fig. 3.13).

An initial investigation of the sewerage system was carried out using the traditional Transport and Road Research Laboratory method but before committing the expenditure, the Anglian Water Authority asked Howard Humphreys, Consulting Civil Engineers, to reassess the situation using the most recently developed methods.

Sewerage structure

The plan of the sewerage system draining 500 ha is shown in Fig. 3.13 with indications of its present unsatisfactory performance. The *Sewerage rehabilitation manual* (WAA–WRc, 1983) recommends an ordered appraisal of sewerage problems as follows

(1) Upgrading of sewer records.
(2) Identification of critical sewers.
(3) Assessment of structural condition.
(4) Assessment of hydraulic performance.
(5) Assessment of overall performance.

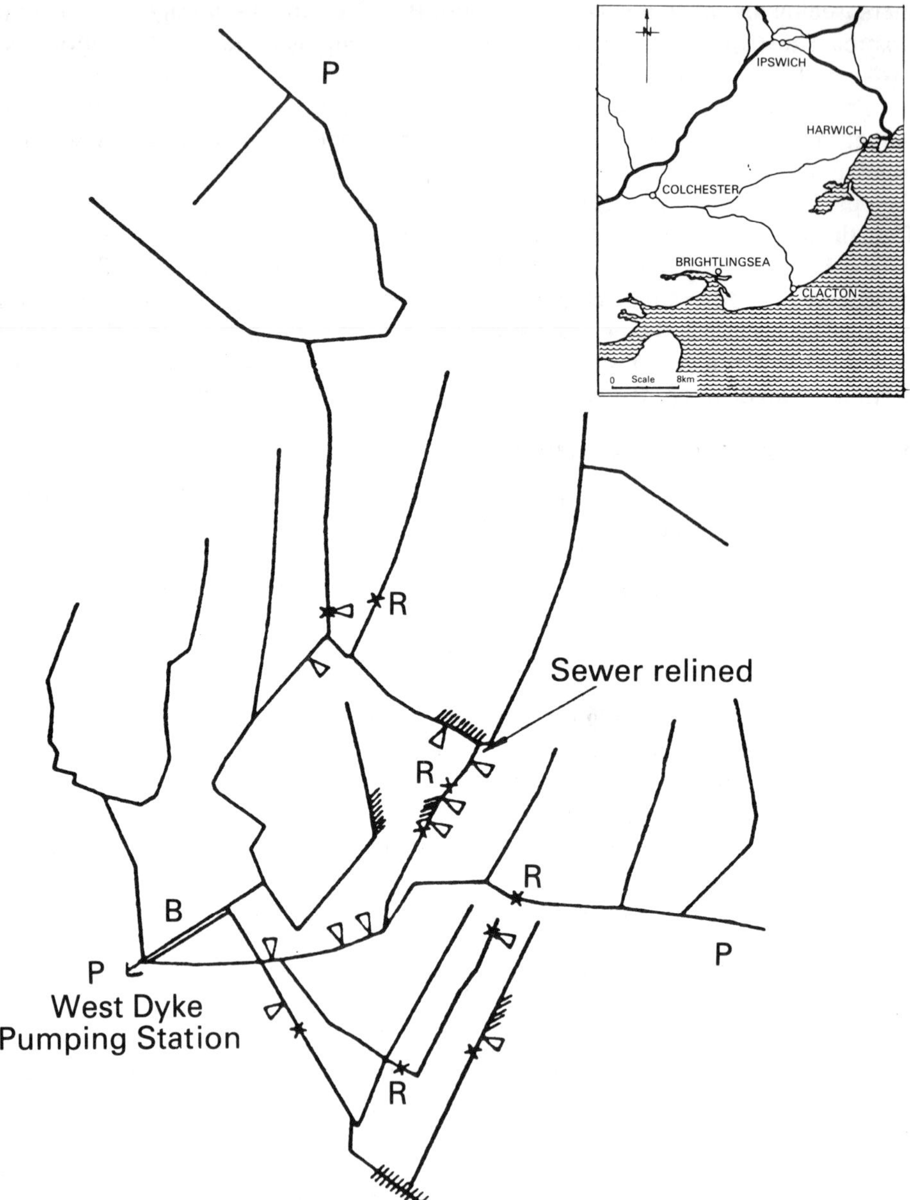

Fig. 3.13 — Brightlingsea sewerage network: /////, observed flooding; ◁, major defects; R, repaired sewers; ×, collapsed; OP, rain gauges.

The Brightlingsea sewer records together with the survey of the system for the first study were checked and considered adequate for the new investigation.

Structural assessment was carried out using closed circuit television for the trunk sewers. Then for the hydraulic assessment the consultants adopted the Wallingford Procedure (NWC–DOE, 1981), a suite of computer programs for the design and analysis of urban storm drainage networks.

Wallingford Procedure (WASSP)

The package contains the generation of specific rainfall profiles for any area of the UK, the modelling of overland flow from paved and unpaved areas and an improved method of pipe flow analysis including the modelling of surcharged flows. Ancillary works within the sewer system can also be modelled.

In Brightlingsea, the combined and surface water systems were modelled together since surcharging and overflowing of the surface water system drained to the combined sewers. Within the catchment area contributing areas to either of the systems were identified. In the upper part of the catchment the sandy subsoil was found to have different percentage runoff from the alluvial clays of the lower part of the town. The model was verified using data from three rain gauges on the boundaries of the study area (Fig. 3.13) and from flow monitors at strategic locations in the sewers, nine in the combined system and three in the surface water system. Unfortunately, only one good storm occurred during the study period. A sample of the observed and simulated hydrographs for this storm at point B is shown in Fig. 3.14. The proving of the model using the flow measurements was helped by

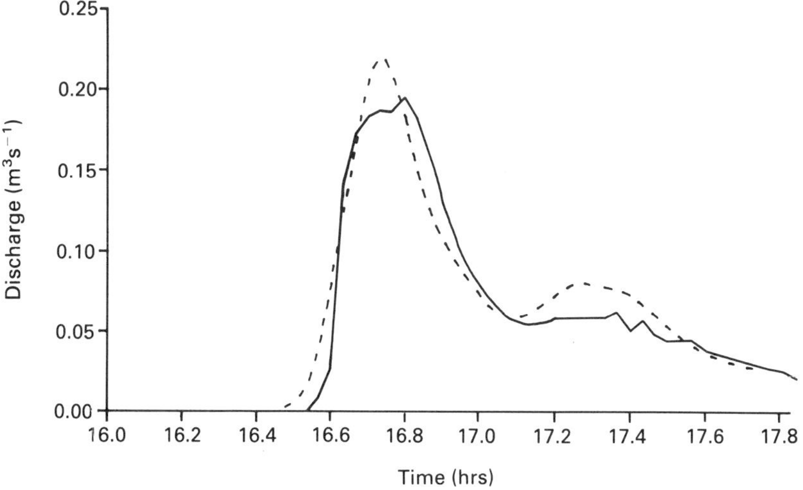

Fig. 3.14 — Discharge hydrographs at gauge B for 5 December 1986 storm: ———, observed, ----, simulated.

subdividing the model into six submodels. Five of these represented the upstream parts of the catchment where flooding was frequently reported and these caused few problems of validation. The lower submodel contained the area with numerous storm overflows and proved to be more difficult to verify.

The hydraulic performance of the system was assessed by running the verified WASSP model with different simulated storm rainfalls. First, the runs of the 1 year return period storms of durations 15, 30, 60 and 90 min produced the most severe flooding from the 30 min storm. Then taking a fixed 30 min duration, the storms of varying return periods 2, 5, 10, 20 and 30 years were simulated in the model. The frequency of storm occurrences to give the onset of surcharge and then the onset of flooding was designated for each critical pipe length. It is a property of the model procedure that the storm of a given return period will produce the flood flow of equal frequency. The overall results showed that the system extensively surcharged with a 1 year storm but with limited flooding.

Conclusions

Taking into account the hydraulic performance of the system and the structural condition of the sewers, a rehabilitation scheme was recommended. This was considerably less ambitious than the previous proposals with an estimated cost of 0.4 times the original. The saving is principally due to the differing strategy of the methods. The traditional Transport and Road Research Laboratory model only identifies inadequate pipe full capacities and immediately recommends a larger pipe size whereas WASSP allows the surcharge capacity of the pipes to be considered, assesses flood flows and simulates storm overflows.

4

Transport drainage

INTRODUCTION

The means of transport for people and commodities have increasingly become the concern of the civil engineer particularly since the Industrial Revolution. For the water engineer, the age of the canal posed many problems of water supply and control especially across undulating terrain. However, for land-based vehicles, the engineer has to be concerned with drainage for the maintenance of the routeways and the assurance of freedom of passage.

There are two major aspects of transport drainage. The most important is the safety of the structure of the routeway whether it is road or rail. Considerable care is afforded to the planning of the route, taking note of suitable gradients and minimizing the need for costly bridges and embankments. This aspect of routeway design was well appreciated by the Romans in the construction of their famous roads needing solid foundations for their heavy chariots, although minimizing distances seemed to be their prime concern.

The lines to be followed by railways were carefully planned, moderate gradients being essential and since they tended to follow river valleys absolute levels were important. Particular attention was paid to ensure that the railways were sited above flood levels and it is only the severest of flood events that today overtops a railway. Only now, after many centuries of random road development, with the advent of the internal combustion engine and the escalation of motorized transport, is the civil engineer concerned with the planning of new modern roads. For both roads and railways, the capacity of culverts and bridges over watercourses is of prime concern and it is in this aspect of ensuring structural safety that the calculations of the hydrologist are needed.

The second aspect of transport drainage stems from the direct onset of flood water from heavy rainfall and the nature of the routeway surface. Railway lines are laid on firmly compacted but freely draining foundations but heavy rainstorms may soon make a road impassable if provision has not been made for the adequate drainage of the tarmac or concrete surface. Accurately designed roadway gradients and efficient gullies must be built into every motorway to ensure that high rainfall intensities do not inhibit safe vehicular passage. Experiments with different surfaces have been made to reduce the dangers of splash and spray but the key to the problem is careful surface laying to encourage rapid runoff and to avoid pondage.

The drainage problems of impervious surfaces associated with modes of transport are extended to car-parks now covering large areas but more importantly to the vast tarmac expanses of airports. The runways used for take-off and landing of aircraft need to be kept clear of standing water even from the heaviest rainstorms. In addition, the perimeter tracks and standing bays cover considerable areas and together with the runways would provide total immediate runoff required to be accommodated in gullies and drains.

The vital hydrometric data necessary for the determination of design criteria are long records of short-duration rainfalls. Autographic charts from rainfall recorders form the basis of these records but intensity measurements are now being collected on solid state event recorders. For the direct runoff from impervious surfaces, the Rational method for peak discharges, the Transport and Road Research Laboratory method and more recently the Wallingford Procedure incorporating the Rational and Hydrograph methods may be used in the design of drainage systems (NWC–DOE, 1981). Stream and river flows to be accommodated by culverts and bridges crossing the routeway may be derived by any of the accepted rainfall-runoff methods depending on available data. More often, probabilities of occurrence and return periods of events are estimated from the short-duration rainfall records and runoff derived by the unit hydrograph method or conceptual catchment models. The design capacities are usually determined by the importance of the routeway and the resources of the client.

The limited examples demonstrate the hydrological studies undertaken for the design of new roads in underdeveloped countries and the reappraisal of the flood situations along an established railroad in Australia.

4.1 TIMOR ROAD SCHEME

LOCATION The Province of Timor administered by Portugal is the eastern part of the island of Timor in the Indonesian archipelago.

SOURCES Hall, M. J. (1974) *Rainfall depth–duration–frequency relationships for the Province of Timor.* Report to Planning & Transport Research & Computation Co. Ltd.

Hall, M. J. (1975) Rainfall depth–duration–frequency relationships for drainage design in developing countries. *Proceedings of a Seminar on Highway Design in Developing Countries, Warwick.* Planning & Transport Research & Computation Co. Ltd.

PROBLEM The Timor road scheme involved the design and construction of three roads in the Province of Timor (Fig. 4.1). The roads were to connect the main town of Dili with Baucau, Same and Maliana. The

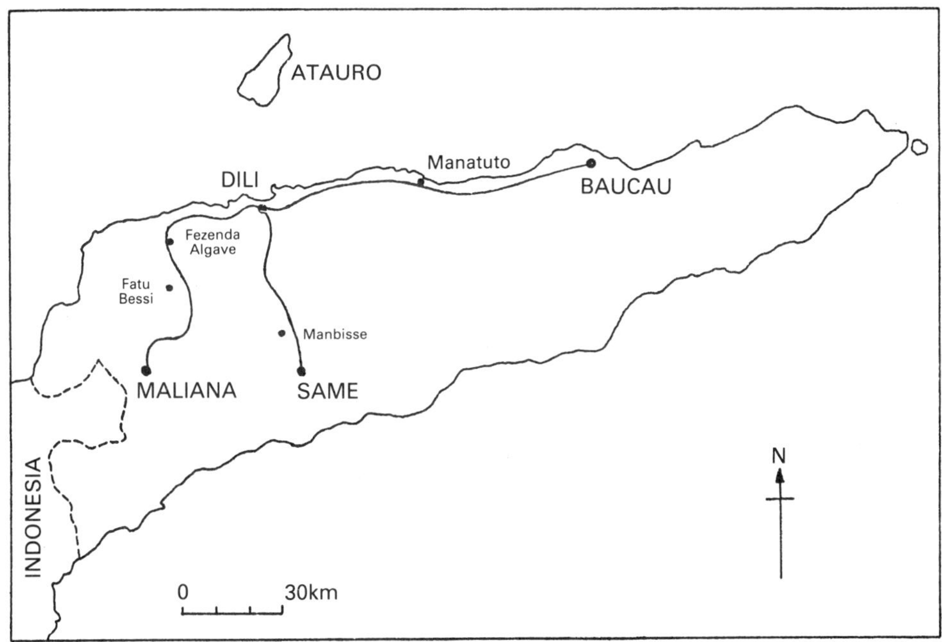

Fig. 4.1 — Province for Timor, showing road routes and selected rain gauge stations.

hydrometric information available for drainage design was minimal and a first study tackled the problem of defining rainfall depth–duration–frequency relationships fundamental to drainage design procedures.

Investigations

The island of Timor at latitude 9°S lies at the southern fringe of the equatorial climatic region. The topography is extremely rugged; although the island is less than 80 km in width, the mountain backbone rises to 2920 m above sea level. The heavy rainfalls result in rapid runoff and the rivers with high bed slopes carve deep valleys and deposit extensive alluvial fans along the coastal plain. There were 32 rain gauge stations providing daily rainfall measurements and of these 30 had more than 10 years of records at the time of the investigations. Pertinent information from a selection of the stations located near the planned roads is shown in Table 4.1. The 18 years of annual maximum observer day rainfalls at Dili were considered sufficient to use the Gumbel EV1 frequency distribution to make estimates of daily falls for a range of recurrence intervals. The frequency distribution fitted by the method of moments together with the 95% confidence limits are shown in Fig. 4.2. The annual values have been plotted using the simple relationship $T=(n+1)/m$ where T is the recurrence interval, m is the rank of the rainfall values in descending order of magnitude in the sample of n years. The high variability of the annual maximum observer day rainfalls is also indicated at the other stations by the high standard deviations. A closer study of the Dili record comparing statistics derived for different

Table 4.1 — Selected rainfall information in Timor Province

Stations	Altitude (m)	Years of record	Annual Maximum observer day rainfall (mm)		Annual total thunderstorms	
			Mean	Standard deviation	Mean	Standard deviation
Dili	4	18	75.7	28.9	68	33
Manatuto	4	21	54.2	33.8	4	8
Baucau	527	22	93.5	28.5	47	29
Fazenda Algarve	916	23	102.0	34.9	47	19
Fatie Bassi	1125	21	118.2	25.6	84	29
Maliana	278	20	113.8	43.9	46	26
Maubisse	1432	18	71.8	24.6	11	11
Same	544	16	160.6	61.7	32	15

periods, resulted in Table 4.2. Differences of over 8% are seen in the values derived from:

$$X_T = X_{\text{bar}} + KS \qquad (4.1)$$

where X_{bar} and S are the mean and standard deviation, T is the recurrence interval and K is the EV1 frequency factor 0.72, 2.04 and 3.14 for 5, 25 and 100 years respectively.

In the design of road drainage works, rainfalls of shorter durations are required and the only information available was for the 12 year period at Dili. Owing to the sampling variability of the data it was considered inappropriate to use Dili reduction factors to derive shorter-duration rainfalls from the longer record of annual maximum observer day values.

Following this unsatisfactory consideration of the local records, a second method of determining rainfall depth–duration–frequency relationships was adopted. Estimates obtained from generalized relationships based on a large number of long rainfall records can be produced more easily and are probably more reliable than those obtained from short periods of local data. This assumes that heavy short-period rainfalls result from similar meteorological conditions such as convective cells often associated with tropical cyclones. Generalized rainfall depth–duration–frequency relationships were derived by Bell (1969) relating to an index value of rainfall P_T^t, the rainfall depth in a duration of t min with a recurrence interval of T years. Thus, knowing the 2 year, 60 min rainfall P_2^{60}, other values of P_T^t can be found from

$$P_T^t = (0.35 \ln T + 0.76)(0.54 t^{0.25} - 0.5) P_2^{60} \qquad (4.2)$$

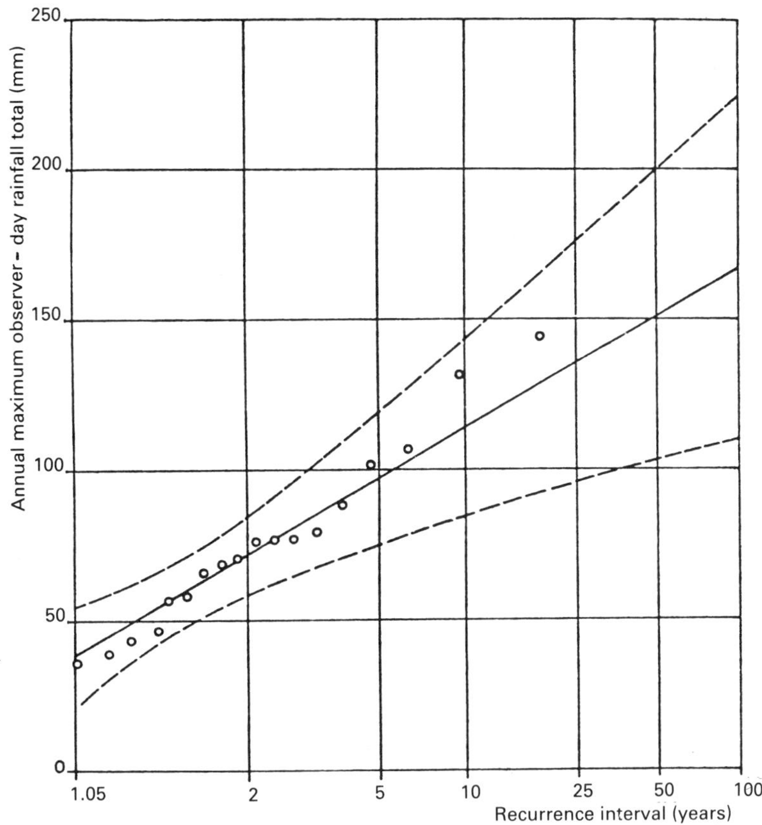

Fig. 4.2 — Distribution of annual maximum observer day rainfalls for Dili, 1955–1972, showing 95% confidence limits.

Table 4.2 — Annual maximum observer day rainfalls at Dili

Recurrence intervals (years)	5	25	100
Day rainfall, 1955–1966 (mm)	105.9	145.8	179.0
Day rainfall, 1955–1972 (mm)	96.5	134.7	166.4

for 2 years $< T <$ 100 years and 5 min $< t <$ 120 min. Where there are no short-duration data, the index P_2^{60} can be obtained from the empirical relationship

$$P_2^{60} = 0.17MN^{0.33} \qquad \text{for } 0<M<51 \text{ and } 1<N<80$$

or

$$P_2^{60} = 0.617M^{0.67}N^{0.33} \qquad \text{for } 51<M<114 \text{ and } 1<N<80 \quad (4.3)$$

where M (mm) is the mean annual maximum observer day rainfall and N is the mean annual number of thunderstorm days. Using equation (4.2), the ratios of P_T^t to P_2^{60} can be obtained for a range of durations and recurrence intervals. These reduction factors to obtain a required rainfall over a selected duration for a specific recurrence interval from the 2 year 1 h rainfall are given in Table 4.3. The P_2^{60} value obtained for

Table 4.3 — Reduction factors for 2 year, 1 h rainfall

T (years)	Reduction factor for the following durations t			
	$t = 2$ h	$t = 1$ h	$t = 30$ min	$t = 10$ min
2	1.29	1.00	0.77	0.46
5	1.70	1.33	1.01	0.61
10	2.02	1.57	1.20	0.72
25	2.43	1.89	1.44	0.87
100	3.05	2.38	1.81	1.09

Dili from equation (4.3) was 45.1 mm and when the shorter duration falls were obtained using the Table 4.3 reduction factors, the values overestimated the rainfall depths based on the Dili record. However, this is acceptable for drainage design purposes at the higher durations and lower recurrence intervals.

Conclusions

The method of synthesizing rainfall depth–duration–frequency relationships using the generalized equations was recommended since it was based on much longer records although derived from other countries. Its application to the higher altitudes along the roadway routes needed further study. Nevertheless, as a first step in calculating the 10 year, 2 h rainfall at Maliana for example, the procedure would be as follows

(1) Table 4.1: mean annual maximum observer day rainfall is 113.8 mm.
(2) Table 4.1: mean annual thunderstorm days is 46.
(3) Equation (4.3): $M = 113.8$, $N = 46$; then the 2 year, 1 h rainfall is 52.1 mm.
(4) Table 4.3: reduction factor for 10 year, 2 h rainfall is 2.02.
(5) The 10 year, 2 h rainfall at Maliana is therefore estimated to be $2.02 \times 52.1 =$ 105.2 mm.

The short duration rainfall values estimated from the mean annual maximum daily rainfalls and mean annual number of thunderstorm days provided initial guidance to the engineers charged with the task of designing adequate road drainage and selecting culvert and bridge dimensions for safe roadway construction.

4.2 RAILROAD DESIGN STORMS

LOCATION The Mount Newman to Port Hedland railroad in the north of Western Australia.

SOURCE Netchaef, P. (1983) Design storm criteria for a long railroad in a cyclone area. *Hydrology and Water Resources Symposium*, NCP 83/13. Institution of Engineers, Australia, pp. 210–214.

PROBLEM The Mount Newman to Port Hedland 426 km railroad serving the Pilbarra iron ore region of Western Australia occasionally suffers flooding from heavy cyclonic rainfall (Fig. 4.3). The mining company

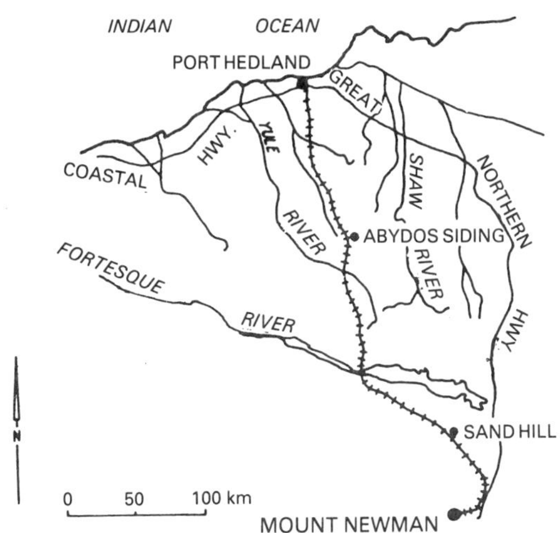

Fig. 4.3 — The Mount Newman railroad.

asked BHP Engineering, Sydney, to review the original design flood criteria adopted for the 35 bridges and 600 culverts along the line. Individual structures along the railroad were originally designed to withstand overtopping once every 10 years. As a result it had been assumed that the railroad was safe from flooding for this frequency. However, considering the length of the railroad, it was suspected that independent hydrological regimes would cause rail access to be cut by flooding more frequently than indicated by this return period. The first stage in the investigations was therefore a thorough study of the cyclones of Western Australia in order to adopt a regional approach to determining the design rainfall for the railroad as a whole.

Cyclones

The northern parts of Western Australia have been affected by over 160 tropical cyclones since records by the Bureau of Meteorology began in 1909. Most of the cyclones developed over the tropical seas north of the continent and moved in a general southwesterly direction before turning southwards to encroach on the land areas. The highest 24 h rainfall measured in the region was 747 mm but over a catchment draining to the railroad 400 mm was the maximum recorded. From earlier studies, two strong correlations have been found between rainfall and topography (with air dew point) and between total precipitation and the pressure change within a storm. By investigating incidence patterns incorporating both these relationships, a 'region of equal cyclone chance' was defined.

Three topographical zones were identified (Fig. 4.4).

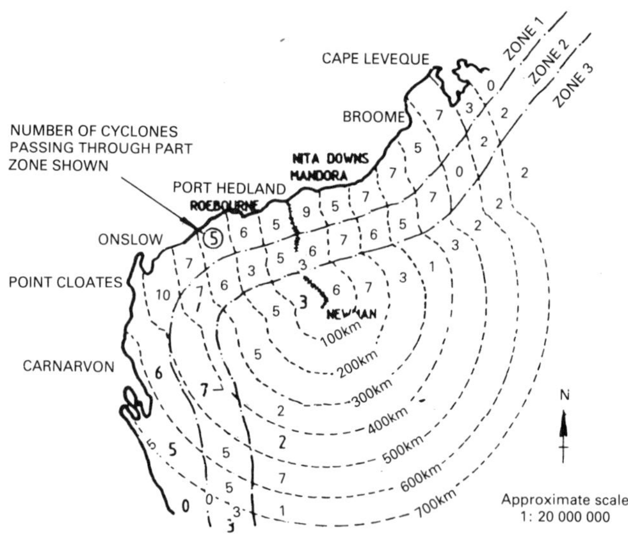

Fig. 4.4 — Area zones in the region of the railroad and the cyclone passages therein: zone 1, flat coastal zone; zone 2, elevated middle zone; zone 3, desert inland zone.

(1) A flat coastal zone.
(2) An elevated middle zone.
(3) An inland zone, mainly desert.

Equidistant lines from the railroad increasing by 100 km were drawn on the map and the number of cyclones (1909–1975) passing through the areal subdivisions were noted. No significant differences in cyclonic incidence could be found within several hundred kilometres either side of the railroad. The proportion of cyclones passing with similar random frequency in the three zones were 40% in zone 1, 34% in zone 2 and 26% in zone 3, indicating the expected tendency for cyclones to decay inland.

The rainfall distribution in the cyclones was investigated by plotting all maximum total storm rainfall isohyets for cyclones (1909–1980) on a map. These isohyets grouped into two main areas: round Cape Leveque trending to the northeast and round Port Hedland mainly trending to the southeast. The cyclones affecting the northerly area produced nearly twice the total rainfall of those near the railroad and this was clearly demonstrated on a plot of the 300 mm isohyets with much fewer occurrences round Port Hedland. Fig. 4.5 showing plots of the 200 mm isohyets

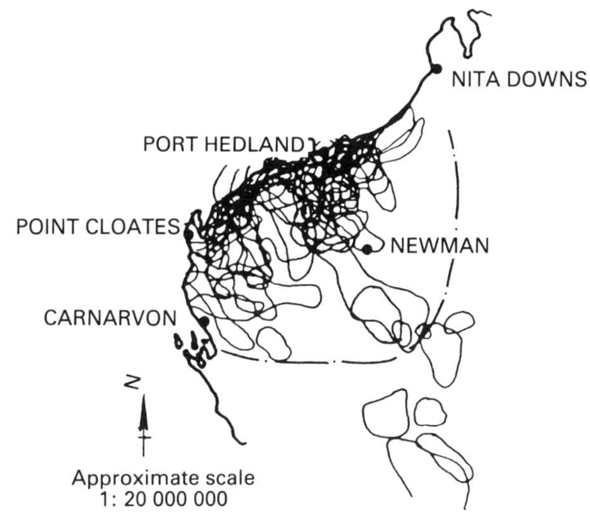

Fig. 4.5 — 200 mm isohyets and 'region of equal cyclone chance'.

indicates that 200 mm cyclones penetrate far inland. From assessing the cyclone tracks and the rainfall patterns a line was drawn defining a 'region of equal cyclone chance'.

Regional rainfall

Although a storm rainfall frequency analysis made for each of the zones showed that for given frequency the total rainfall decreases inland, it was decided to take together all the cyclones within the region of equal cyclone chance to obtain an exceedence probability plot. The lower probabilities of exceedence plotted well on a straight line which also contained the point representing the 50 year return period total storm rainfall over 72 h duration for Port Hedland. Since most rainfall measurements are made over 24 h durations, the regional frequency analysis storm totals were converted to 24 h values by applying a factor of 0.84, estimated after consideration of point rainfall ratios for differing durations at Port Hedland, Mount Newman and Abydos. The final results are shown in Fig. 4.6, where the regional total storm and

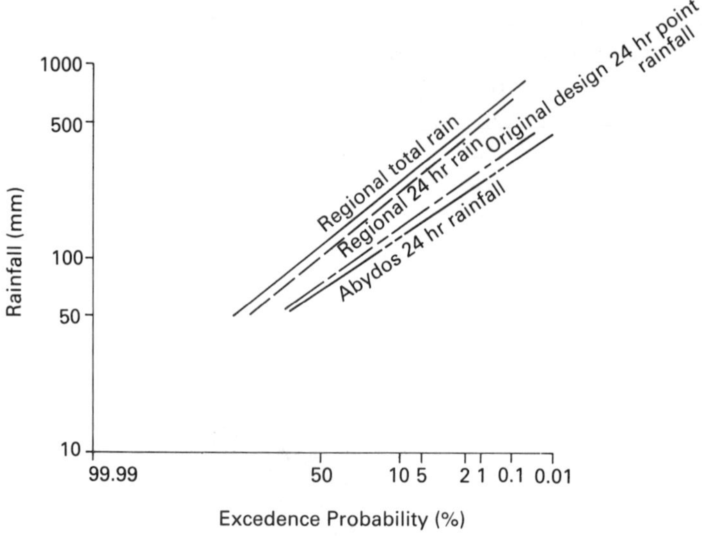

Fig. 4.6 — Comparison of rainfall design criteria.

regional 24 h rainfalls are compared with the original point rainfall design criteria used for the railroad. It can be seen that the point rainfall with return period of 1 in 50 years is equivalent to a rainfall with a return period of 1 in 10 years for the region or generally a ratio of 5 to 1. Hence it can be expected that storms affecting the railroad as a whole may be up to five times more frequent than indicated by using the point rainfall approach.

Conclusion

The regional approach to defining the storm rainfall frequencies was a great improvement on the application of point rainfall values when attempting to evaluate flood frequencies along the length of the railroad. Further studies were made to calculate the associated design floods over the numerous catchments draining to bridge and culvert sites. The rainfall temporal pattern was determined first and then linear and non-linear catchment models were used for predicting the required design floods with a return period of 1 in 50 years, the design life of the railway.

4.3 TAPPITA–TOBLI–BLAY ROAD

LOCATION The Tappita–Tobli–Blay road is an important line of communication in northeastern Liberia, west Africa.

SOURCE Howard Humphreys & Partners (1984) *Tappita–Tobli–Blay Customs Road, Liberia, hydrological studies*, Final and Interim Report to Techsult & Company Ltd.

PROBLEM The 50 km road from Tappita to Blay Customs runs cross-country near the Liberian–Ivory Coast border. It crosses one major river, the Cestos, and in making its way via tributary valleys has to pass over numerous small streams. The location of the route is shown in Fig. 4.7. Drainage channels across the line of the road are provided by 32 corrugated pipe culverts (single or double) and seven double or triple concrete box culverts. In addition, there are five bridges across the larger watercourses. Serious flooding of the road has been caused by inadequate or deformed pipe culverts or at sites where the capacity of the pipes has been reduced by accumulated sediments. A reappraisal of the drainage facilities for the road was required and Howard Humphreys & Partners were commissioned to undertake the necessary hydrological studies.

Hydrometric data

A 2 week visit to the project area enabled the problems to be identified, the hydrometric sites to be inspected and basic maps and data to be obtained.

The area lies within the equatorial climatic region with average temperatures over 25°C, high humidity and rainfall all the year round. However, of the average annual rainfall of over 5000 mm, at least 85% falls in the main wet season May–October. The vegetation is mainly secondary logged equatorial forest with restricted clearings around isolated villages.

Daily rainfall measured by a gauge at Sawolo near the Castos River (Fig. 4.7) provided 2 years of records (1982 and 1983) but, for a longer record, daily values from Sanniquellie, a gauge north of the project area were available from 1972 to 1981. However, for durations less than 24 h, data from Ghana, Ivory Coast, Benin and Sierra Leone had to be consulted.

River flow records were available from two river gauging stations. On the Gwehn Creek near Tappita, mean monthly and daily minimum and maximum water levels were obtained from August 1978 to January 1984 and daily water levels for 1982 and 1983. On the Cestos River, the gauging station provided annual maximum mean daily flows (1962–1983). Both rating equations and tables were provided for each gauging station. In addition, copies of the autographic charts at the Cestos River station for the maximum flood events in 1978, 1979 and 1983 were produced.

Rainfall analysis

The maximum daily rainfall recorded at Sanniquellie was 107 mm equivalent to 4.46 mm h^{-1} if continuous rainfall is assumed. To obtain some idea of the occurrence of rainfall intensities, a Gumbel analysis was made of the 10 years of annual maximum daily data converted by a factor of 1.13 to 24 h period data (WMO, 1973), resulting in the values in Table 4.4. However, tropical rainstorms rarely last for 24 h and rainfall intensities for short durations are absolutely necessary for assessing flood flows from small rapidly responding catchments. Autographic records from the neighbouring countries were analysed and rainfall intensity–duration relationships obtained for 5,

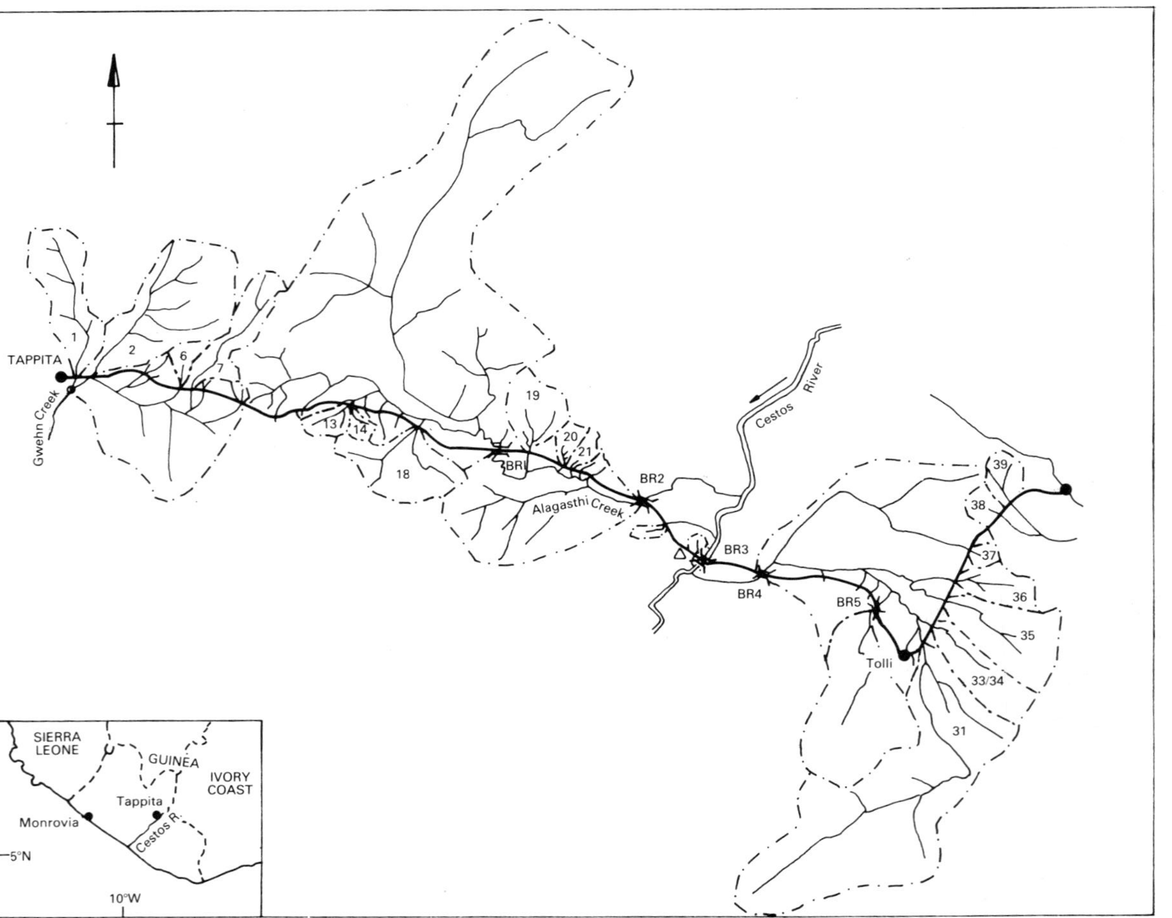

Fig. 4.7 — Tappita–Tobli–Blay Road: ———, rivers; ———, road; ·—·—, catchment boundaries;][, bridges; the numbers are the culvert catchments greater than 1 km^2; ◑, river gauge; △, rain gauge.

Table 4.4 — Average rainfall intensities

Return Period (years)	Average rainfall intensity (mm h^{-1})		
	24 h	48 h	72 h
2	3.81	2.44	1.87
5	4.50	3.03	2.39
10	4.95	3.42	2.72
20	5.39	3.79	3.04

10 and 20 year return periods. The results from the Sierra Leone data conformed well with data for four short-duration high-intensity storms presented by the road designers and connected with the intensities derived from the daily data. Thus with due regard to the Sierra Leone results, rainfall intensity–duration–frequency curves were derived for the Tappita–Blay area for durations from 0.1 to 72 h. Equations fitted to the three return period curves gave the following relationships:

$$I \text{ (5 years)} = \frac{28}{(D+0.2)^{0.575}}$$

$$I \text{ (10 years)} = \frac{30}{(D+0.2)^{0.560}}$$

$$I \text{ (20 years)} = \frac{32}{(D+0.2)^{0.550}}$$

where I (mm h^{-1}) is the intensity and D (h) the duration.

Flood estimates

The catchments of the streams draining to the 39 pipes and culverts are shown in Fig. 4.7. Their areas range from 0.15 km^2 (catchment 30) to 58.72 km^2 (catchment 31).

Catchments less than 1 km²

From previous experience in Liberia, it had been found that the Rational formula using appropriate coefficients was satisfactory for estimating peak flows only for very small catchments. There were 21 catchments of less than 1 km^2 draining across the road and, applying the rainfall intensities for the three return periods, the assumed corresponding flood peaks were calculated using the formula

$$Q_p = C_A C_R i A \times 0.278 \text{ m}^3 \text{ s}^{-1}$$

where C_A=0.60 is an attenuation coefficient, C_R=0.50, 0.55 and 0.60 for 5, 10 and 20 year return periods respectively is a runoff coefficient, i (mm h^{-1}) is the rainfall intensity for the time of concentration (assumed 0.5 h for areas less than 5 km^2 and 0.75 h for areas 0.5–1.0 km^2) and A (km^2) is the catchment area. The results are given in Table 4.5.

Table 4.5 — Flood peaks for catchment areas less than 1 km^2

Catchment	Area (km^2)	Q_p ($m^3\,s^{-1}$) for the following return periods		
		5 years	10 years	20 years
3	0.48	1.38	1.61	1.87
4	0.27	0.77	0.91	1.05
5	0.17	0.49	0.57	0.66
8	0.36	1.03	1.21	1.40
9	0.46	1.32	1.55	1.79
10	0.18	0.52	0.60	0.70
11	0.18	0.52	0.60	0.70
12	0.35	1.00	1.18	1.36
15	0.33	0.95	1.11	1.29
16	0.96	2.31	2.72	3.16
17	0.36	1.03	1.21	1.40
22	0.50	1.43	1.68	1.95
23	0.38	1.09	1.28	1.48
24	0.77	1.85	2.18	2.54
25	0.73	1.76	2.07	2.41
26	0.18	0.52	0.60	0.70
27	0.16	0.46	0.54	0.62
28	0.32	0.92	1.08	1.25
29	0.16	0.46	0.54	0.62
30	0.15	0.43	0.50	0.58
32	0.23	0.66	0.77	0.90

Catchments greater than 1 km²

For the 18 larger catchments draining to pipes or culverts and the areas draining to the four bridges over tributaries, a rainfall-runoff model developed by Fiddes (1977) for east Africa was used to estimate the peak flows. The components of this model include an initial land phase system whereby rain falling on the estimated contribution area is converted to runoff through a simple linear equation and this runoff translated into mean runoff rate from which peak discharge is obtained. Estimated catchment data and short-duration rainfall intensities are required and the model assumes that the return period of the rainfall is preserved in the predicted peak discharge.

Some of the parameters applicable in east Africa were modified to suit the conditions in Liberia. The modified model procedure is described in the following stages.

(1) The contributing area C_A for each catchment is assessed from three component coefficients: C_S, soil type and average land slope; C_L, land use and vegetation cover; C_W, antecedent conditions.

$$C_A = C_S C_L C_W.$$

(2) Catchment lag time K and initial retention Y were assessed from slope and land use characteristics over several model runs.

(3) The base time T_B of the flood hydrograph was estimated from

$$T_B = T_p + 2.3K + T_A$$

where T_p is the duration of high rainfall intensity and T_A the flood wave attenuation time. T_A is derived from

$$T_A = \frac{0.028}{Q^{0.25}} \frac{L}{S^{0.5}}$$

where L (km) is the length of the main stream, Q ($m^3\,s^{-1}$) the average flow during base time T_B and S the average slope along the main stream. Since T_A requires the unknown Q, a first approximation is made with $T_A = 0$ and subsequent iterations are made to provide a final estimate of T_A.

(4) Point rainfall for duration T_B and selected return periods are taken from the rainfall intensity–duration relationships and areal values calculated using the areal reduction factor ARF:

$$\text{ARF} = 1 - 0.044A^{0.275}$$

with catchment area A in square kilometres (Rodier & Auvray, 1965)

(5) The runoff volume RO during T_B is calculated from

$$\text{RO} = C_A(P - Y)A\ 10^3\ m^3$$

where P (mm) is the storm rainfall and Y the initial retention taken as 0. The average flow Q is given by

$$Q = \frac{0.93\text{RO}}{T_B \times 3600}\ m^3\,s^{-1}.$$

(6) Steps (3)–(5) are repeated until Q is within 5% of the previous estimate.

(7) The peak discharge Q_p is then given by

$$Q_p = FQ$$

with $F = 2.3$ for lag times greater than 1 h and $F = 2.8$ for shorter lag times (Rodier & Auvray, 1965)

A sample printout from a model run is given for bridge 2 in Table 4.6.

Table 4.6 — Tappita-Tobli-Blay road, Liberia: modified Fiddes model to predict flood flows, Alagashi Creek Bridge 2

Zone Parameters: rainfall intensity equations

5 year, $I = 28.00/(D+0.20)^{0.575}$

10 year, $I = 30.00/(D+0.20)^{0.560}$

20 year, $I = 32.00/(D+0.20)^{0.550}$

Catchment Parameters

Catchment area (km^2) 208.10

Main stream length (km) 29.50

Main stream slope 0.002 60

Initial retention 0

Contributing area 0.380

Catchment wetness 1.000

Catchment land use 0.350

Catchment lag time 8.000

Contributing area coefficient 0.133

Return period (years)	ARF	Rainfall (mm)	Base time (h)	Peak flow ($m^3 s^{-1}$)
5	0.8083	91.9	27.3	55.4
10	0.8082	103.3	27.2	62.4
20	0.8080	113.7	27.1	68.9

A full listing of the model results is given in Table 4.7. Although it was not possible to calculate standard errors of these estimates, it was considered that they were accurate to ± 10%. The model results were compared with observed data for flood events on the Gwehn Creek and found to be within an acceptable range of likely values.

Cestos River

The flow records at bridge 3 over the Cestos River (catchment area, 1770 km^2) for the period 1962–1983 were treated to the conventional probability analysis. The annual maximum MDFs were checked against the peak flows on the autographic charts and the differences found to be negligible; so the analysis of the maximum MDFs was considered adequate to assess magnitude of the flow peaks. The data ranging from

Table 4.7 — Flood peaks for catchments greater than 1 km^2

Catchment	Area (km^2)	Q_p ($m^3\ s^{-1}$) for the following return periods		
		5 years	10 years	20 years
1	11.97	9.6	10.6	11.6
2	22.43	15.4	17.2	18.9
6	1.83	2.5	2.7	3.0
7	7.69	7.7	8.5	9.3
13	2.09	2.2	2.5	2.7
14	1.01	1.1	1.2	1.3
18	11.32	14.1	15.4	16.7
19	6.63	4.2	4.6	5.0
20	1.70	1.6	1.7	1.8
21	1.37	1.4	1.5	1.6
31	58.72	20.5	23.0	25.4
33/34	7.49	6.8	7.5	8.1
35	12.07	8.2	9.0	9.8
36	4.90	2.4	2.7	2.9
37	1.19	2.0	2.1	2.2
38	3.04	2.7	2.9	3.1
39	3.00	2.2	2.4	2.6
Bridge				
1	165.60	47.3	53.1	58.5
2	208.10	55.4	62.4	68.9
4	166.20	36.2	40.7	44.8
5	22.60	17.9	19.7	21.4

126 to 237 $m^3\ s^{-1}$ over the 21 years were found to be best fitted by the Pearson type III distribution. The 20 year flood peak was estimated to be 243 $m^3\ s^{-1}$ with the 95% confidence limits $\pm 21\ m^3\ s^{-1}$.

Conclusions

Following the completion of the hydrological studies, the consultants were able to advise on the remedial measures necessary to inhibit further flooding of the roadway. All the box culverts and bridges were able to pass the 20 year flood and only needed attention where there had been structural failure or where there was evidence of scouring of the river bed. Details of pipes to pass the 20 year flood flows at 30 crossings were given and a new concrete box culvert recommended for catchment 33–34.

5

Flood mitigation

INTRODUCTION

Rivers and streams are the natural drainage channels leading surplus water from the land towards the sea, thereby forming the final link in the hydrological cycle. Only in some special areas of the world do rivers lead into basins of inland drainage. The variations in river flow have a distinctive effect on the formation of the river channels. In times of low flow with reduced velocities the river may begin to block the channel with deposits of sediment and then seek to find another course. With the arrival of high discharges and increased velocities, the river is capable of more work; it scours the river bed and, should the volume be great enough, it will overtop the usual river banks and spread out over the adjoining land. The task of the land drainage engineer is to ensure that watercourses maintain permanent channels for as much as practicable of the range in variation of river flows.

The mitigation of floods and the entrainment of river channels are specially important in densely peopled areas. In underdeveloped parts of the world, the natural behaviour of rivers causes little harm and, if an odd bridge is washed away in a flood, then it is replaced with another perhaps bigger and better but with no detriment to anyone. However, flood water breaking out of the normal river chennel and inundating surrounding land can do untold damage in a congested developed area with sometimes the ultimate disaster, loss of life. It is therefore in cities and towns that engineers have the most challenging projects for the alleviation and perhaps total prevention of flood catastrophes.

Flood protection schemes are undertaken at the direction of the local community. It may take several flood events before the democratic wheels of local government begin to turn effectively; a serious catastrophic flood, however, often gives a sense of urgency and speeds the process of investigation and design. It is becoming more common in practice now to make assessments of the damage done during a major flood and to balance the cost of an alleviation scheme with the benefits that would accrue from the protection afforded. A considerable contribution to damage assessment has been made in the UK by Penning-Rowsell and his team of collaborators (Penning-Rowsell & Chatterton, 1977).

The size and importance of a flood event and its location influence the scale of engineering expertise allocated to the job. The magnitude of a scheme can range from the annoying occasional flooding of a few houses in a small rural village, a project given to the most junior engineer to tackle by himself, to the extensive inundations of an industrial or expensive residential area by a rare flood, a project which can employ a team of engineers and support staff for several years.

The initial hydrological studies may be carried out by the lone engineer, the design office may have its own hydrologists or the work may be put out to contract. In all circumstances, the first task is to determine the magnitude of the worst flood and assess to the probability of its recurrence. There is usually a lack of hydrometric data upon which to base necessary analyses and recourse is often made to empirical formulae developed by experienced engineers over the years. However, The Flood Studies Report (NERC, 1975) is used increasingly in the UK to provide necessary design criteria for flood alleviation schemes. Comparable aids to practice include in the USA, the guidelines of the Interagency Advisory Committee on Water Data (1982) and, in Australia, the manual edited by Pilgrim, (1987).

Once the magnitude and return period of the offending flood have been assessed, then the design engineer generally produces a selection of solutions to the flooding problem based on a range of flood discharges. These are costed and compared with the possible damage costs saved by each proposed scheme, before presentation to the promoting authority. It is then in the hands of the local government, central government or any other interested agency such as the World Bank for developing countries to decide on the scheme that can be afforded and to find the required finance.

The case study examples range from the consequences of a flash flood in an arid area to the control of a river of continental proportions.

5.1 EXETER FLOOD PROTECTION

LOCATION The City of Exeter is on the River Exe in the County of Devon, England.

SOURCES Sir M. MacDonald & Partners (1961) *Preliminary report on the flooding in Exeter City during Autumn 1960 with proposals for flood protection,* Report to the Devon River Board.

South West Water, Peninsula House, Rydon Lane, Exeter.

PROBLEM During the exceptionally wet autumn of 1960, the low-lying parts of the City of Exeter, prone to flooding since the sixteenth century, were seriously affected by inundations from the swollen waters of the River Exe. There were three particularly heavy widespread storms, two of which caused the flooding of 1000–1200 houses and business premises, some to a depth of 2 m. Roads, railways, culverts, sewers and underground telephone communications were also damaged. The extent of the flooded area in the city is shown in Fig. 5.1 but both upstream and downstream of the built-up

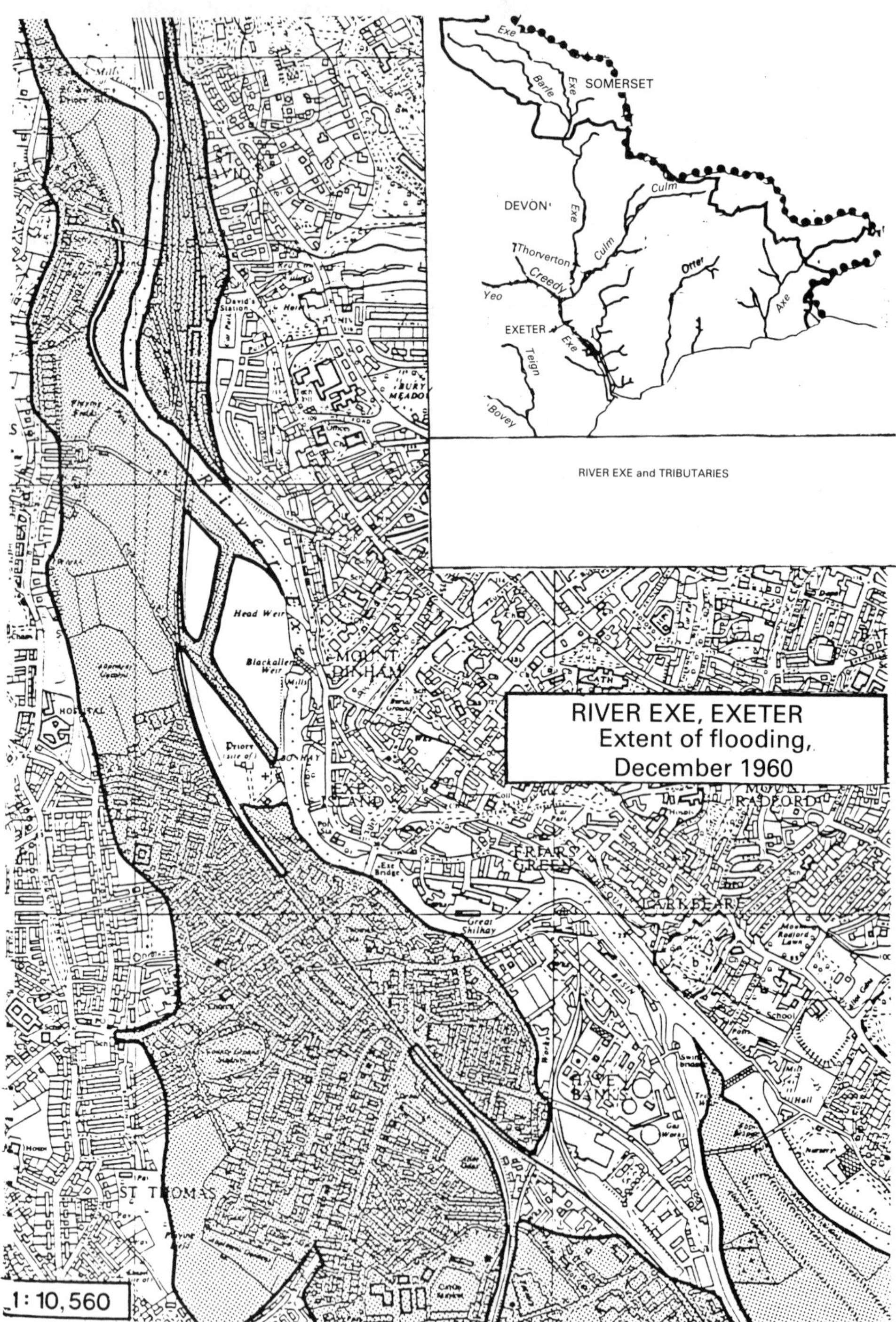

Fig. 5.1 — Exeter flooding of River Exe: ●●●, water authority boundary.

areas the river had overtopped its banks and covered the flood plain. Following this serious event, the authorities decided that a flood alleviation scheme should be undertaken and the consulting engineers, Sir M. MacDonald & Partners, were asked to investigate and advise.

Investigation

The main problem was to determine the magnitude of the peak flow through the city for the worst event in the December 1960 and if possible to derive the flood hydrograph to give the temporal pattern of the flood and hence an estimate of the total volume of discharge. The latter is required for assessing means of containing such a flood and controlling its discharge down the channel.

At this time, in 1960, there was only one river gauging station on the Lower Exe, at Thorverton (Fig. 5.1) and none on the two tributaries, the Culm and Creedy. Although there were many reliable daily rain gauges in the catchment, there were no recording gauges to give the timing and duration of the major storms. (Three rainfall recorders were operational outside the catchment.) Information from the Meteorological Office indicated that there was complete soil saturation in southwest England by the last week in September so that thereafter runoff would be immediate from the steep-sided valley of the Upper Exe. An assumption of 100% runoff from storm rainfall would therefore be reasonable.

Hydrological approach

The problem was tackled first by using these limited available records to determine the runoff from rainfall for the three outstanding storms on 30 September, 26 October and 3 December 1960. The major contributing flows to the discharge through Exeter were defined by the Upper Exe, the Culm and the Creedy. Mean 24 h flows, assuming 100% runoff in 24 h, calculated from the three catchment areal rainfalls derived by the Thiessen method are given in Table 5.1. It will be noted that the tributaries contributed greatly to the first two storms but the Upper Exe contribution dominated the flow in the December storm.

To ascertain the timing of the peak flows, the period of concentration of each catchment and the records of rainfall recorders were required. Attempts were made to determine the period of concentration for the Upper Exe first from records from an adjacent rain recorder related to the flood peak at Thorverton and then using an empirical formula but no consistent answers were obtained. Relating the total volume of rainfall to the total flood volumes recorded at the gauging station proved useless and there was no satisfactory relationship obtained between peak rainfalls and peak river levels or discharges. Discrepancies were partly due to considerable errors in the stage–discharge table for the Thorverton gauging station. There had been scouring of the river bed causing irregularities and the peak levels were well above bankful capacity and beyond the range of the measured discharges. Following many unsuccessful trials at rainfall-runoff relationships, it remained to compare the daily mean runoffs derived from the daily rainfalls assuming 100% runoff with the peak river levels and dubious discharges (Table 5.2). It is highly unlikely to have peak discharges less that the derived daily mean runoff values although the measured surface runoff could have extended beyond 24 h. However, the undervalued peaks

Table 5.1 — Runoff from rainfall

Catchment	Area (km^2)	Storm date	Areal rain (mm)	Rainvolume (Mm^3)	Equivalent mean 24 h Q ($m^3\ s^{-1}$)
Upper Exe	624	30 September 1960	50.0	31.20	361
		26 October 1960	53.8	33.57	389
		3 December 1960	65.0	40.56	469
Culm	277	30 September 1960	51.6	14.29	165
		26 October 1960	50.3	13.93	161
		3 December 1960	29.5	8.17	94
Creedy	269	30 September 1960	53.8	14.47	168
		26 October 1960	40.1	10.79	125
		3 December 1960	30.5	8.20	95

Table 5.2 — Daily runoff versus peak discharges at Thorverton

Date	Areal rain (mm)	Mean 24 h flow $m^3\ s^{-1}$)	Maximum stage m	Peak Q ($m^3\ s^{-1}$)
30 September 1960	50.0	361	2.58	215
26 October 1960	53.8	389	3.16	357
3 December 1960	65.0	469	3.64	510

are more likely to be due to the unsatisfactory extrapolation of the stage–discharge relationship. Even the December storm might be expected to have produced a higher peak at Thorverton.

At this stage, failing satisfactory results from these analyses, it was deemed appropriate to apply the Institution of Civil Engineers normal maximum flood (NMF) curve usually used for small catchment areas to be impounded (Table 5.3). If it is assumed that the three contributing catchments peaked at the same time, a maximum of 1170 $m^3\ s^{-1}$ would result from the whole catchment as the flood approached Exeter. From historic rainfall records, this has never occurred and indeed the incidence of storms over the area suggests that the timing of the separate catchment peaks is never likely to coincide.

Further investigations into the estimated speed of translation of the peaks in the three subcatchments led to the conclusion that the arrival times of the three peaks to

Table 5.3 — Normal maximum flood

Catchment	Area (km^2)	Normal maximum peak flow ($m^3\ s^{-1}$)
Upper Exe	624	736
Culm	277	453
Creedy	269	453

the River Creedy confluence fell within 24 h. Thus the daily mean flood flows would overlap and there would be three individual contributing peaks.

A final hydrological appraisal of the possible flood discharge on 3 December 1960 is given in Table 5.4. In the event, this maximum flow, say 700 $m^3\ s^{-1}$, did not reach

Table 5.4 — The maximum flood, 3 December 1960

	Maximum flood discharge ($m^3\ s^{-1}$)
Upper Exe mean flow	469
Culm mean flow	94
Creedy mean flow	95
Total mean flow	658
Add peak from Upper Exe (510–469, Table 5.2)	41
Maximum peak flow below Creedy confluence	699

the Creedy confluence since a considerable area upstream had been flooded and an appreciable volume of water was retained on the floodplain. Nevertheless, a damaging flood flow penetrated the low-lying parts of Exeter.

Hydraulic approach.

As soon as possible after the flood event, the maximum levels reached by the river flow were identified along the river banks in the city. In some places, the peak levels were indicated by 'wrack marks', the deposition of debris, and in others by noted levels on walls and buildings. These were surveyed in the reaches 2, 3 and 4 (Fig. 5.2) and cross-sections were surveyed at convenient points along the reaches where the flow path could be identified. From these measurements and the resulting drawings, mean values of the slopes, cross-sectional areas and wetted perimeters were

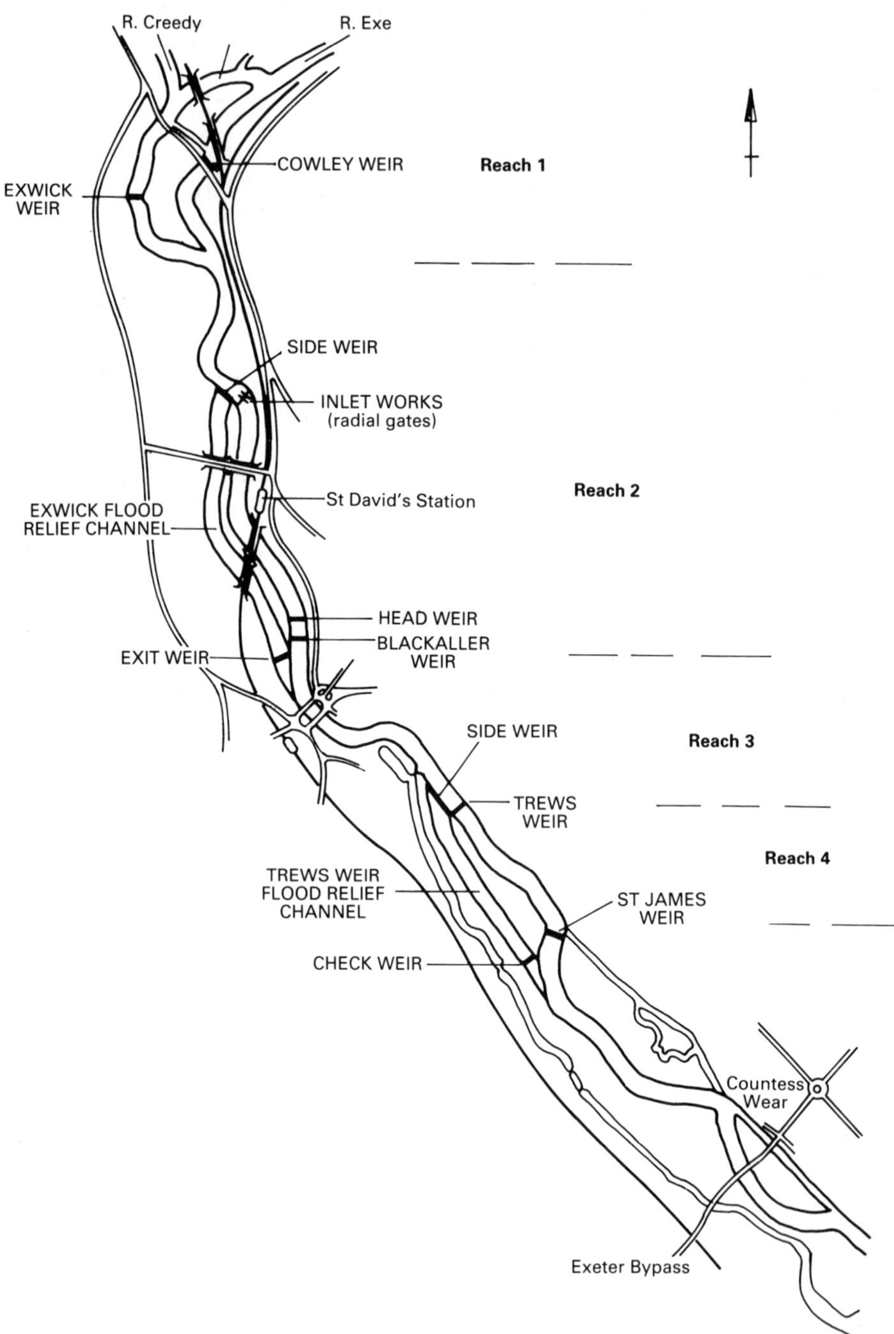

Fig. 5.2 — Exeter flood alleviation scheme.

evaluated for each reach with the results in Table 5.5. Then Manning's equation for discharge Q which is

$$Q=AV=\frac{AR^{2/3}S^{1/2}}{n}$$

was used to compute the peak discharge in each reach. The value of n, the roughness factor to be used in the equation, needed careful assessment of the flow paths and a mean value of 0.025 was taken. The computations are shown in Table 5.6. The large differences in these results can be explained by the irregularities in the flow paths of the flood. If the estimated flood flow of 700 $m^3\,s^{-1}$ did reach the Creedy confluence, then about 100 $m^3\,s^{-1}$ spilled over the flood plain in reach 2. Some of this returned to the main river channel in reach 3 which was further augmented by the runoff from the City of Exeter. The discharge value for reach 4 was not reliable owing to the unknown behaviour of St James Weir during the flood and possible tidal effects from further downstream.

Conclusions

As a result of the investigations based on such minimal hydrometric data, it was estimated that a flood of approximately 700 $m^3\,s^{-1}$ affected the City of Exeter on 3 December 1960 and that a flood of such magnitude could happen at any time. The Devon River Board (the responsible body in 1960) proceeded with a flood alleviation scheme which provides for the safe passage of 700 $m^3\,s^{-1}$ through the City of Exeter. With the collaboration of the city and other interested parties, relief channels with new control weirs have been built and larger bridges constructed where needed (Fig.5.2). It has recently been estimated that the 700 $m^3\,s^{-1}$ represents a flood magnitude with a return period of once in 50 years.

Table 5.5 — Hydraulic characteristics

Reach	Slope S	Mean area $A(m^2)$	Mean wetted perimeter (m)	Hydraulic mean depth $R(m)$
2	1.16	176.5	45.4	3.89
3	1.02	202.5	47.9	4.23
4	1.39	195.1	49.7	3.93

Table 5.6 — Calculated reach discharges

Reach	$R^{2/3}$	$S^{1/2}$	V $m^3\,s^{-1}$)	A (m^2)	Q ($m^3\,s^{-1}$)
2	2.47	0.0341	3.37	176.5	595
3	2.62	0.0319	3.34	202.5	676
4	2.49	0.0373	3.72	195.1	726

5.2 KUANTAN FLOOD MITIGATION

LOCATION Kuantan is on the east coast of the State of Pahang in Peninsular Malaysia.

SOURCE Binnie Dan Rakan, M. (1977) *Water resource development in the Kuantan region,* Report to the Government of Malaysia.

PROBLEM The extensive urban development of the Kuantan area is expected to aggravate the flooding of the lowland reaches of the Sungai Kuantan (Fig. 5.3). In conjunction with investigations into water resources for increased demands, special studies were carried out on the causes and incidence of flooding. The results of these hydrological studies provided the planners and design engineers with the bases for suitable flood mitigation schemes. Assessment of previous flood damage and expected increases in future damage together with the estimates of costs of flood protection schemes enabled the developers to select the most practical and economic solutions.

Flood areas

Flooding is nearly an annual problem in the Sungai Kuantan Basin and can affect three distinct zones.

(1) Kuantan urban area north of the river estuary where floods may be caused by the interaction of high river discharges, tidal action and internal drainage problems.
(2) Lowlands at the confluence of the Sungai Kuantan and Sungai Belat, the floodplain which may be inundated for a week with depths of 2–3 m. The extent of a major event is shown in Fig. 5.3.
(3) Narrow stretches of the floodplains along the river valleys.

Hydrology

A general account of the rainfall and runoff of the Sungai Kuantan Basin is given in the study of the water resources for Kuantan (see section 7.1). The heaviest recorded rainfalls in the region occurred in December 1926 at Jeran Kuantan. On the 27 December, 630 mm was recorded and the total for the month was 1956 mm. The variable timing of the floods in the vulnerable areas is demonstrated by the event of November 1975. The whole river basin was affected by a severe rainstorm. The daily totals were 396 mm in Kuantan and 350 mm at Sungai Lembing Town. The hourly distribution of the Kuantan rainfall is shown in Table 5.7. The flooding in Kuantan Town due to the incapacity of the local storm drainage occurred during or soon after the peak rainfall at 11.00 hours. The river did not reach its peak in the lower reaches

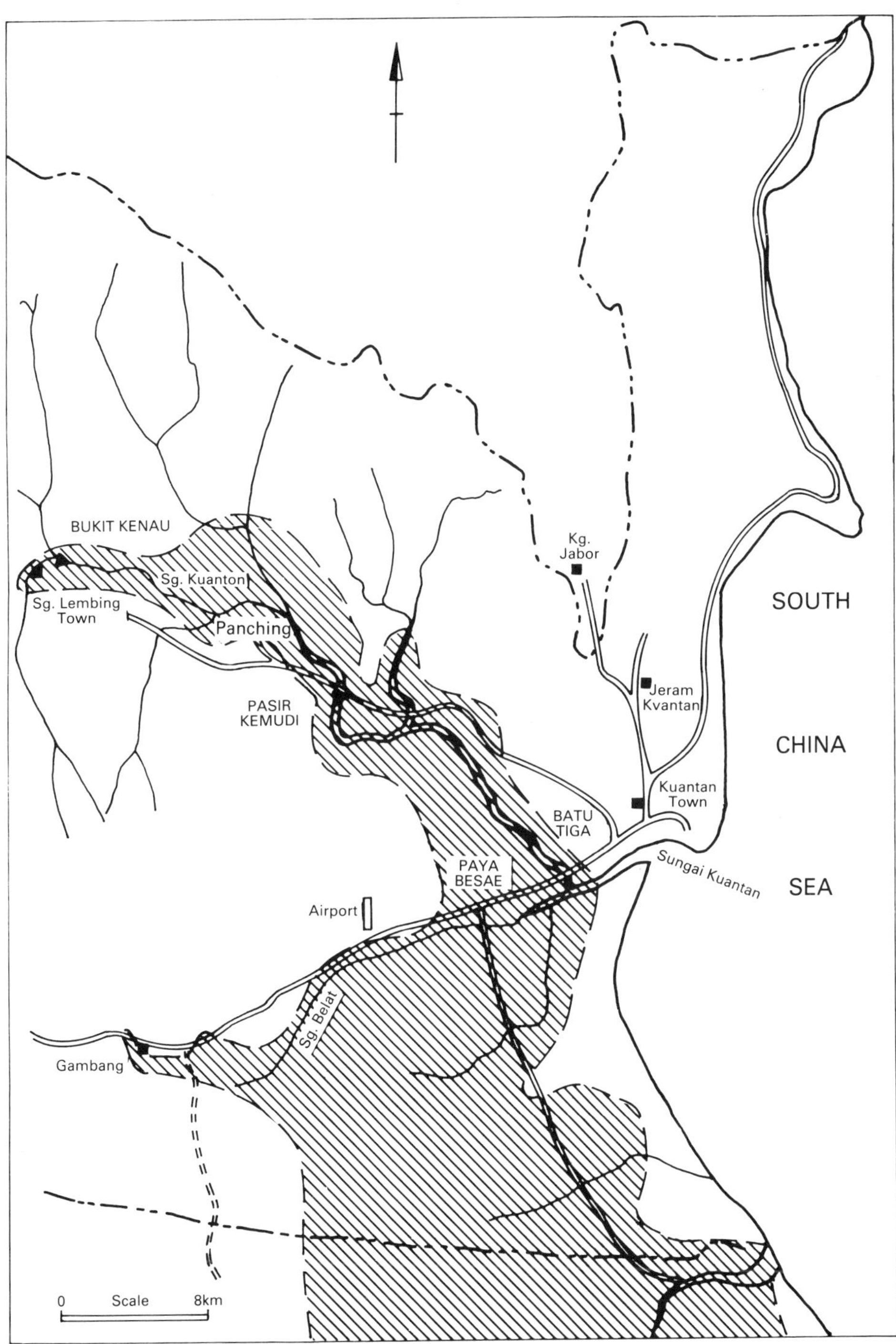

Fig. 5.3 — Kuantan flooded area, 1970–1971.

Table 5.7 — Hourly rainfall, Kuantan, 24 November 1975

Time at end of hourly interval (hours)	10.0	11.0	12.0	13.0	14.0	15.0	16.0	17.0	18.0	19.0
Rainfall (mm)	20.1	62.0	46.0	45.0	40.1	50.0	27.9	40.9	18.0	20.1

until the afternoon of the following day by which time the local flood waters had subsided.

Owing to the paucity of quantitative data, no previous flood magnitude–frequency studies had been made. Therefore peak water levels recorded or marked on buildings were assembled and the flood profiles for the major events plotted from Sungai Lembing down to the lowest road bridge. These were related to the river gauging station at Pasir Kemudi and, by affording return periods to the notable historic events in comparison with records in neighbouring catchments, a stage–return period relationship was constructed (Fig. 5.4). An event of the severity of 1926 would cause considerable flooding in Kuantan Town. A check on the incidence and ranking of 3, 5 and 7 day rainfall totals at four rain gauge stations gave good agreement with the severe flood events. The flood of January 1971 (shown in Fig. 5.3) was the third largest recorded on the Sungai Kuantan with an estimated return period of 1 in 20 years. Fourty-four villages in the catchment area were under 2–3 m of water for up to 14 days. The town centre of Kuantan was not affected by this major river flood. Information on the damage and related costs were collected by various organizations and, from these, estimates of the total flood damage costs were made.

Most of the flood discharge measurements available relate to the gauging station at Bukit Kenau. Since 1973, a series of measured peak flows ranging from 564 to below 280 $m^3 s^{-1}$ have enabled the rating curve to be extended and thereby estimates of peak discharges and flood volumes of notable events to be made. The mean annual peak discharge was estimated to be at least 700 $m^3 s^{-1}$ and the corresponding daily flow 450 $m^3 s^{-1}$. These values were supported by regional studies but comparisons between the response of forested catchments and of developed catchments suggested that downstream runoff would be greater. In addition, flood volumes could be enhanced by higher rainfall intensities nearer the coast. A mean annual daily flood flow for the whole catchment was estimated to be 1200 $m^3 s^{-1}$.

From a plot of annual flood water level hydrographs at Pasir Kemudi, it was assumed that at 7.5 m serious flooding would probably prevail downstream for at least 3 days during major floods. These events were greater than a 5 year flood at Pasir Kemudi and if it were assumed that such a flood would be contained within the channel then the storage required to prevent serious damage from a T year flood would be the difference between the T year, 3 day flood volume and the corresponding 5 year value.

A series of flood volume indices were derived from the long records of a neighbouring catchment relating T-year floods to 1 day mean annual flood volume for a range of durations. These were used to derive 3 day annual flood volumes for selected return periods for the Sungai Kuantan Basin and for strategic flood control

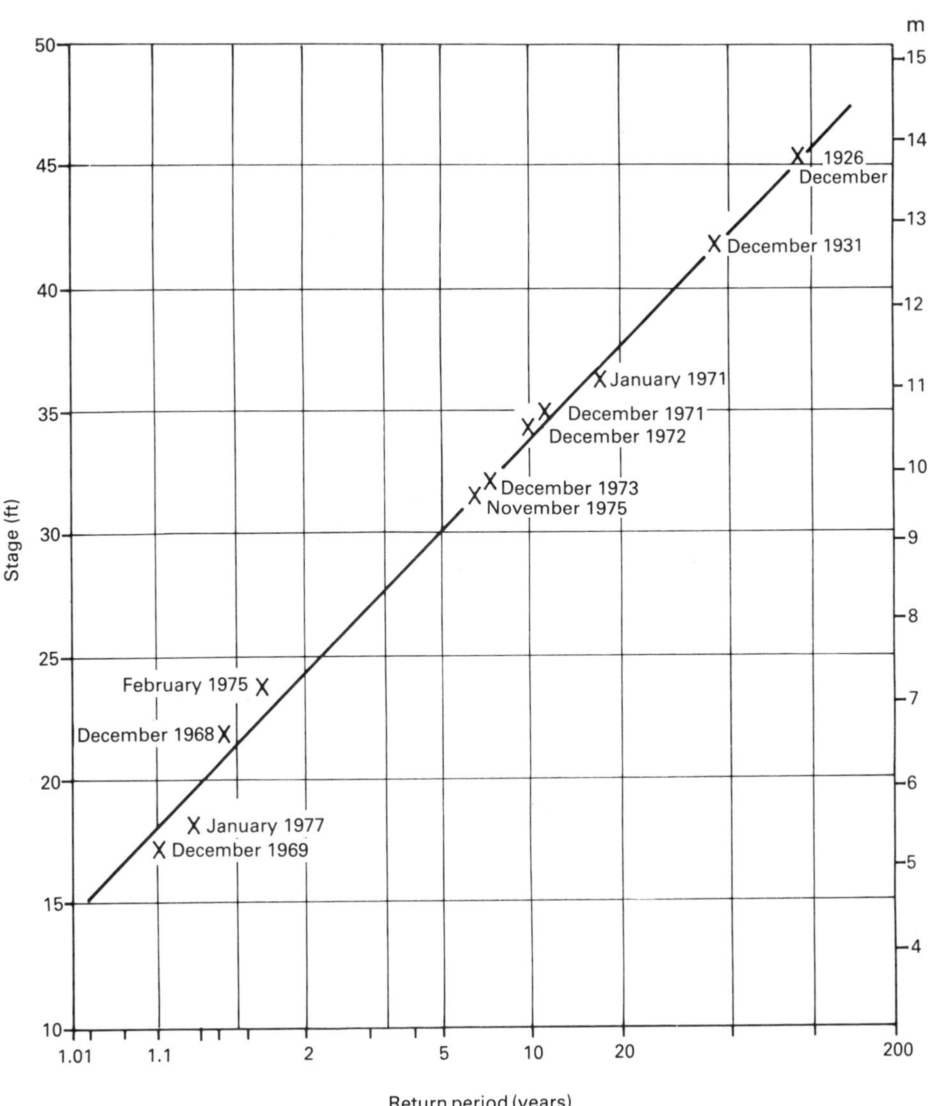

Fig. 5.4 — Annual flood peak stage–return period, Sungai Kuantan at Pasir Kemudi.

points, three proposed reservoir sites in the headwaters and a flood channel offtake on the Sungai Belat (Table 5.8). It will be noted that, although the true runoff rate from the suggested forested upland reservoir catchments would be less than the overall basin value used for the calculations, the higher estimated combined storages for the 20 year return period would still be insufficient to prevent serious flooding near Kuantan Town. The storages in Table 5.8 are, however, certainly underestimated since, for the higher floods, the period of inundation of the floodplains is usually much longer than 3 days. Considerations of the operational problems of the

Table 5.8 — Estimated 3 day annual flood volumes in the Sungai Kuantan basin

Site	Area (km^2)	Estimated 1 day volume (Mm^3)	Estimated 3 day volume (Mm^3) for the following return periods and 3 day/1day ratios			
			2.33 2.35	5.0 3.31	20 4.84	50 years 5.79
Kuantan Town	1658	103.68	244	343	502	600
Dam site 1	166	10.38	24	34	50	60
Dam site 2	130	8.13	19	27	39	47
Dam site 3	111	6.94	16	23	34	40
Sungai Belat offtake	101	6.32	15	21	31	37

impounding reservoirs, which from cost–benefit assessments would necessarily have to be multipurpose serving water supply, irrigation and power generation, were such that their effectiveness for flood mitigation would be minimal.

Mitigation schemes

The following methods for the alleviation of the lowland flooding were studied and assessed.

(1) Reduction of flood peaks by upland regulating reservoirs.
(2) River channel improvements.
(3) Provision of embankments.
(4) Release of flows to lowland storage.
(5) Tidal control barrage.

The most economical scheme involved the construction of embankments to protect the development areas of Paya Besar and Batu Tiga from all but the largest flows. This would convert the two areas into polders. The defined flood path along the rivers would be cleared of shrubs and mangroves and the main road would have to be raised. Preliminary calculations indicated that embankments 3–4.5 m high would be required along the Sungai Kuantan to contain a 10 year flood and, for a 20 year flood, the embankments would need to be 4–5 m high. The cost of flood protection to the Paya Besar and Batu Tiga development areas would be justified by the benefits obtained in alleviating flood damage and more detailed studies were recommended to be given priority.

5.3 NEWTON ABBOT FLOOD ALLEVIATION

LOCATION Newton Abbot is in south Devon, UK, at the confluence of the River Lemon and the River Teign.

SOURCES South West Water Authority (1980) *River Lemon, Newton Abbot,* Report on the analysis of flooding and alternative proposals for flood alleviation.

South West Water (1983) *The derivation of a SWW flood growth curve,* Internal Report.

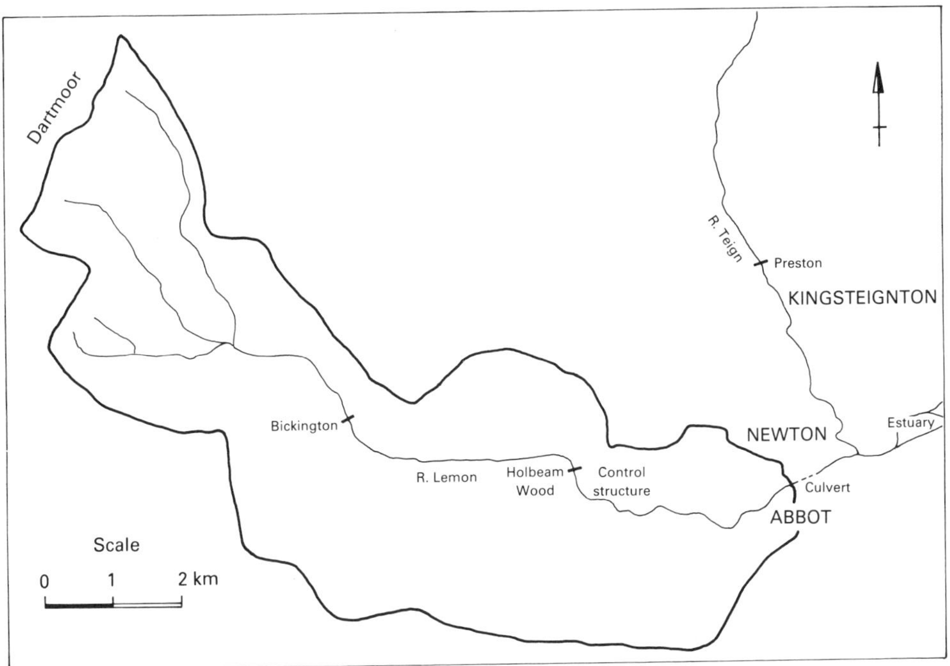

Fig. 5.5 — River Lemon catchment.

PROBLEM Newton Abbot, a town of 20 000 population, developed at the lowest crossing point of the tidal River Teign where a small tributary, the River Lemon, was originally available for a freshwater supply (Fig. 5.5). Expanding encroachment of the floodplains of both rivers has gradually increased the vulnerability of the town to flooding from the rivers and the Teign estuary. There have been nine serious flood events in Newton Abbot since 1894 of which those of 1894, 1900, 1938 and 1979 have caused widespread inundations and disruption. The extent of flooding depends on the cause of the extreme event. The December 1979 flood resulted from high river flows in the Lemon combined with estuarial flooding.

Investigations

There were no gauged flow records available for the River Lemon but an interrogated river level gauge at Bickington provided information on the sequence of levels during the 1979 floods. From these levels, wrack marks and noted flood levels in Newton Abbot, hydraulic analysis resulted in peak estimates of 30 $m^3\ s^{-1}$ at Bickington and 60 $m^3\ s^{-1}$ in Newton Abbot. To evaluate the frequency of occurrence of this major flood, the historic records were considered. The main hindrance to flood flows in Newton Abbot is a culverted length of the river which from hydraulic calculations was estimated to have a maximum capacity of 40 $m^3\ s^{-1}$. The open channel capacity was approximately 35 $m^3\ s^{-1}$. With serious flooding occurring at say 45 $m^3\ s^{-1}$ four times in the past 85 years, a return period of 21 years may be assigned to this discharge. Similarly, nine events of less serious flooding occurring at flows equal or greater than 35 $m^3\ s^{-1}$ would result in an approximate return period of 9 years. An estimate of the mean annual flood (24 $m^3\ s^{-1}$) was made by a modified method from The Flood Studies Report (NERC, 1975). These frequency points fitted The Flood Studies Report growth curve ($Q_T/\overline{Q}$ versus T) for southwest England (region 8) which was then extrapolated to give an estimated 60 $m^3\ s^{-1}$ peak flow for a return period of 100 years.

Alleviation scheme

On the basis of these investigations, the design engineers proposed several schemes for accommodating the 100 year flood of 60 $m^3\ s^{-1}$ and, after consideration of the separate cost–benefits evaluated by the 'Penning–Rowsell' method (Penning-Rowsell & Chatterton, 1977), a flood control structure across the River Lemon at Holbeam Wood was built. From a synthesized inflow hydrograph (peak, 47.4 $m^3\ s^{-1}$) based on the Bickington estimated discharge hydrograph, the storage required from a fixed orifice structure was 465 000 m^3 to give a peak outflow of 20 $m^3\ s^{-1}$ downstream, thus effecting a 27.4 $m^3\ s^{-1}$ reduction in the flood flow. A detailed site survey at Holbeam Wood revealed an available storage of only 400 000 m^3 and therefore a variable orifice (radial gate) had to be used to reduce storage required to that available. The maximum flow in Newton Abbot would then be 32.6 $m^3\ s^{-1}$ for the 100 year event. No further works were necessary to contain this and it could be accommodated by the existing culvert but minor repair works to the channel were recommended to cater for 40 $m^3\ s^{-1}$.

Addendum

As follows from most completed schemes incorporating control works, the question of operating procedures arose. The necessary further hydrological investigations revealed that the original estimate of 60 $m^3\ s^{-1}$ for the 100 year flood was exaggerated and also the town centre culvert capacity was overestimated. A wider study of the flood records of catchments in the region with reliable river gauging stations lay the foundations for a new frequency growth curve for the South West Water area (see source cited above). Relating the behaviour of the Lemon Catchment to the similar Teign catchment and employing the new growth curve, a reappraisal of the December 1979 flood was made which resulted in the threshold for serious flooding at Newton Abbot for the 21 years return period from historic floods to be 35.6 $m^3\ s^{-1}$ (instead of 45 $m^3\ s^{-1}$) and the 100 year event peak flow to be 53.9 $m^3\ s^{-1}$.

The storage capacity of the Holbeam Wood Flood Control structure was checked and The Flood Studies Report unit hydrograph method adapted to provide release rules for operating the radial gate in times of flood flows.

The Newton Abbot Flood Alleviation Scheme was designed and carried out using conservative estimated discharges based on limited available data. The subsequent later calculations confirmed that the completed scheme would contain a 100 year flood event but it has been further recommended that the hydrological variabilities of the catchment be monitored by the installation of a crump weir gauging station at Newton Abbot upstream of the culvert which also requires further study to establish a more realistic roughness coefficient for the hydraulic calculations.

5.4 THE WADI ADAI FLOOD

LOCATION The Wadi Adai flows from the Hajar Mountains of Oman to the sea west of Muscat.

SOURCES Bell, N. C. (1981) *Flood hydrology of northern Oman,* M. Sc. Dissertation. Imperial College, London.

Wheater, H. S., & Bell, N. C. (1983) Northern Oman flood study. *Proc. Inst. Civ. Eng.* Part 2, **75** 453–473.

PROBLEM On 3 May 1981, a violent storm swept across northern Oman resulting in loss of life and extensive damage in the capital area round Muscat (Fig. 5.6). In particular, the flow from the Wadi Adai breached the

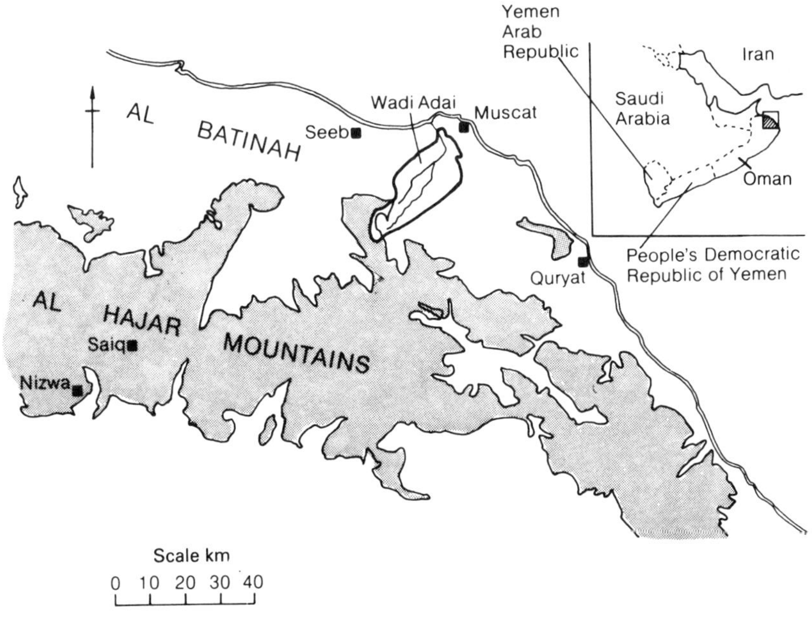

Fig. 5.6 — Wadi Adai location: ⛰, land above 600 m; ■, major towns.

coastal highway, flooded houses to a depth of 1.5 m and severed communications with the Police Headquarters. A hydrological study of the event was undertaken by hydrologists of Imperial College in association with Jack & Letman Associates (consulting structural engineers) on behalf of the Royal Oman Police with the object of making recommendations for flood protection works.

Wadi Adai

Wadi Adai is one of several major wadis draining from the inland chain of steep rocky mountains (jebel) composed of igneous formations and limestone massifs. Across the northern coastal plain of alluvial sands and gravels, the wadi channels are indistinct and may be realigned during high flows. The catchment of the Wadi Adai to Wattayah (380 km^2) (Fig. 5.7) consists in its upper reaches of gravel hills

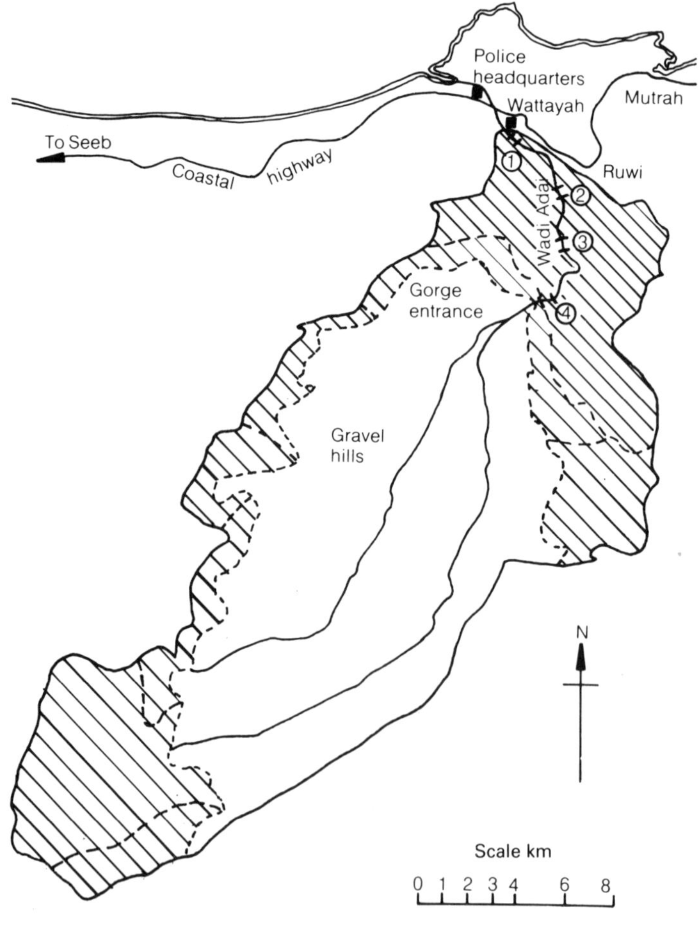

Fig. 5.7 — Wadi Adai catchment: ▨, hard rock.

surrounded by steeply sloping areas of bare rocks. Flow from this area (321 km^2) is concentrated through a gorge beginning 10 km upstream of Wattayah and tributaries from a further rocky 49 km^2 contribute to the flow in this reach. The channel slopes in the upper jebel ring range from 0.13 to 0.50, in the two channels 50 m wide in the gravel area the overall slope is 0.014 and in the gorge, the wadi channel 100 m wide with steep sides has a bed slope of 0.0055.

Hydrology

The hydrological studies may be subdivided into considerations of the cause and possible frequency of occurrence of the damaging storm and the evaluation of the resultant peak flood discharge of the ungauged Wadi Adai.

Storm rainfall

The incidence of rainfall in arid and semiarid climatic regions is highly irregular. The variability of the rainfall amounts within a distinct seasonal pattern is seen in the summary data from the longest record at Muscat where the average annual total is 106 mm (Table 5.9). It will be noted that a heavy rainfall in May at the end of the

Table 5.9 — Muscat rainfall, 1893–1959

Month	Jan	Feb	Mar	Apr	May	June	July	Aug	Sept	Oct	Nov	Dec
Mean rainfall (mm)	31.2	19.1	13.1	8.0	0.38	1.31	0.96	0.45	0.0	2.32	7.15	22.0
Standard devi ation (mm)	38.9	25.1	18.9	20.3	1.42	8.28	4.93	2.09	0.0	7.62	15.1	35.1
Maximum rain fall (mm)	143.0	98.6	70.4	98.3	8.89	64.0	37.1	14.7	0.0	44.5	77.2	171.2
Average number of rain days	2.03	1.39	1.15	0.73	0.05	0.08	0.10	0.07	0.0	0.13	0.51	1.60
Maximum daily rainfall (mm)	78.7	57.0	57.2	51.3	8.9	61.5	30.0	10.4	0.0	36.8	53.3	57.2

winter season is an unusual occurrence. The 1981 storm resulted from the eastward penetration of a depression originating over the Mediterranean Sea and an associated secondary low pressure system developing over the Arabian Gulf. The merging of the two lows with high air temperatures and humidity and subsequent instability caused heavy thunderstorms and strong winds during the night of 2 May. Rainfall measurements are organized by four authorities and the assemblage of data to provide a comprehensive picture of the rainfall distribution was a major undertaking. The highest storm total over two rainfall days was 140 mm with three other measurements over 100 mm, all recorded along the coastal plain. From a plotted isohyetal map, the storm rainfall over the Wadi Adai catchment was estimated to be 89 mm.

The temporal storm pattern was obtained by averaging the recorded profiles at seven autographic stations and applying the percentage rainfalls to the catchment areal rainfall (Table 5.10). Indications of the frequency of occurrence of this unusual

Table 5.10 — Storm profile over Wadai Adai catchment

Time from storm onset	Cumulative amount (%)	Cumulative rain (mm)	Intensity for duration ($mm\ h^{-1}$)
15 min	41	36	144
30 min	60	53	106
1 h	68	60	60
3 h	86	76	25
6 h	96	85	14
9 h	100	89	10

event were obtained by comparison with the results of a Gumbel (EV1) analysis of the annual maximum daily falls at Muscat and a composite station year series of data from a number of stations in northern Oman. Considered as a 24 h storm, the return period would lie between 60 and 160 years but, taking into account its short duration, comparison with a derived intensity–duration–frequency diagram gave a return period of 100–300 years for the storm.

Storm flow

Flow measurements of the seasonal wadis are sparse and the only record available was a single-stage hydrograph in a wadi only partially affected by the storm with no stage–discharge calibration. First, detailed surveys of four reaches in the Wadi Adai gorge (Fig. 5.7) were carried out starting 1 week after the storm and completed within 2 months, to determine the peak discharge using the slope–area method (Chow, 1959). After careful consideration, a value of Manning's n of 0.032 was adopted. The results are summarized in Table 5.11. The peak discharge adopted for the wadi at Wattayah was 1150 $m^3\ s^{-1}$.

To provide information on the volume and shape of the flood hydrograph for design purposes an attempt was made to synthesize the Wadi Adai storm hydrograph. A physically based rainfall-runoff model with the model parameters determined by measurable catchment characteristics was used. Since flood runoff in semiarid areas is mainly generated by overland flow, a numerical model of overland flow and channel routing was adopted to simulate the wadi response. In the chosen model, a computer package KINEROS (Kibler & Woolhiser, 1970) applies the kinematic wave approximations to the solving of the Saint Venant equations of gradually varied unsteady flow for both overland and channel flow. The catchment was modelled by a series of rectangular planes and trapezoidal channels relating to

Table 5.11 — Wadi Adai peak discharges

Reach	Adopted discharge ($m^3 s^{-1}$)	Average velocity ($m s^{-1}$)	Between reaches	
			Distance (km)	Travel (min)
4	1350	4.5		
			3.9	15.5
3	1160	3.9		
			1.8	7.7
2	1250	3.9		
			2.9	12.9
1	1160	3.6		

the varied parts of the catchment with Manning's n= 0.016 for hard-rock areas 0.025 for the gravel hills and 0.035 for channel sections except in the gorge where 0.032 was applied. The model was run with the derived storm profile. Considerable research was made into the possible rainfall losses and a figure of 8 mm was estimated for the initial loss with an ensuing 0.75 runoff coefficient for the jebel areas and 0.50 for the gravel hills. The resulting hydrograph had a peak of 1163 $m^3 s^{-1}$ from the total rainfall of 89 mm with an overall runoff coefficient of 0.56. Several model runs were made to test the sensitivity of the parameters but the adopted simulation fitted the eye-witness reports of the timing of the peak.

Conclusion

The high degree of variability of hydrological phenomena in semiarid areas compounded with the lack of reliable rainfall and stream flow measurements makes for extreme difficulty is assessing water quantities. The lack of vegetation results in rapid responses to intense rainfalls on rocky surfaces and in the transport of loose material to join the sediment load in flooded wadis. The Wadi Adai flood study results were based on the available rainfall measurements, the detailed survey of the channel and the experience of hydrologists in other semiarid areas of the world.

5.5 YORK FLOOD DEFENCES

LOCATION The City of York is at the confluence of the tributary, the Foss, with the main River Ouse, about 70 km upstream from the Humber estuary.

SOURCES Mott, Hay & Anderson with Sir M. Madonald & Partners, Consulting Engineers (1982) *River Foss flood alleviation,* Draft Report.

Yorkshire Water Authority (1982) *River Ouse flood frequencies, York and Cawood.*

Yorkshire Water, Rivers Division (1985) *York flood defences.*

PROBLEM The City of York has a long history of flooding from records as far back as the thirteenth century. In 1982, over 540 residential and industrial properties were affected and the commercial and industrial life of the city was at a standstill for 3 days (Fig. 5.8). Although there have been several flood studies, an appraisal of the flood frequencies of the River Ouse at York was made by the Yorkshire Water Authority (see source cited above) as a necessary pre-requisite to the design of the defence schemes.

Investigation

The catchment area of the River Ouse above York is 3315 km^2 (Fig. 5.9). The principal tributaries are the Swale, Ure and Nidd draining from the Pennines where the annual rainfall exceeds 2000 mm. Snow on the mountains is common in winter and rapid thaws accompanied by heavy rain are often the cause of flooding. Low river gradients in the Vale of York attenuate flood flows so that peak flows are sustained and recessions are extended. Within the city, there are several overflows in the combined sewerage system which result in secondary flooding with high river levels.

Owing to the complexity of the catchment drainage, the flood frequency study was confined to analyses of the river records. A river gauging station on the Ouse at Skelton, 2.5 km upstream of the city centre, provides level data from 1956 but doubtful discharges at higher levels. At the York Guildhall, water levels have been recorded since 1885 and at the Viking Hotel from 1976. The critical level at which property flooding commences is 8.93 m AOD (N) at the Guildhall. The 1982 peak level was 10.12 m after 3 days of continuously rising levels.

The annual maximum series of Guildhall levels was analysed using the Gumbel EV1 distribution for extreme values. The Weibull plotting position was adopted and the theoretical distribution calculated by the method of moments. Two series of the data were treated, the whole record of 98 years (1885–1982) and the last 48 years (1935–1982) to check whether there was any evidence that high levels were occurring more frequently. From the two reasonably well-fitting theoretical curves, the data in Table 5.12 were calculated.

The frequency table in Table 5.13 gives the number of floods above selected levels for various periods of record. While peak levels are the critical variable, their temporal relationships are dependent on constant channel sections. Knowledge of the frequency distribution of the annual maximum discharges would be more valuable in designing remedial channel works. The inadequacy of the stage–discharge relationship at Skelton precludes this and unfortunately the Guildhall section is also not calibrated. Extensive studies in connection with the Foss alleviation scheme found no physical reason for the apparent increased trend in high flows in the Ouse mostly seen during the period 1900–1935 and it was considered that the later period 1935–1982 provided the most stable base for the statistical analyses. However, the variable compilation of the level measurements suggested inhomogeneity in the series and the records needed to be treated with caution. These studies found that the shorter data series of the Ouse annual maximum water levels were fitted better by the normal distribution than by the EV1 distribution. From this theoretical

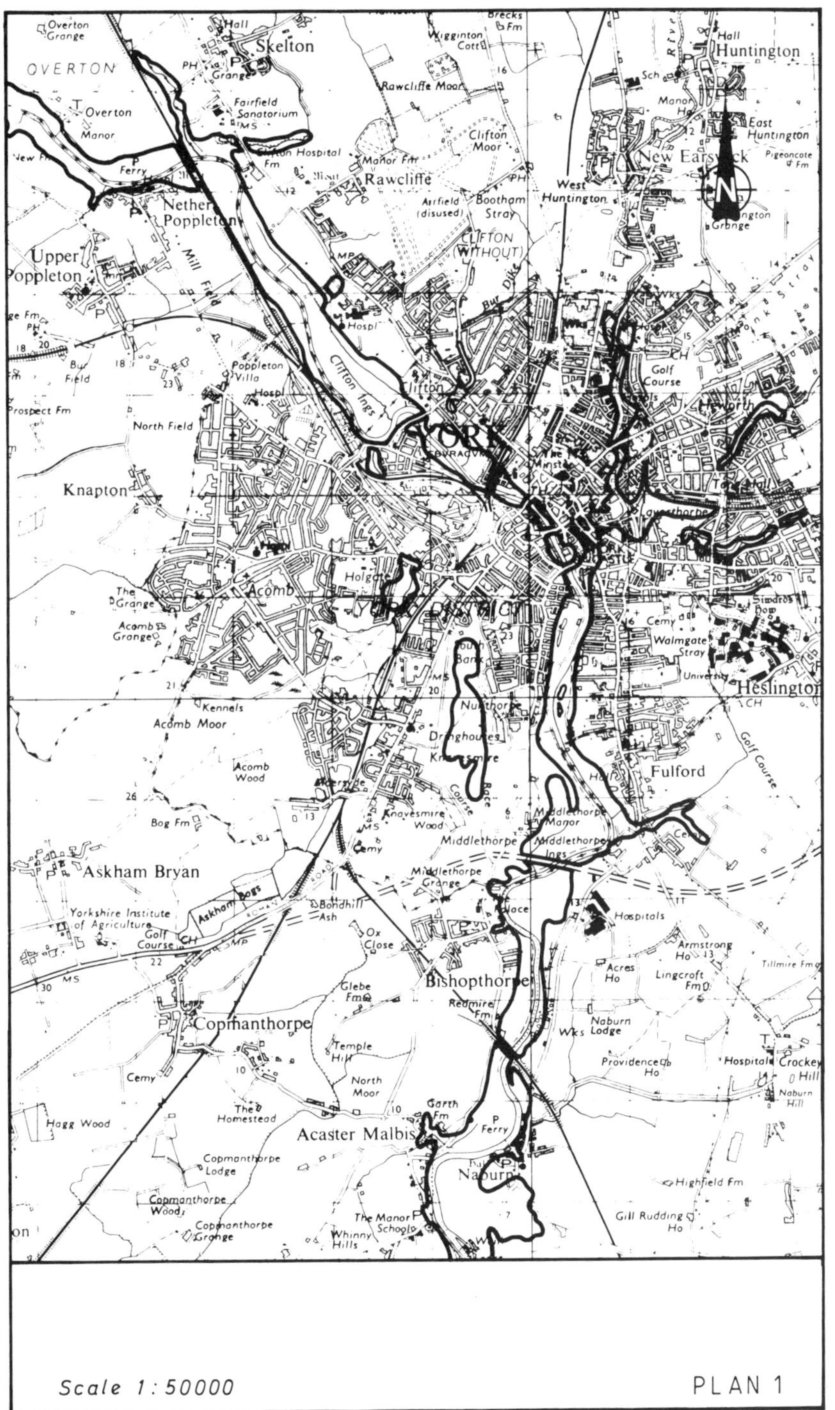

Fig. 5.8 — York, flooded areas, January 1982.

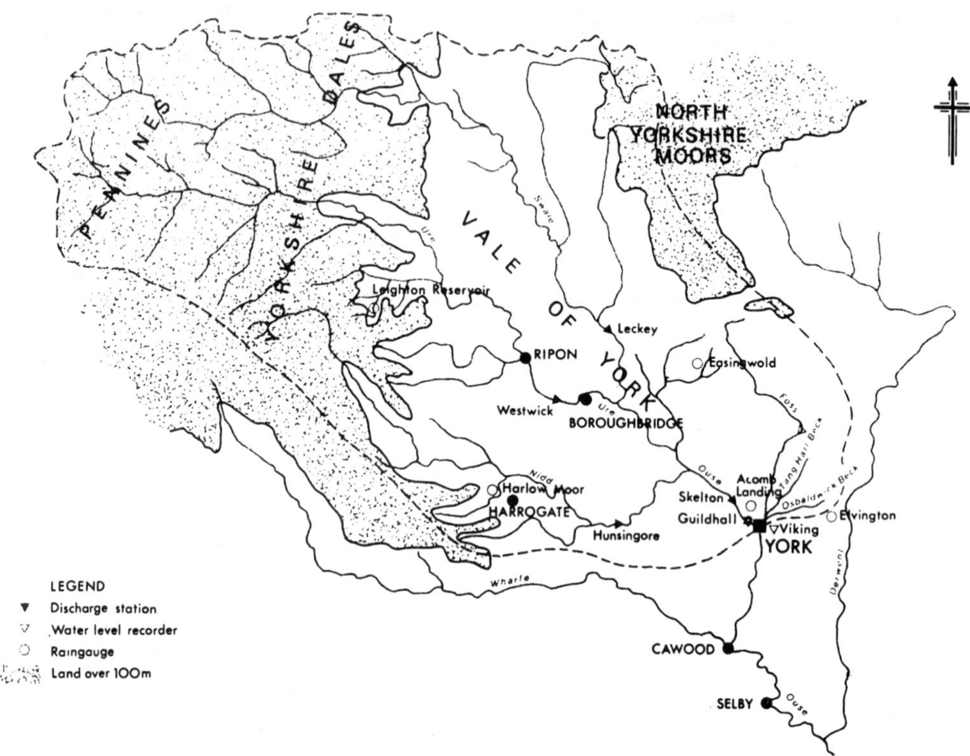

Fig. 5.9 — River Ouse catchment.

Table 5.12 — Data on flood levels at York Guildhall

Level records	1885–1982	1935–1982
Return period of 1982 peak (10.12 m) (years)	56	35 years
Peak level of 100 year flood (m)	10.40	10.63

extrapolation, the 100 year return period gives a flood level of 10.16 m at the Guildhall, much less than the 10.63 m from the EV1 distribution and deemed to be more reliable. In obtaining different answers from the distributions, the engineer must exercise judgement in using the most appropriate value.

Since the flooding along the Foss tributary results mainly from the backwater influences from the River Ouse, a special detailed study was made of the probability of the coincidence of peak flows in the Foss with high Ouse water levels. Flows in the Foss were synthesized using The Flood Studies Report (NERC, 1975) unit hydro-

Table 5.13 — Number of floods above given levels

Level (m AOD)	Number of floods in last 98 years	20 years	5 years
⩾8.80	44	19	8
⩾8.95	25	14	7
⩾9.25	25	14	7
⩾9.55	4	2	2
⩾9.85	2	1	1

graph method and local rainfall data, with the unit hydrograph derived from catchment characteristics. The synthesized Foss peak flows were used to derive partial series conditional on recorded coincident levels in the Ouse exceeding threshold values. Four such partial series were derived for Ouse levels of more than 7, 7.5, 8 and 8.5 m and return periods of peak Foss flows evaluated. For example, the 100 year event conditional on an Ouse level greater that 7.5 m had a flow of 30 $m^3 s^{-1}$ in the Foss.

The hydrological studies provided a broad range of criteria from which the engineers could select design levels and discharges, in conjunction with cost–benefit considerations, to form the basis of flood alleviation schemes.

Conclusion

The flood defences for York have been divided into 10 separate schemes. Several of these are concerned with raising existing flood banks and building new concrete walls to protect housing and industrial properties. The two most costly schemes are of special interest. The first is the enlargement of the capacity of the natural washlands upstream of the city to store temporarily more of the high flood flows. The storage capacity of Clifton Ings (Fig. 5.8) has been increased from 0.75 to 2.30 Mm^3 of flood water by raising and extending the flood banks to 5.1 km in length and installing control intakc and outfall structures to regulate the filling and draining of the temporary storage. The most expensive scheme concerns the River Foss with its additional catchment area of 125 km^2. The proposed scheme recommends a short length of retaining wall along the river, a turnover barrier gate to separate the flow of the Foss from the Ouse and a large pumping station (capacity, 30 $m^3 s^{-1}$) to lift the Foss flow into the Ouse. In addition, the combined sewerage system is to be upgraded.

5.6 FLOOD PROTECTION OF THE YELLOW RIVER

LOCATION The Yellow River (Huanghe) is the great river of northern China.

SOURCE Gong Shiyang & Wu Zhiyao (1983) *Flood prevention on the lower reaches of the Yellow River,* Natural Resources/Water Series No. 11, *Flood damage and control in China.* Department of Technical Cooperation for Development, UN, pp. 45–52.

PROBLEM The Yellow River rises on the Qinghai Plateau, north of Tibet, flows for 5464 km to the Bohai Sea and has a total catchment area of 752 000 km^2 (Fig. 5.10). The lower course of the river downstream of Zhengzhou (785 km) flows between dykes (levees) with the river bed above the general level of the North China Plain owing to sediment deposition. During the past 2000 years the river has breached the dykes more than 1500 times and changed its course 26 times. A particularly serious flood this century in 1933 with a peak discharge of 22 000 $m^3 s^{-1}$ at Sanmenxia caused 54 dyke breaches and flooded 10 000 km^2 inhabited by 3.6 million people. During recent years special studies have been made of the floods and of the sediment load in order to plan a strategy for containing the flood flows and improving the flood defence works.

Hydrometric studies

Located in eastern Asia, the Yellow River catchment experiences a monsoon climate with seasonal precipitation diminishing inland from an annual average of around 600 mm near the river mouth to less than 200 mm in the source regions. The low winter precipitation is all snowfall; the wettest months are July and August which can bring heavy rainstorms. There are four periods of high water each year in the lower reaches of the river. The ice runs in February, when the ice floes from the thaw near Kaifeng pile up downstream against the still frozen stretches near Jinan and cause a rise in river levels and hence a threat to the main dykes. The spring floods in March and April result from the snow and ice melting in the upper reaches but the peak discharge and total runoff are relatively small and are no threat to the dykes in the lower reaches. It is the summer floods from the July and August rainstorms and autumn floods from continuous wet weather in September and October which can be the most disastrous.

In addition to the quantitative data available from gauging stations established this century, the valuable historical information from many centuries regarding rainfall and floods is used to assess river flow characteristics. The hydrometric station established at Sanmenxia in 1919 and the station at Huayuankou where the river enters the plains are the key gauging stations in the network run by the Yellow River Conservancy Commission of 474 hydrometric stations and 53 gauging stations in the catchment which also has 1400 rain gauge stations.

Floods

During the summer flood season, the catchment area above Tuoketuo provides a baseflow of 2000–3000 $m^3 s^{-1}$ to the floods generated by the rainstorms in the three areas of the middle reaches (Fig. 5.10).

(1) Tuoketuo down to Longmen, of area 111 691 km^2 with tributary catchments Wuding, Fenhe and others.

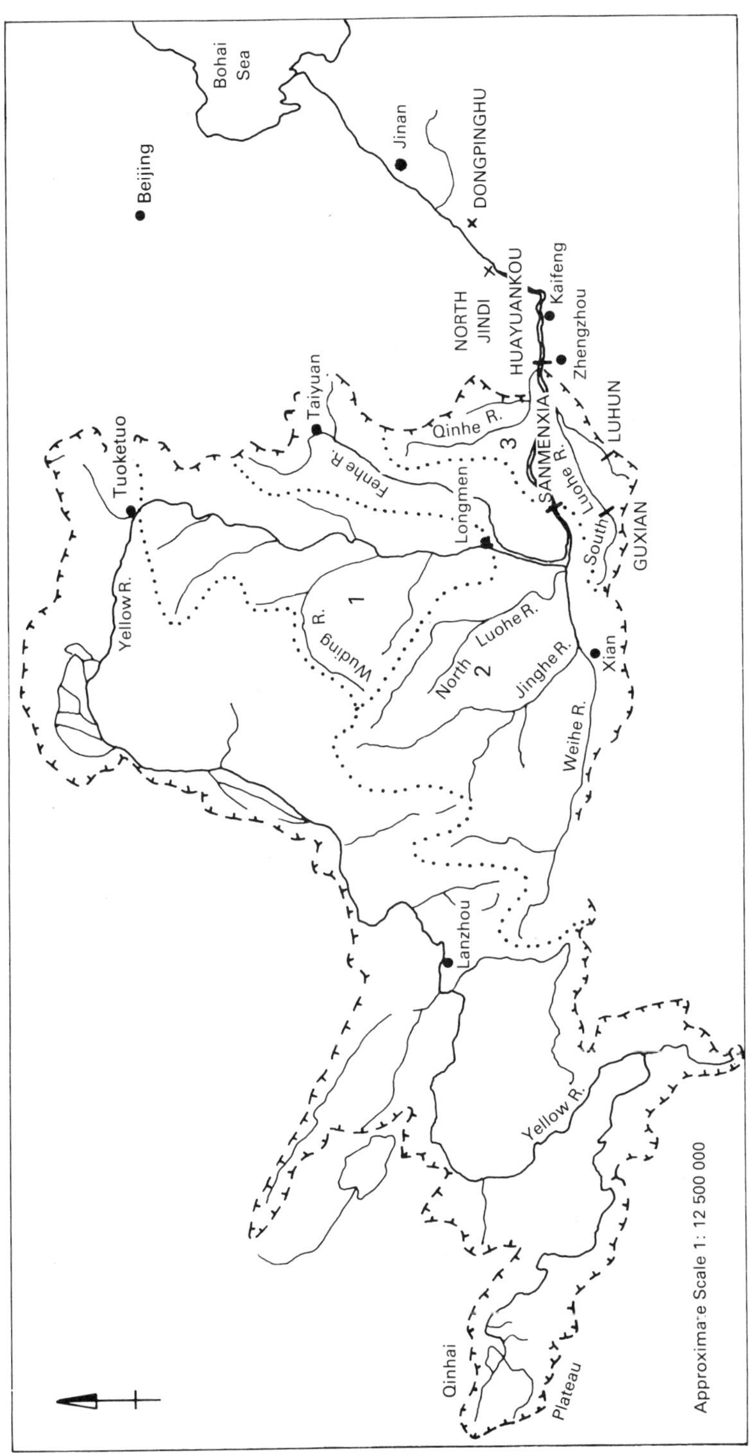

Fig. 5.10 — Sketch map of the Yellow River Basin: 1, 2 and 3, middle reach areas; ●, town.

(2) Longmen to Sanmenxia, of area 161 671 km^2 with tributaries Luohe (northern), Jinghe and Weihe.
(3) Sanmenxia to Huayuankou, of area 41 687 km^2 with tributaries Luohe (southern), Yihe and Qinhe.

Examples of two major floods having their origins in different areas are shown in Table 5.14. Hydrographs of these floods are seen in Fig. 5.11. Floods having their

Table 5.14 — Observed floods at two stations

Main area of flood origin	Year	Sanmenxia			Huayuankou		
		Peak flow ($m^3 s^{-1}$)	Flood volume ($10^9 m^3$)		Peak flow ($m^3 s^{-1}$)	Flood volume ($10^9 m^3$)	
			5 days	12 days		5 days	12 days
1 and 2	1933	22 000	5.18	9.18	20 400	5.96	10.05
3	1958	6 400	2.65	5.15	22 300	5.63	8.68

origins in areas 1 and 2 have a high sediment content eroded from the loess plateaus while area 3 floods concentrate rapidly giving a short forecast time and hence constitute a serious threat in the lower reaches. Fortunately the two types of flood peak do not often coincide but the Sanmenxia 5 and 12 day discharge volumes generally provide over 50% of the corresponding Huayuankou total discharges (Table 5.15). The duration of each flood is usually about 12 days but most of the discharge occurs in 5 days and the volume of flood flow greater than 10 000 m^3 s^{-1} is not large.

Sediment

The sediment load in the flood flows provides the greatest problem in the work of flood protection in the lower reaches. Severe erosion in the loess plateau of the middle reaches produces on average 1.6 billion tons year^{-1} carried to the danger areas. Some 400 million tons are deposited in the channel, some 800 million tons are deposited at the mouth of the river and the remainder is carried out to sea. The delta of the river grows out at a rate of 0.3–0.4 km year^{-1}. At the Huayuankou hydrometric station for a flood discharge of 3000 m^3 s^{-1}, the corresponding stage was 2 m higher in 1978 than it was in 1950. Most of the sediment is carried in the flood flows and it varies considerably within a year and from year to year. The alternation of intensive scouring and channel deposition radically changes the flow regime and can quickly threaten vulnerable sections of the dyke. These large-scale modifications to the river bed and floodplain terrace areas within the dykes results in continuous anxiety for the safety of the flood defences.

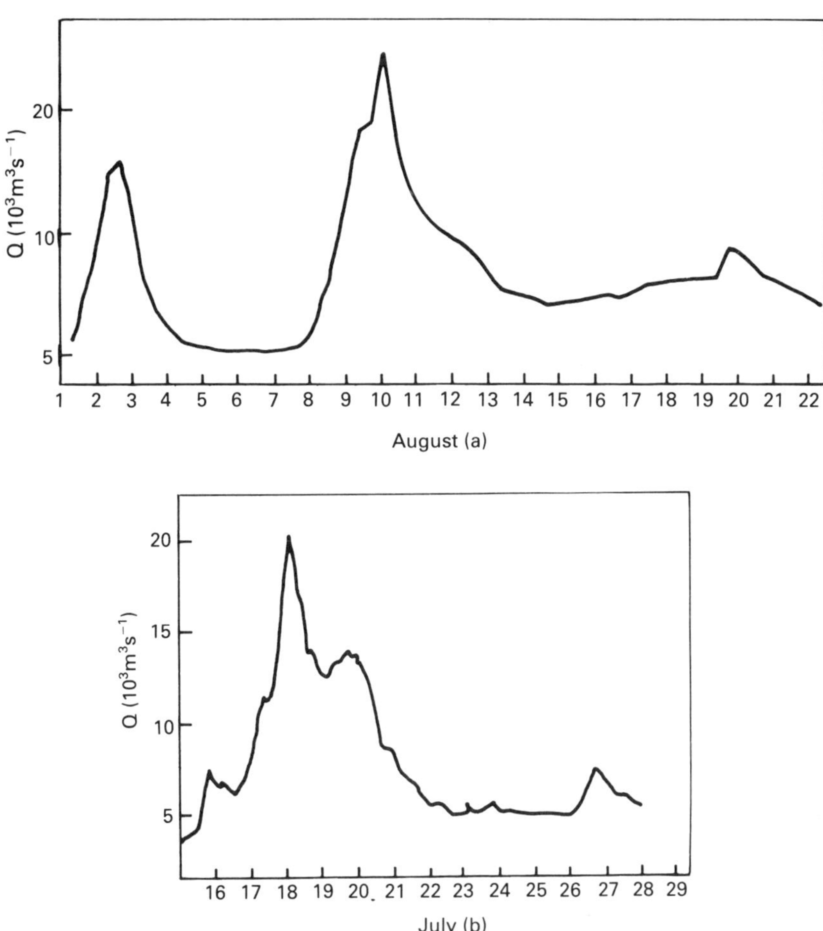

Fig. 5.11 — Flood hydrographs on the Yellow River: (a) Sanmenxia, 1933; (b) Huayuankou, 1958.

Flood protection works

A policy of retaining the floods upstream and controlling their discharge downstream, diverting the floods and detaining them on either bank if necessary, has been adopted to manage the floods in the lower reaches. The remedial works proposed are as follows.

(1) *Dykes* The dykes along both banks totalling 1369.5 km in length are to be made safe for peak discharges up to 22 000 m^3 s^{-1}.
(2) *Training works* These are carried out along vulnerable stretches where the dykes and the banks of the floodplain are open to direct scour by the current and spur dykes or short groins are built to divert the flow.
(3) *Detention basins* Of the naturally flooded areas along the lower reaches, two are to be developed into controlled detention basins. The design capacity for the

Table 5.15 — Flood parameters at Huayuankou

	Number of years	Mean value	Coefficient of variation	Frequency, 1%	Frequency, 0.1%
Peak flow ($m^3 s^{-1}$)	39	9780	0.50	27 400	38 700
Discharge volume in 5 days ($10^9 m^3$)	51	2.65	0.49	7.13	9.84
Discharge volume in 12 days ($10^9 m^3$)	51	5.29	0.41	12.2	15.8
Discharge volume over 10 000 $m^3 s^{-1}$ ($10^9 m^3$)				3.3	6.2

diversion works is 10 000 $m^3 s^{-1}$ and the effective storage for both the Dongpinghu (right bank) and the north Jindi (left bank) detention areas will be 2.0 $10^9 m^3$.

(4) *Middle-reach reservoirs* Three reservoirs are to retain the peak flood waters generated in the three areas experiencing the summer storms. The Sanmenxia Reservoir built on the main river in 1960, originally primarily for water storage and power generation, has been modified to operate effectively for flood control. Bottom outlets in the dam allow the passage of flood water and sediment. In a major flood above Sanmenxia the outflow will be restricted to 15 000 $m^3 s^{-1}$ and, when a flood arises in area 3, the reservoir will be operated to hold 3–4 $10^9 m^3$ of water, thus relieving the pressure on the lower reaches. Two dams in area 3 on the catchments of the Yihe and Luohe have a joint capacity of 600 Mm^3 available to retain the 100 year flood and to reduce the peak flow at Huayuankou by 2000–6000 $m^3 s^{-1}$ The details of the three reservoirs are given in Table 5.16.

Table 5.16 — Particulars of the middle-reach reservoirs

Reservoir	Sanmenxia	Luhun	Guxian
Location	Main River	Yihe	Luohe (southern)
Catchment area (km^2)	688 421	3492	5370
Dam	Gravity concrete	Earth	Gravity concrete
Dam height (m)	106	55	118
Total storage ($10^9 m^3$)	36.0	1.29	1.16
Flood capacity ($10^9 m^3$)	6.04	0.64	0.46
Outlet capacity ($m^3 s^{-1}$)	15 000	16 150	12 000

Conclusion

The remedial works carried out in the past 30 years have ensured success in protecting the main dykes but the task of controlling the flood waters and the sediment is enormous and will continue for many years. Modern methods of planning and management based on sound hydrometry together with the employment of the latest engineering techniques and equipment will help in this vital work. In conjunction with the engineering aspects of the project, concern for the safety and well-being of the millions of local inhabitants plays a large part in the design and operation of the remedial works.

5.7 CONON VALLEY FLOOD PROTECTION SCHEME

LOCATION The River Conon and its tributaries drain from the north-west highlands of Scotland and flow eastwards into the Cromarty Firth at Dingwall.

SOURCES Guganesharajah, K., Thorn, D.G., Evans, T.E., & Harpin, R. (1985) Mathematical modelling of the River Conon. *The Hydraulics of Floods and Flood Control Conference.* British Hydromechanics Research Association, Paper H3.

Johnson, F. G., Jarvis, R. M., & Reynolds, G. (1981) Use made of The Flood Studies Report for reservoir operation in hydroelectric schemes. *Flood Studies Report — five years on.* Thomas Telford, pp. 85–90.

Sir M. MacDonald & Partners (1984) *Conon Valley Flood Protection Scheme,* Final Report to the Highland Regional Council, Department of Water and Sewerage.

PROBLEM The embankments of the lower reaches of the River Conon were breached during high flood flows in 1962, 1966 and 1983. The flooding led to extensive damage to agricultural land, to the flooding of main roads and in 1983 to the drowning of a local inhabitant. Concern was expressed that, if the embankments failed near Conon Bridge, then serious urban flooding would occur. Local opinion was that the hydropower development in the catchment and the transfer of flows from neighbouring catchments were also partly responsible for the worsening floods. Sir M. MacDonald & Partners were retained to study the flood problem of the lower Conon and advise.

Investigation

The Conon catchment area, 1250 km^2, ranges in elevation from a 1110 m peak to sea level at the Cromarty Firth (Fig. 5.12). The contributing main rivers flow through natural lochs, lochs used by the North of Scotland Hydroelectric Board and man made reservoirs. The development of hydropower generation was accelerated in 1945 and, in the following 15 years, seven power stations, seven main dams, 32.2 km of tunnels and 24.1 km of aqueducts were constructed.

The hydrometric records available for the study were 29 rain gauges (20 read

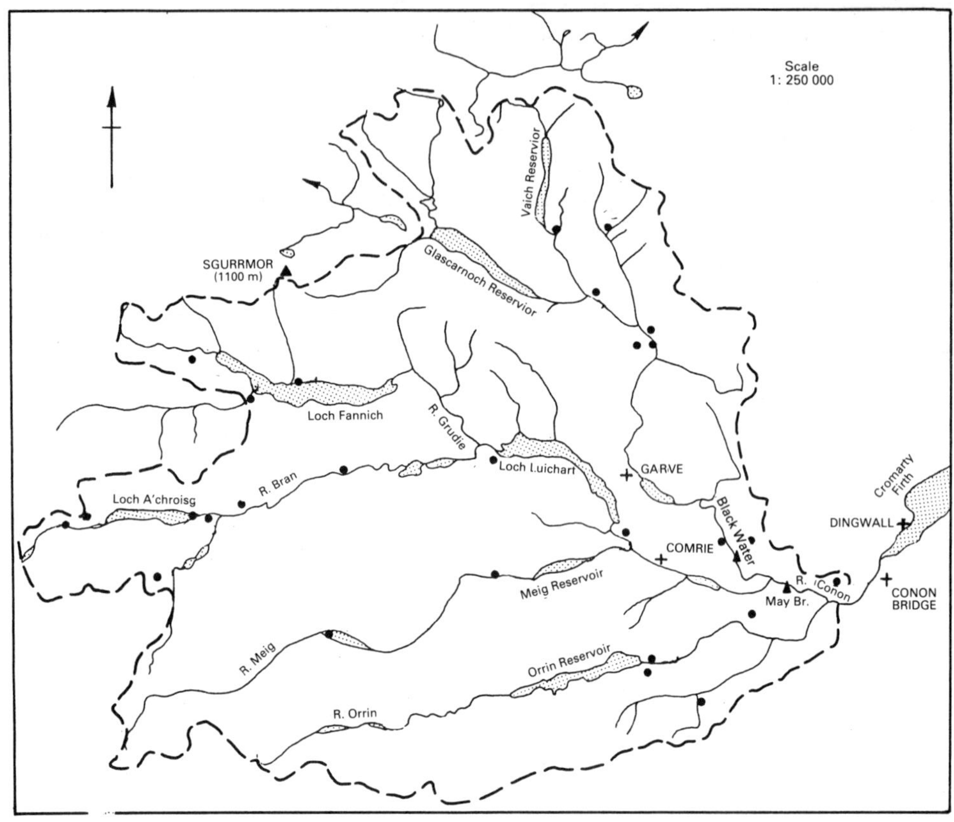

Fig. 5.12 — Conon catchment hydrometric network: ———, rivers; ◌, lochs or reservoirs; ---, catchment boundary; ●, rain gauges; ▲, water level gauges.

daily and nine monthly), two river gauging stations (Conon and Blackwater) (Fig. 5.12) and two tidal stations on the north bank of Cromarty Firth. In addition, flow data were available from the North of Scotland Hydroelectric Board stations.

The analytical studies were subdivided into the hydrological modelling to investigate the floods in the catchment for return periods of between 10 and 100 years and the hydraulic modelling of the lower Conon reaches under steady and transient flow conditions to determine critical water levels.

Hydrological modelling

The North of Scotland Hydroelectric Board catchment model based on The Flood Studies Report method (NERC, 1975) but incorporating North of Scotland Hydroelectric Board experience in multireservoired catchments reported by Johnson *et al.* (see source cited above) was used in the hydrological studies. The Conon catchment was subdivided into 19 subcatchments (Fig. 5.13) and, for each subarea, a unit hydrograph was obtained. For the reservoirs, spillway discharges, area–level curves,

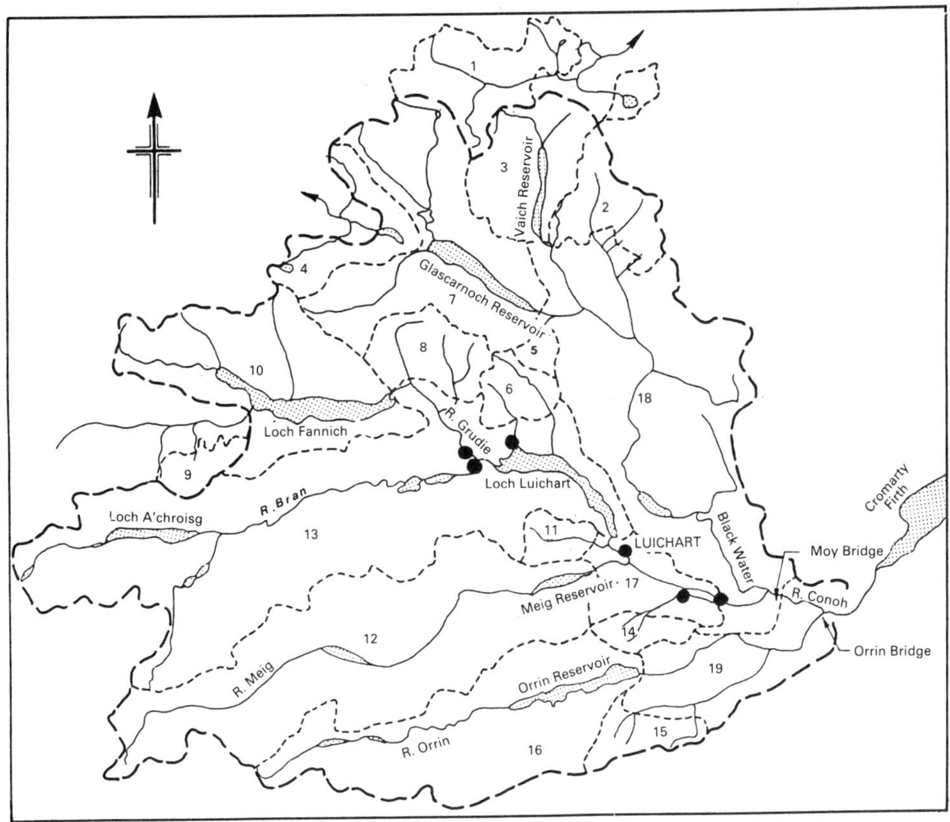

Fig. 5.13 — Conon catchment model subcatchments: ——, river; ◌, loch or reservoir; – – –, catchment boundary; - - - subcatchment boundaries; ●, power stations. The subcatchments are as follows: 1, Glen Beag and Crom Loch; 2, Strath Rannoch; 3, Vaich; 4, Droma; 5, Allt Guibhuis; 6, Coire Mhuilion; 7, Glaxarnoch; 8, Allt Dearg; 9, Strath Chrombuill; 10, Fannich; 11, Glen Marksie; 12, Meig; 13, Luichart, 14, Allt Na Fainich; 15, Allt Gowrie; 16, Orrin; 17, Torr Achilty; 18, Black Water; 19, Orrin Residual.

initial reservoir levels and details of flow diversions had to be incorporated where pertinent. The model was run for the following ranges of variables: reservoir levels for 50%, 40%, 30%, 20% and 10% exceedences; rainfall depths for 10, 25, 50 and 100 year return periods; storm durations of 24, 36, 48 and 72 h; nine winter storm profiles from 5.6% to 94.4% exceedence probabilities. The results for the peak flows at Moy Bridge showed that the critical storm duration was about 48 h and this was adopted as the design storm duration. To determine the return period of the Moy Bridge peak flows, the combined probability analysis of the rainfall probabilities and reservoir level probabilities was necessary. The resulting Gumbel plot, most reliable in the return period range 30–85 years, indicated that the 1983 flood of approximately 694 $m^3 s^{-1}$ had a return period of about once in 48 years.

A special study of the operation of the Luichart and Meig Reservoirs was made to establish the effects of uprating the Luichart generators on flooding in the Conon valley and found a significant reduction in the peak flows at Moy Bridge was possible.

The value of an analysis of the annual maximum peak flows recorded at Moy Bridge is governed by the quality of the record and owing to the changes in the catchment, the annual maximum series is not homogeneous. Statistical tests indicated no trend or periodicity but the reservoir modifications and suspect data made the Gumbel (EV1) distribution strictly invalid. However, a joint probability curve was constructed from the North of Scotland Hydroelectric Board catchment model results and a Gumbel fit to the recorded points omitting the outliers of 1983 and 1966. Based on this analysis the 1983 peak would have a return period of approximately 50 years, with the bankful flow of 467 $m^3 s^{-1}$ having a return period of around 9 years.

Several design flood hydrographs were simulated with the North of Scotland Hydroelectric Board catchment model with the peak flows adjusted to correspond with the frequency curve values and taking into consideration the historic flood effects on the river embankments. The following flood events were used for input into the hydraulic model of the lower reaches of the river (Table 5.17).

Table 5.17 — Derived floods used in hydraulic analysis

Flood event	Approximate peak flow at Moy Bridge ($m^3 s^{-1}$)	
	Present Luichart, 26 MW	Future Luichart, 34 MW
1 in 25 years	570	520
1 in 48 years	665	620
1 in 100 years	770	720
1966	1080	1055
1983	695	650
1981	380	320

1983 Luichart capacity, 30 MW.

Hydraulic modelling

The model HYDRO is based on a four-point implicit finite difference solution of the Saint Venant equations for open channel flow. Newton–Raphson iteration is used to obtain the solution for the whole river system at a given time step. The model can accommodate loops in the drainage system and can use large time steps and irregular lengths of channel reaches. It is generalized to cater for natural or man-made drainage systems of varying physical features such as irregular geometry, varying roughness coefficient, lateral inflows, onstream and offstream reservoirs, off-channel storage and the impedance of bridges, culverts, weirs and orifices in any reaches of the channel network.

The lower Conon reaches modelled by HYDRO are shown in Fig. 5.14 and the schematic layout is also given showing the out-of-bank storages due to the 1983 flood (Fig.5.15). Additional surveying work was needed to provide river reach lengths and

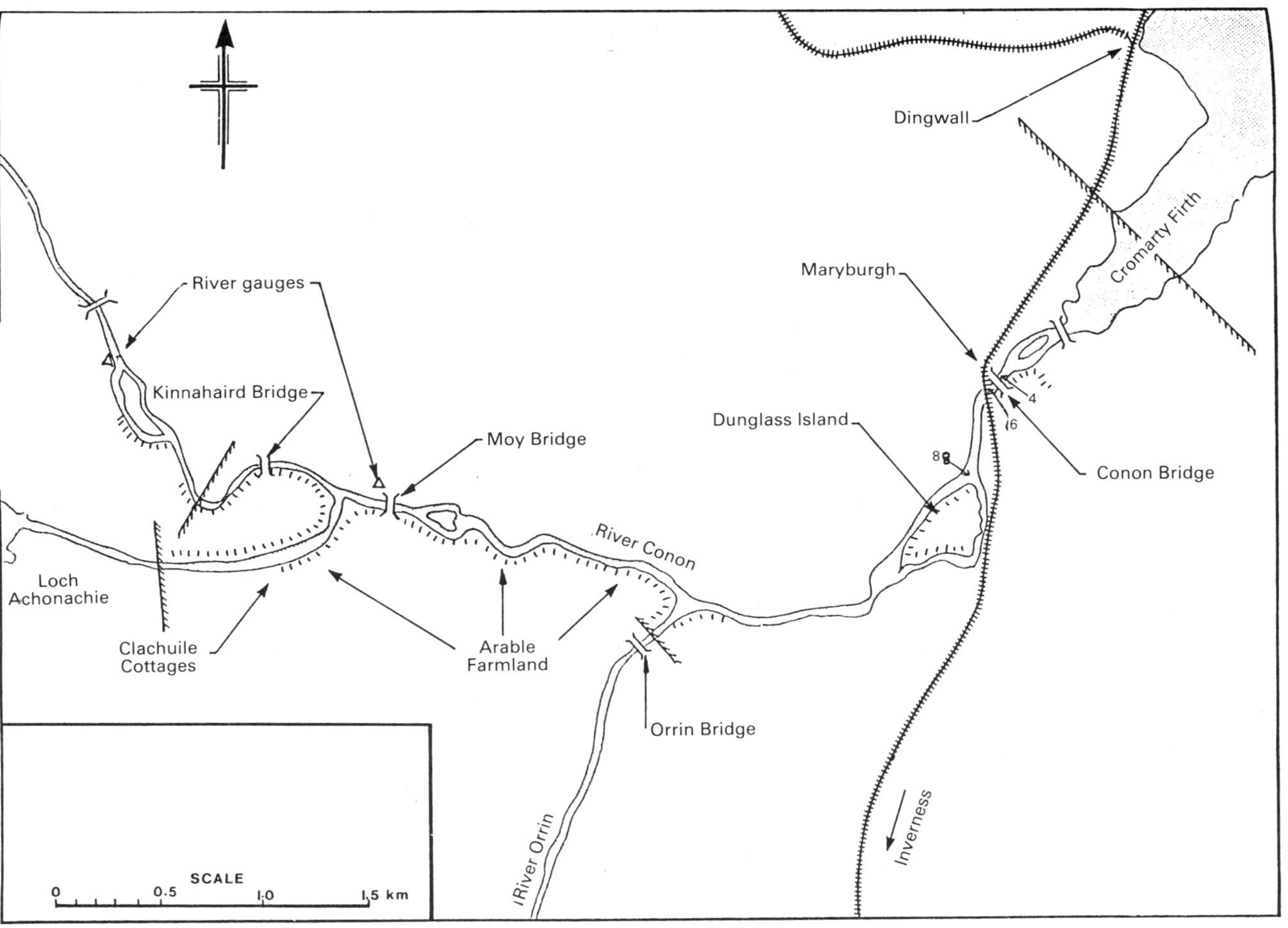

Fig. 5.14 — Conon river model area: ///////, model boundary; ''''''', existing embankments;)(, bridges; +++++, railways.

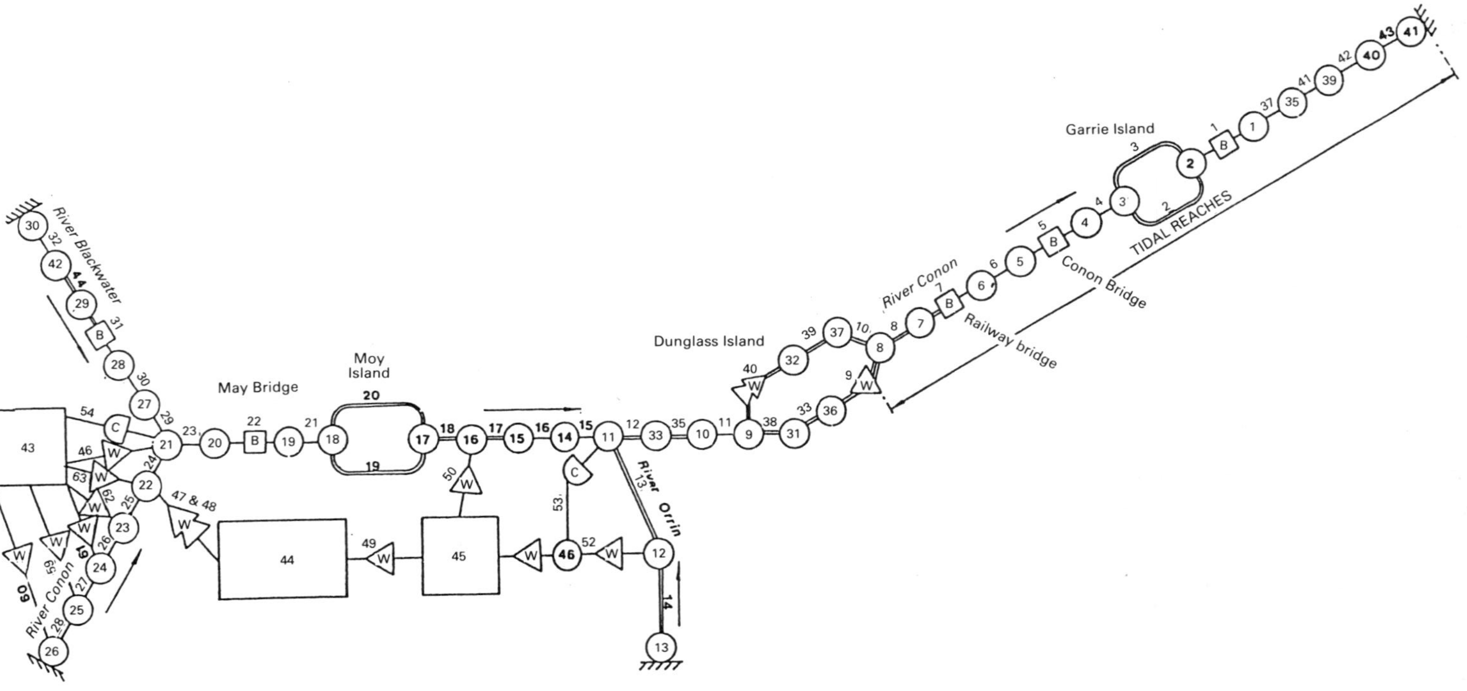

Fig. 5.15 — Conon river model schematic layout for December 1983 flood event: —○—, nodes; —○—, boundary nodes; —B—, bridges; ○—○, channel reaches; ○═○, channel reaches with floodplain; —W—, weirs with bidirectional flow; —W—compound weirs; —C—, culvert; □, storage cells.

cross-sections and area–depth characteristics for the floodplains. Seven data files needed to be compiled for a specific model run. The model was calibrated using six historic floods, the observed water levels at Moy Bridge and the peak levels at Conon Bridge.

Before presenting a selection of the results from HYDRO, it is pertinent to outline the objectives of the flood alleviation scheme.

(1) Protection at Conon Bridge against the 1 in 100 year flood.
(2) Reduction and/or elimination of flood surges across the trunk road and of main damage costs resulting from breaching of flood embankments.
(3) Reduced frequency of flooding of agricultural land.
(4) Reduced flood levels at Moy Bridge.

Of several schemes considered, two options were selected for presentation.

Option A This achieves each objective for events up to the magnitude of the 1983 floods.

Option B This achieves all but the last objective for events up to the magnitude of the North of Scotland Hydroelectric Board 1 in 48 year hydrograph.

Options A and B differ in the extent of works proposed for the lower Conon valley. Option A which includes some controlled use of washlands provides a higher standard of protection than B at the danger points on the river but is more expensive. This is demonstrated in Table 5.18 which gives resulting levels at three nodes 8, 6 and

Table 5.18 — Water levels at Conon Bridge

	Water level (m AOD)			Q_{max} at node 6 ($m^3\ s^{-1}$)
	Node 8	Node 6	Node 4	
Historic events				
1966	7.35	6.27	5.69	1337
1983	5.96	5.55	5.41	656
1981	4.62	3.91	3.75	391
Design events-option A				
100 year flood	5.90	5.37	5.20	723
25 year flood	5.39	4.79	4.61	558
1966	7.26	6.04	5.50	1338
1983	5.43	4.86	4.69	560
1981	4.59	3.88	3.72	386
Design events-option B				
100 year flood	5.99	5.52	5.37	739
25 year flood	5.46	4.87	4.68	582
1966	7.33	6.26	5.68	1337
1983	5.50	4.99	4.83	587
1981	4.62	3.91	3.75	391

4 (shown in Fig. 5.15) for 13 runs of the model using the three single historic events and then together with the design flood inputs assuming protection works under options A and B.

Conclusion

From the model runs, the consultants used the detailed level results for the critical river reaches to design the remedial and defensive works, with detailed recommendations for the containing embankments. A summary of the costs and an economic analysis were also presented.

A rather unique and satisfactory engineering solution to alleviating the flood problem was found. Lengths of embankment were slightly lowered at crucial locations and protected using geotextile material to prevent breaching, the cause of most of the damage. The agricultural land could then be used to act as a washland storage area with controlled flooding. The uprating of the Luichart turbines enabled the frequency of controlled flooding (1 in 17 years) to be reduced from 1 in 9 years currently experienced even though the embankment was lowered. Without the unsteady-state modelling, the positioning and the levels of the lowered embankments would have been very difficult to evaluate.

5.8 SANTA CRUZ BASIN DESIGN FLOODS

LOCATION The Santa Cruz River flows by Tucson, Arizona, into the Gila River, a tributary of the Colorado River.

SOURCE Ponce, V. M., Osmolski, Z., & Smutzer, D. (1985) Large basin deterministic hydrology: a case study. *J. Hydraul. Eng., Am. Soc. Civ. Eng.* **111** No 9, 1227–1245.

PROBLEM Prolonged and widespread rainfall in September 1983 over southeastern Arizona resulted in new record flows in the Santa Cruz River at Tucson. Severe flooding and bank erosion along the main river and tributaries led to damage to bridges, failure of bank protection and destruction of property, with damage estimates totalling 500 million dollars. Although many flood studies have been made in the arid and semiarid regions of the USA, the local authorities considered that a fresh appraisal of the surface water hydrology of the Santa Cruz Basin would be beneficial. The Pima County Department of Transportation and Flood Control District, Tucson, supported this study of the application of the deterministic approach to evaluate the hydrology of a large complex catchment in semiarid climatic conditions.

Santa Cruz Basin

The Santa Cruz River rises in the mountains in southeastern Arizona (2860 m above sea level) takes a looped course through part of Mexico before settling in a northerly direction to Tucson (Fig. 5.16). The drainage basin to the river gauging station at Cortaro (9073 km^2) is characterized by isolated mountain blocks separated by broad

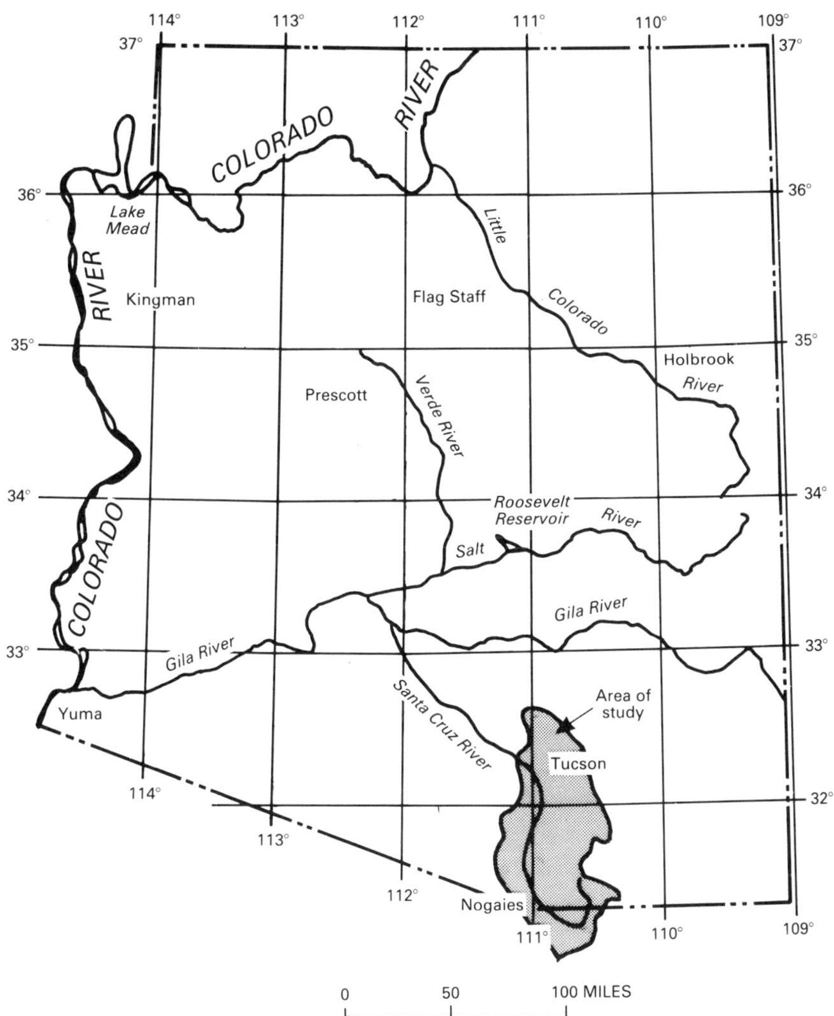

Fig. 5.16 — Santa Cruz Basin location.

alluvial valleys. The variable precipitation, short intense thunderstorms in summer and more widespread frontal rains in winter, results in an average annual rainfall of 356 mm. Stream flow occurs mainly in direct response to rainfall. In the upper basin, channels are dry for long periods. Stream beds are permeable and stream flow is lost to the subsurface water and groundwater. With sparse vegetation, there is a rapid runoff response to rainfall but flood peaks are reduced by the stream bed infiltration. Particulars of the US Geological Survey river gauging stations (Fig. 5.17) are given in Table 5.19. The 1983 event resulted from 152–203 mm of rain falling in the 6 day period 28 September–3 October, which contributed to a new record annual total of 635 mm in southeastern Arizona.

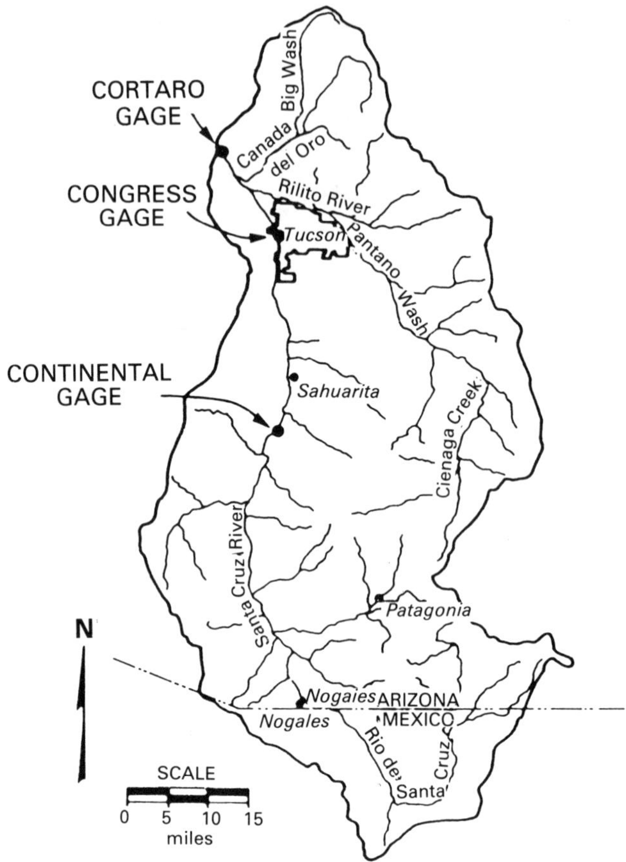

Fig. 5.17 — Stream configuration and gauging stations.

Table 5.19 — Santa Cruz River gauging stations

Station	Catchment area (km^2)	Record (years)	Peak flow ($m^3\ s^{-1}$)	
			Pre-1983	1983 estimate
Cortaro	9073	44	651	1840
Congress	5755	80	671	1491
Continental	4356	37	750	1273

Basin model

A comprehensive computer-based model system was used to simulate the rainfall-runoff processes in the complex configurations of subcatchment areas. The Santa Cruz Basin was divided into 60 upland subcatchments contributing flow hydrographs and 119 reach subcatchments contributing lateral inflow into well-defined river catchment reaches. For generating the subcatchment hydrographs, the unit hydrograph method of the Soil Conservation Service (USDA, 1972) was applied. A unit hydrograph for each subdivision, derived from the dimensionless unit hydrograph formula and the relevant subcatchment characteristics was convoluted with the effective storm profile to give a contributary hydrograph at each outlet.

For the river reaches, river channel routing applied the Muskingum–Cunge method (Ponce & Yevjevich, 1978) based on local flows, channel cross-sections and gradients. The channel routing module can vary its parameters in time as a function of the local flows and also account for channel losses. Each channel reach was subdivided into two or three subreaches with their own geometric data. The basin model also included a simple reservoir routing module.

A composite deterministic model of such complexity requires a large amount of physical data but, once this has been assembled, the spatial variability of the hydrology over a large basin can be well represented.

Data requirements

The subcatchment areas, hydraulic lengths and slopes of the streams were evaluated from the US Geological Survey topographic maps. Channel cross-sections were compiled from several agencies and extra measurements were made at remote locations. Runoff parameters for the different subcatchments were obtained from soil and vegetation map information using Soil Conservation Service methods. Since the aim of the project was to estimate extreme flood flows, baseflows were considered negligible and assumed to be zero. Three reservoirs in the basin were assumed to be at crest level for running the model. The important measures of stream bed infiltration were gleaned from research reports and maps of water table depths helped to assess initial rates at 0.305, 0.61 and 1.22 m day^{-1} for low-, medium- and high-level flows, respectively.

Model calibration

The model was calibrated with the data from the October 1983 flood. Runoff coefficients and stream bed infiltration rates were adjusted to match calculated with recorded flows. A time interval of 7.5 min was found to be satisfactory for the smaller subcatchments and a matching maximum subreach length of 1.61 km was taken for subdivisions of the channel reaches. The model automatically generated additional cross-sections related to the input sections when required. The most reliable records of the flood at the Continental and Congress river gauges were used and a decrease in runoff coefficients to a range 62–85 was found to be necessary. Infiltration rates were increased by a factor of 1.6 upstream of Continental, by 2.2 between Continental and Congress and by 5.2 between Congress and Cortaro. Thus the stream bed losses ranged from 0.49 to 6.34 m day^{-1}. Peak flows obtained in the calibration runs were 2374 m^3 s^{-1} at Cortaro, 1610 m^3 s^{-1} at Congress and 1259 m^3 s^{-1} Continental.

Storm simulations

Since floods can be caused by the widespread winter rains and the short summer storms, flood hydrographs were simulated with the two types of rainfall input. A general storm assumed coverage of the whole basin with rainfall depth uniformly distributed in time. Local storms which cover approximately 1050 km^2 were centred at three strategic points to produce peak flows at the gauging stations. The Soil Conservation Service standard dimensionless temporal storm was applied to produce hyetograms for these simulated summer storms.

The estimated time of concentration for the whole basin for a 100 year frequency flood event was estimated to be 24 h. Therefore storm durations of 24, 48 and 96 h were chosen and the point rainfall depths of 100 year frequency for these durations, 116.8, 152.4 and 170.2 mm respectively, were obtained from the National Weather Service publications. These point values were reduced by depth–area factors from published National Weather Service curves. A summary of results from the storm simulations using the 100 year frequency rainfall data is given in Table 5.20.

Table 5.20 — Simulated peak flows

	Peak flow ($m^3\ s^{-1}$) for the following duration		
	24 h	48 h	96 h
General storms			
Cortaro	1573	1407	555
Congress	1290	1160	504
Continental	1661	1132	453
Local storms			
Cortaro	1078	1353	1002
Congress	1658	1896	1324
Continental	1174	1336	1016

Conclusion

As a result of the deterministic model study of the Santa Cruz Basin, the Pima County authority set new statutory discharges for the Santa Cruz River. Regulatory discharges for floodplain delineation and management and design discharges for channel and bridge improvements are given in Table 5.21. The adopted discharges

Table 5.21 — Santa Cruz design discharges

Station	Design discharge ($m^3 s^{-1}$) Regulatory	Design
Continental	1270	1560
Congress	1700	1980
Cortaro	1980	2260

were influenced by the simulated 24 h general storm peak at Continental, the 48 h local storms peak at Congress and the calibration peak at Cortaro.

6

Reservoir spillways

INTRODUCTION

The most spectacular constructions associated with water engineering are the impounding dams built across river valleys to hold back in storage the natural river flows. The dams and their resultant reservoirs vary in size according to location on a river system and to the purpose which they have to fulfil. Similarly the dam structures can have a variety of designs depending on the construction material of masonry, concrete or earth and rockfill chosen usually with regard to availability, suitable foundations and cost. Whatever the design, a gravity or buttress dam or an earth embankment, the structure must have a spillway built to pass extreme flood discharges which would otherwise endanger it. The evaluation of the design flood discharge is one of the most challenging tasks that face the engineering hydrologist. In this chapter are described a selection of different hydrological studies that have been undertaken prior to the design of some notable dams.

Although dams have been built for many centuries, the increase in engineering skills and construction capabilities in the last hundred years has led to developments in design considerations. To assess extreme floods, engineers have habitually turned to the collection of historic flood records and these naturally have a range of origins and are of variable reliability. For many rivers to be impounded, there are no flood records and recourse must be made to relating extreme events from other catchments to the design site. Thus were developed empirical flood formulae for flood discharge in terms of catchment area and other effective measurable variables. In the USA, the Creager formula based on hundreds of 'unusual flood peaks' in the various states plus some from foreign countries came from an enveloping curve of the logarithmically plotted points of peak discharge against area of catchment. A coefficient in the formula is dependent on climatic and catchment characteristics (Creager *et al.*, 1964). The Creager formula is still applied worldwide to give a supporting value to results from other estimation methods.

In the UK, Richards (1955) produced a sequence of formulae with more detailed considerations of the effects of rainfall intensity, duration, and distribution and catchment characteristics such as area, length and slope. Further guidance for UK engineers was provided by a Committee of the Institution of Civil Engineers in their Interim Report of 1933, which was revised and reprinted following the accumulation of further records (ICE, 1960). The Interim Report was particularly concerned with the safety of impounded reservoirs. On plots of the runoff intensities against catchment area for upland catchment areas up to 100 km^2 and for larger areas similar in character, envelope curves were drawn and these were taken to define the 'normal maximum flood', NMF. Values of the NMF over a requisite area were widely used in the UK as a first assessment of an extreme event likely to occur on a river to be impounded. It was also suggested that due regard be paid to a possible catastrophic flood which could have extreme rates of runoff at least twice those of NMF. In a sequel to the Interim Report (Allard *et al.*, 1960), attention was drawn to the increasing application of statistical techniques to long series of accurate discharge measurements and the subsequent ability to ascribe return periods to specific flood discharges, i.e. to obtain the chance of their occurrence on average once in a certain number of years.

The methods of flood frequency analysis have been favoured especially in the USA which has extensive samples of long flow records covering the country. The choice of frequency distribution used in the analysis is important since widely differing results can be obtained from different distributions. To encourage uniformity in practice, Benson (1968) investigated the validity of six different distributions over 10 representative long-term records. His findings suggested that the log Pearson type 3 distribution be recommended as the basic formula for application in the USA. The difficulties of estimating probabilities of occurrence from small data samples and the methods of obtaining unbiased estimates of the population probabilities have been discussed further by the Interagency Advisory Committee on Water Data (1982).

The rapid growth of deterministic hydrology has encouraged the scientific study of the causes of extreme floods. The structure of severe rainstorms, rapid snow melt, an initial saturated catchment and the impervious nature of steep rocky catchments — all have their part to play in the concentration of runoff into a serious river flood situation. Particular attention has been paid to the maximizing of rainfall from rainstorms and this hydrometeorological approach to the evaluation of maximum rainfall and hence extreme floods was introduced in the 1930s (ASCE, 1973). The worst floods may be expected from the probable maximum precipitation (PMP) which can be estimated by the hydrometeorologists. The PMP varies across the world according to climate and precipitation regime and can be evaluated specifically for particular catchment areas. The probable maximum flood (PMF) can be determined from the PMP by taking the worst catchment conditions, minimizing rainfall losses and using high runoff coefficients. The PMP is converted to a flood hydrograph using a deterministic catchment model such as the catchment unit hydrograph derived from major historic storms. While no return period can be afforded to the PMF, an arbitrary return period of 10 000 years can be assigned to a PMF when assessing the risk of occurrence for economic analysis of the spillway design (ASCE, 1973). The 10 000 year flood is equated to 0.5 PMF in the UK (ICE, 1978) and in the USA further considerations of the difficulties in estimating floods with very low probabilities have been published (Office of Water Data Coordination, 1986).

Current engineering practice, in the UK and elsewhere, is benefiting greatly from the results of The Flood Studies Report (NERC, 1975), an exhaustive appraisal of the flood records of the British Isles undertaken by a special team at the Institute of Hydrology. Following many discussion meetings and consultations with practising engineers, an engineering guide was published by the Institution of Civil Engineers (ICE, 1978). This gives recommended standards of safety for different categories of reservoirs and outlines the derivation of the design flood closely following the methods of the Flood Studies Report which incorporates the techniques of statistical and deterministic hydrology. A comparable exhaustive guide to flood estimation, the 3rd edition of *Australian Rainfall and Runoff*, has recently been published for practising engineers in Australia (Pilgrim, 1987).

The case studies demonstrate many of the hydrological techniques applied to dams impounding a great range of catchment areas.

6.1 MANGLA DAM

LOCATION Mangla Dam is on the River Jhelum, a major tributary of the River Indus, in Pakistan.

SOURCES Binnie & Partners (1968) Mangla. Reprinted from (1967) *Proc. Inst. Civ. Eng.* **38** 337–576; (1968) *Proc. Inst. Civ. Eng.* **41** 113–203.

Binnie, G. M., & Mansell-Moullin, M. (1966) The estimated probable maximum storm and flood on the Jhelum River. In: *River flood hydrology, Proceedings of the Symposium at the Institution of Civil Engineers*, 18 *March* 1965. pp. 189–210.

PROBLEM The construction of the Mangla Dam was part of the Indus Basin project to regulate the flow of the western rivers, Rivers Indus, Jhelum and Chenab, to feed the eastern rivers, Rivers Ravi, Beas and Sutlej, whose headwaters by treaty were to be diverted to India. In the first stage of the Mangla project, the retained reservoir area was to be 259 km^2, with the gross reservoir capacity 7253 Mm^3 and with the earthfill dam 2560 m long earthfill dam at a maximum height above the core trench of 138 m.

A major hydrological problem was the determination of the design flood flow for the dam spillway. It was considered that 'there must not be the slightest risk' (Binnie & Mansell-Moullin, 1966) of the dam being overtopped by an extreme flood discharge. The disruption of irrigation supplies and hydroelectric power, loss of life and damage resulting from a dam failure would constitute a national disaster. Therefore, the design capacity of the spillways was based on the estimated PMF.

Investigations

The spillway design studies carried out in 1958 and 1959 adopted the hydrometeorological approach using the concept of the PMP related to the most adverse catchment conditions to give an estimate of PMF under current climatic conditions.

The catchment area above the dam site is 33 300 km^2 with the altitude ranging from the peaks of the Great Himalaya over 5000 m down to about 300 m at Mangla.

As seen in Fig. 6.1, there are four main topographical regions within the catchment: the Great Himalaya separated from the Pir Panjal Range by the extensive Vale of Kashmir and the southwestern foothills of the Pir Panjal Range leading down to the plains. The tortuous course of the Jhelum is joined near Muzaffarabad by two tributaries ungauged in 1958, the River Kunhar and River Kishanganga draining from the Himalaya and near Mangla by the River Punch from the Pir Panjal Range.

Probable maximum precipitation

Precipitation occurs in two distinct seasons: December–May, mainly snow to the northeast of the Pir Panjal Range, and June–November, the heavy rainfalls of the summer monsoon coming mainly on the southeasterly winds from the Bay of Bengal. Rapid winter snow melt combined with severe storm rainfalls could possibly give rise to the PMF so that storms occurring in both seasons were studied. Since the precipitation on the catchment is greatly affected by orography, mean isohyetal maps for the two 6-month seasons were first constructed from precipitation gauge records. Storm records were available from 1926–1951 and from 1953 onwards but from a very variable distribution of gauges. Storm intensity records started in 1953. Relationships between the seasonal precipitation and topographical features such as station elevation, slope, orientation to the storms and different precipitation zones were derived to allow areal extension of the limited information. For the winter season, a relationship between mean winter precipitation and distance from the axis of the Pir Panjal Range was used. Precipitation in mountain areas has been found to increase only up to moderate elevations, typically about 2700 m. Records in the Mangla catchment show increasing precipitation up to that elevation, above which no increase was assumed to take place. From the two mean seasonal isohyetal maps, the mean seasonal and mean annual areal precipitation values were obtained for the whole basin and the subcatchments.

The precipitation data and river flow records at Mangla and sites downstream provided the evidence of major storms from 1926. The precipitation data from 14 storms, five in the winter period and nine in the monsoon period, were related to the appropriate mean seasonal precipitation maps. Thus storm isopercental maps giving more uniform distributions which could be more reliably extrapolated in ungauged areas were used in the derivation of the storm isohyetal maps. Graphs of cumulative rainfall against time were prepared for each storm at each station based on rainfall recorder data, observer's notes and synoptic and other studies of storm movements. Then the storm isohyetal maps were used to derive depth–area–duration relationships for each storm for the whole basin and the subcatchments. The meteorological conditions pertaining during each storm were studied. Persisting surface dew points at Sialkot and Rawalpindi situated in the storm-producing airstreams were combined with wind-speed factors based on 600 m pilot-balloon data for Jhelum and synoptic charts to give storm moisture-inflow indices. Radiosonde data for stations between the Bay of Bengal and the Jhelum Basin were studied to examine whether surface dew point data was representative of saturated upper air conditions. The maximum monthly moisture and wind-speed factors were plotted. An enveloping curve defined the variation within each season of the probable maximum moisture-inflow index. The precipitation during each storm was then maximized in the ratio of the maximum probable moisture-inflow index to that prevailing in the storm. Thus the storm

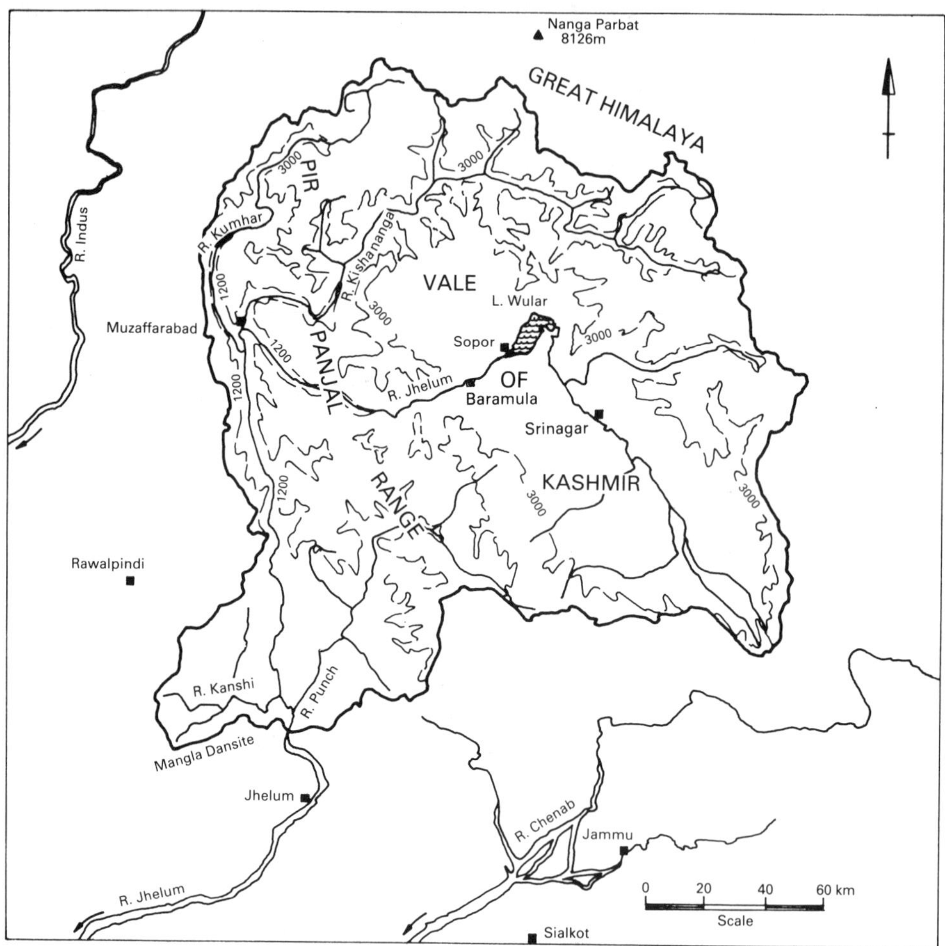

Fig. 6.1 — River Jhelum catchment above Mangla.

depth–area–duration curves were maximized and the worst rainfall sequences of all the maximized storms gave curves for the PMP for selected durations over the required catchment areas. Other investigations supported these results and, from the records at Mangla, it is apparent that the PMP would be caused by a summer monsoon storm from the Bay of Bengal, would be unlikely to exceed 72 h duration and would probably start at night with the heaviest rainfall early in its life.

Probable maximum flood

The PMF, to which no return period can be assigned, is, by definition, the greatest flood which could ever be experienced. It would be due to the worst combination of critical flood-producing conditions which are physically likely to occur in the project basin. A study of the major historic floods recorded at Mangla since 1906 indicated that they were all in the summer monsoon season, the hydrographs featured very rapid rises and falls, the flood durations were short and their timing and shape conformed to a definite pattern. Their principal cause was the rapid runoff from intense monsoon storms on the Punch and Lower Jhelum catchment areas. The floods in the Vale of Kashmir are absorbed in Lake Wular; however, storm runoff from the southern portions of the Kunhar and Kishanganga catchments can result in secondary peaks at Mangla.

There were only two river discharge records available for analysis. At Baramula where the Jhelum leaves the Vale of Kashmir, daily discharges existed over the period 1921–1946 and at Mangla where flood hydrographs could be derived from variable measurements for most major events since 1907. Unrated, mainly short-river-stage records were available for a few other sites. A 3 h unit hydrograph for the combined Lower Jhelum and Punch catchments (downstream of Muzaffarabad) was derived from six summer storms between 1946 and 1958 and synthetic 3 h unit hydrographs were derived from catchment characteristics for the catchments of the Kunhar, Kishanganga and the Jhelum between Sopor and Muzaffarabad. In applying the unit hydrograph method to determine flood flows, an estimate of losses due to evaporation, interception and infiltration must be made. In this connection, antecedent catchment conditions must be assessed and from consideration of the physical characteristics together with the rapid responses demonstrated by the hydrographs, it was decided to adopt a uniform loss rate of 2.54 mm h^{-1}. The 3 h sequences of PMP were reduced by 7.62 mm and the values of rainfall excess were applied to the 3 h unit hydrographs to give the PMF at the outlet of each catchment. The flood hydrographs for the three upper reaches were combined and routed down from Muzaffarabad to Mangla using the Muskingum method.

Results

The PMF at Mangla consists of the following.

(1) Baseflow to include snow melt and groundwater flows. The contribution to major flood flows at Mangla from the Upper Jhelum above Sopor is very small and was allowed for when selecting the baseflow value, 2265 $m^3 s^{-1}$.
(2) An antecedent major storm runoff (4–7 August 1958) occurring 72 h before the start of the PMP.
(3) The three upper catchment floods routed from Muzaffarabad to Mangla.
(4) The storm runoff from the Punch and Lower Jhelum catchments.

The assembled PMF hydrograph at Mangla gave a peak discharge of 73 600 $m^3 s^{-1}$ with a 9 day volume of 9400 Mm^3.

Conclusion

When the consultant's designs were transformed into the project layout for construction, modifications were made to overcome site difficulties. Six possible sites for the main Mangla embankment were investigated until the final decision was made after considerations of cost, storage provided and ease of construction. Even then four locations for the main spillway were considered before placing it on the right flank of the dam with an emergency spillway to be cut beyond the right bank limit of the dam. The accommodation of the PMF was to be made by a combination of spillway discharge and flood storage capacity in the reservoir. After taking into account reduced storage in the reservoir due to sedimentation, the maximum capacity of the main spillway was fixed at $31\,150\,m^3\,s^{-1}$ and of the emergency spillway at $6510\,m^3\,s^{-1}$.

6.2 ARDINGLY DAM

LOCATION Ardingly Dam is on the Shell Brook, a headwater of the River Ouse which rises in the Weald and flows south to the English Channel at Newhaven.

SOURCES Lowing, M. J. (1974) *Flood hydrology at the site of the proposed Ardingly Reservoir*, Report to Rofe, Kennard & Lapworth. Institute of Hydrology.

Rofe, Kennard & Lapworth, Chartered Engineers, Raffety House, 2–4 Sutton Court Road, Sutton, Surrey.

Southern Water, Guildbourne House, Chatsworth Road, Worthing BN11 1LD.

PROBLEM To enhance the water resources of the River Ouse, the Southern Water Authority, in association with the Mid-Sussex Water Company, investigated possible sites for surface storage in the valleys of the headwaters. Rofe, Kennard & Lapworth were asked to undertake a project for a pumped storage reservoir on the Shell Brook near Ardingly with a catchment area of $23.03\,km^2$. The small impounding earth dam 17 m high and crest length 280 m resulted in a reservoir of $0.77\,km^2$ surface area containing a gross volume of $4.77\,Mm^3$. This increased storage in the catchment provided for the regulation of the River Ouse to yield an additional $29.8\,Ml\,day^{-1}$ for abstraction at the Barcombe Treatment Works near Lewes. To assess the size of flood spillway required for the dam, the consultants sought the advice of the Institute of Hydrology.

Investigations

At the time of the Ardingly Reservoir investigations, the major work of The Flood Studies Report team at the Institute of Hydrology was nearing completion and this study was one of the first applications of The Flood Studies Report (NERC, 1975) to a real problem.

There were three useful river flow records in this area of the Upper Ouse catchment from which a unit hydrograph study could be made. A recording rain gauge near Ardingly and a gauging station on the Shell Brook from July 1972 formed the basis of the study but guidance was obtained from a smaller catchment area upstream and the major Ouse station at Goldbridge together with records from Ifield Weir on the River Mole and Weir Wood Reservoir in the River Medway catchment. A further 11 standard daily rain gauges also provided data (Fig. 6.2) and their even distribution meant that areal values of rainfall could be obtained from the simple arithmetic mean. The temporal distribution from the Ardingly recording gauge was assumed to apply over all the catchments.

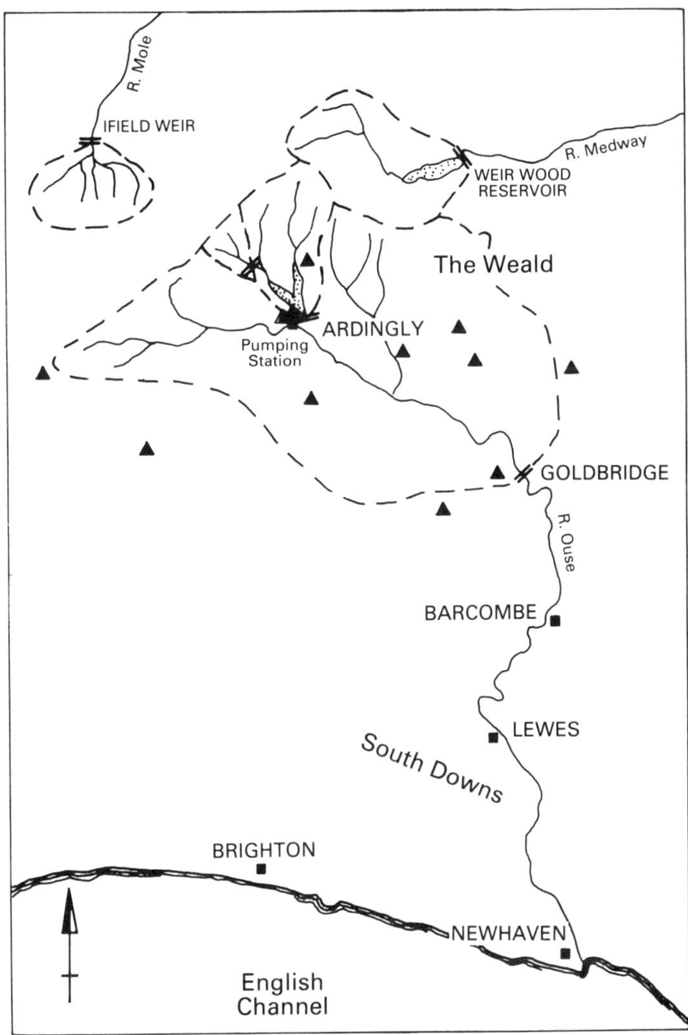

Fig. 6.2 — Ardingly Reservoir location: ▲, rain gauges; ═, river gauges; ----, catchment boundaries.

Unit hydrograph study

From only two suitable storm events, in February 1974 (43.2 mm in 19 h and 24.6 mm in 14 h) unit hydrographs (1 h, 10 mm unit) were derived by the method of least squares and smoothed to remove oscillations. For the Shell Brook catchment (23.03 km^2), the two unit hydrographs had times to peak T_p of 10 and 7 h. Such a difference is not unusual. To test whether local response times showed any tendency to change with storm size, the T_p of 23 previous events recorded at Goldbridge ranging between 13 and 24 h were studied in relation to the volumes of runoff. There was no suggestion of any such tendency. A derived T_p, based on nationwide catchment data, was also compared but it was finally recommended that, for design purposes, the average T_p of the two unit hydrographs for Shell Brook should be used (8.5 h). To determine the percentage runoff Pr in a design storm, The Flood Studies Report, Vol. I, equation (6.40) was used:

$$\mathrm{Pr} = 95.5\mathrm{SOIL} + 12\mathrm{URBAN} + 0.22\,(\mathrm{CWI} - 125) + 0.1(P - 10).$$

On the Shell Brook catchment, URBAN is zero and the SOIL term was taken as the average from all events on the Goldbridge catchment (48%) since the catchment soil types were essentially similar. This gave a prediction equation for the percentage runoff:

$$\mathrm{Pr} = 48 + 0.22\,(\mathrm{CWI} - 125) + 0.1(P - 10) \tag{6.1}$$

where CWI is the catchment wetness index and P the storm rainfall.

To select the design storm, the area particulars were abstracted from The Flood Studies Report maps (Table 6.1). The return period of the storm, its duration D and temporal profile and the catchment antecedent conditions are required to derive a design flow peak of specified return period. From The Flood Studies Report, the calculations proceed as follows:

Table 6.1 — Area particulars

M5-2 day rainfall (mm)	60
r (M5-60 min/M5-2 day)	0.32
SAAR (mm)	900
Estimated maximum 24 h rainfall (mm)	300
Estimated maximum 2 h rainfall (mm)	170

$$D = \left(1.0 + \frac{\mathrm{SAAR}}{1000}\right) T_p$$

With $T_p = 8.5$ h, $D = 16.15$ h. The nearest odd-integer multiple of the data interval is used: $D = 17$. From The Flood Studies Report, Vol. II, Table 2.10, M5-17 h/M5-2 day (with $r = 0.32$) is 0.8. Therefore M5-17 h rain $= 0.8 \times 60 = 48$ mm.

The calculations of the design runoff for a selection of return periods are best shown in tabular form (Table 6.2). The recommended storm profile, the winter 75% (The Flood Studies, Vol. II, Chapter 6) is shown in Fig. 6.3. The predicted

Table 6.2 — Runoff for selected return periods

Peak flow return period (years)	15	50	150	500	1500	5000
Storm return period (The Flood Studies Report, Vol. I, Fig. 6.61)	26	81	200	540	1500	5000
Growth factor, storm MT/M5 (The Flood Sudies Report, Vol. II, Table 2.7)	1.40	1.75	2.10	2.55	3.10	3.95
Growth factor × M5-17 h (mm)	67	84	101	122	149	190
Rain reduced by ARF 0.962 (mm) (The Flood Studies Report, Vol. II, Table 5.2)	64	81	97	117	143	183
Predicted Pr (equation (6.1)) (%)	53	55	56	58	61	65
Design runoff (mm)	34	45	54	68	87	119
Peak flows ($m^3 s^{-1}$) (from design hydrographs)	14.0	18.5	22.0	27.0	34.5	46.5

percentage runoff is applied to each interval during the storm so that each profile of effective rainfall is similar. The profiles are convoluted with the derived unit hydrograph to give the design hydrographs (Fig. 6.3) and a baseflow of 0.7 $m^3 s^{-1}$ should be added to each hydrograph. The corresponding peaks are given in the last row of Table 6.2.

The estimated maximum flood

The previous results give little idea of extreme conditions. The application of the statistical method of peak estimation, correcting the derivation of the mean annual flood, $\overline{Q}$ for Shell Brook from the records of the Goldbridge gauging station, resulted in only small increases in the peak flows for the lower return periods. The growth curve ($Q_T/\overline{Q}$ versus T) for southern England, region 7, allows only limited extrapolation. Therefore estimates of really extreme peaks were made by taking the worst conditions and combining them in the unit hydrograph method. From the estimated maximum 2 and 24 h rainfalls (170 and 300 mm), the 17 h maximum of 265 mm was obtained. A snow melt estimate of 30 mm brought the total precipitation input to 295 mm. A value of CWI = 162 was used in the derivation of percentage runoff.

$$\begin{aligned} \text{Pr} &= 48 + 0.22(162 - 125) + 0.1(295 - 10) \\ &= 85\% \end{aligned}$$

and thus the effective rainfall or runoff equals 85% of 295 which is 253 mm. The unit hydrograph was modified to produce more extreme conditions. The time to peak, originally derived from only two storms, was reduced by a third to 5.7 h. Retaining the storm duration at 17 h, the precipitation input profile was made the peakiest compatible with estimated maximum rainfall falling in any period centred at the middle of the storm. The effective precipitation, each hourly precipitation profile value multiplied by 0.85, convoluted with the new unit hydrograph produced an

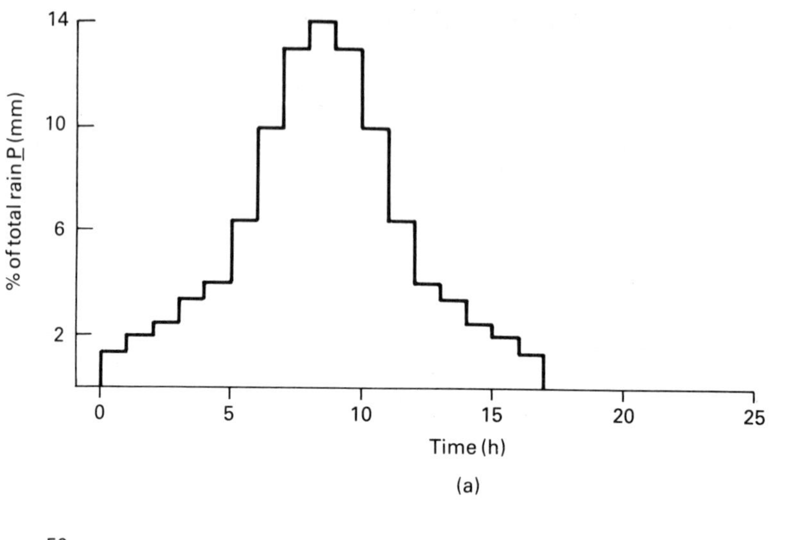

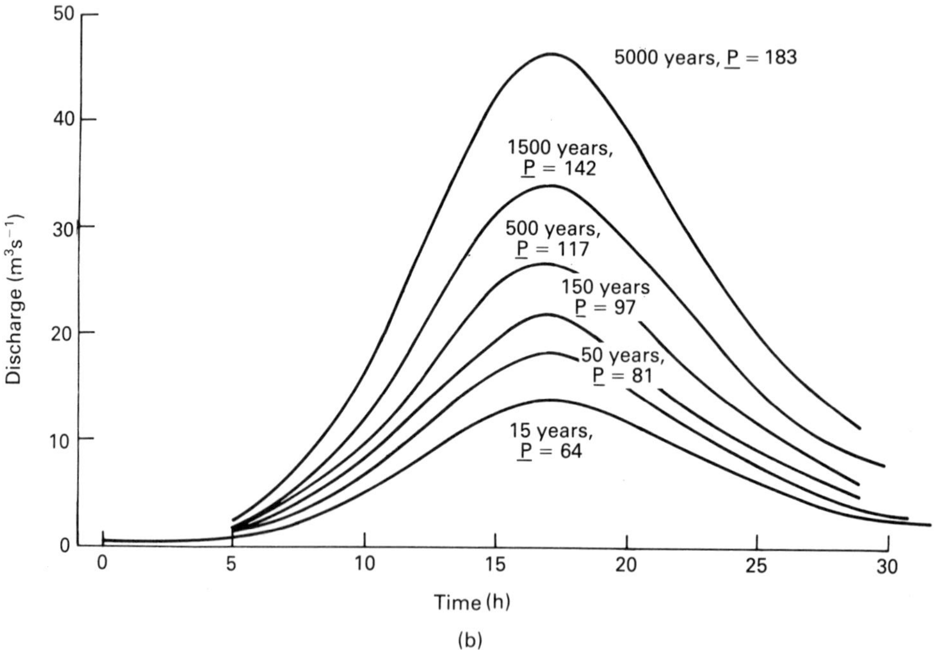

Fig. 6.3 — (a) Winter 75% profile; (b) design hydrographs for selected peak discharge return periods.

estimated maximum design hydrograph with a peak discharge of 156 $m^3 s^{-1}$ (Fig. 6.4).

Conclusion

After considering the special hydrological investigations which had thoroughly tested the response of the Shell Brook catchment under a range of conditions and

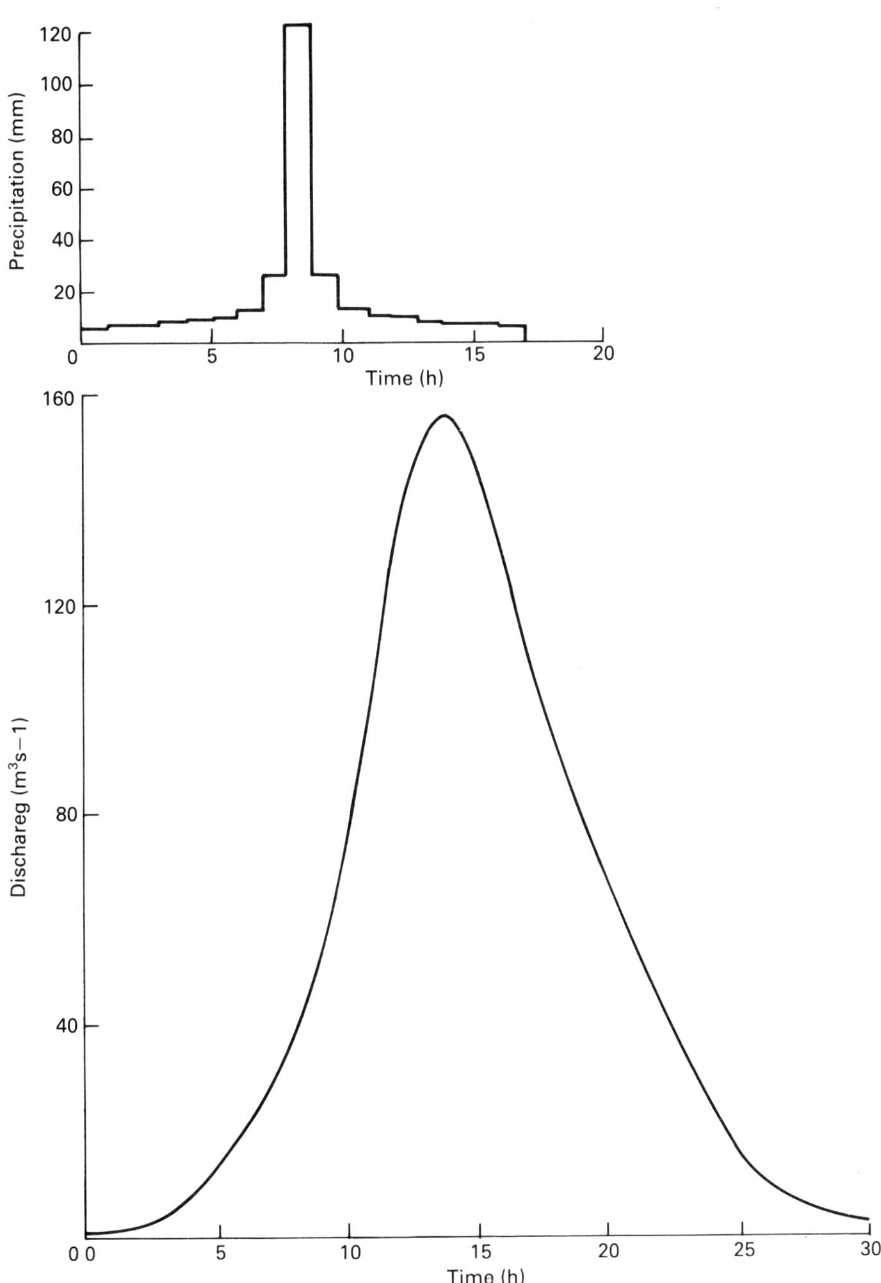

Fig. 6.4 — Shell Brook estimated maximum design hydrograph.

noting the criteria adopted for assessing an extreme flood flow, the design engineers routed the estimated peak discharge of 156 $m^3 s^{-1}$ through the reservoir and obtained a spillway design value of 136 $m^3 s^{-1}$. The dam was completed in 1977.

6.3 WIMBLEBALL DAM

LOCATION Wimbleball Dam is on the River Haddeo, a tributary of the River Exe, in Somerset, England.

SOURCES Bass, K. T., & Isherwood, C. W. (1977) Aesthetics and engineering combine for new reservoir. *Water Power Dam Construct.* January 50–54.

Battersby, D., Bass, K. T., Reader, R. A., & Evans, K. W. (1979) The promotion, design and construction of Wimbleball. *J. Inst. Water Eng. Sci.* **33** No. 5, 399–428.

Rofe, Kennard & Lapworth, Chartered Engineers, Raffety House, 2–4 Sutton Court Road, Sutton, Surrey.

South West Water, Peninsula House, Rydon Lane, Exeter, Devon.

PROBLEM A scheme to provide further water supplies for the West Somerset, North Devon and East Devon Water Boards was investigated by consulting engineers in the mid-1960s. The estimated future demands of 86.6 Ml day^{-1} were to be allocated by abstractions from the River Exe of 46.75 Ml day^{-1} at Pynes intake (including a reserve supply for the South West Water Board area) and 8 Ml day^{-1} at Bolham Weir. A pumped supply of 31.8 Ml day^{-1} was to be sent by pipeline to the Somerset area. After considering in detail four potential sites for a reservoir, the Wimbleball site was found to be the most suitable to meet the demands and the natural requirements of the river (Fig. 6.5). The concrete dam with a crest length of 300 m and maximum height of 50 m has a central buttress section with gravity sections on both flanks and the spillway is incorporated into the northern flank. An acceptable design flood needed to be determined before the spillway details could be finalized.

Haddeo catchment

The catchment area to the dam site is 29 km^2 and no river records were available for the River Haddeo. The yield of the catchment had been estimated from the records of the main gauging station on the Exe at Thorverton. Flood records of this station over 30 km downstream (catchment area, 601 km^2) are of very little help in estimating extreme events in the upper reaches of the river where flood flows usually result from intense local rainstorms on small catchment areas with short times of concentration.

In the equable climate of northwest Europe located in midlatitudes, heavy rainfalls are associated with Atlantic depressions or convective thunderstorms but moderate temperatures and consequently only average atmospheric water contents produce much less serious rainfalls than those experienced in tropical countries. The

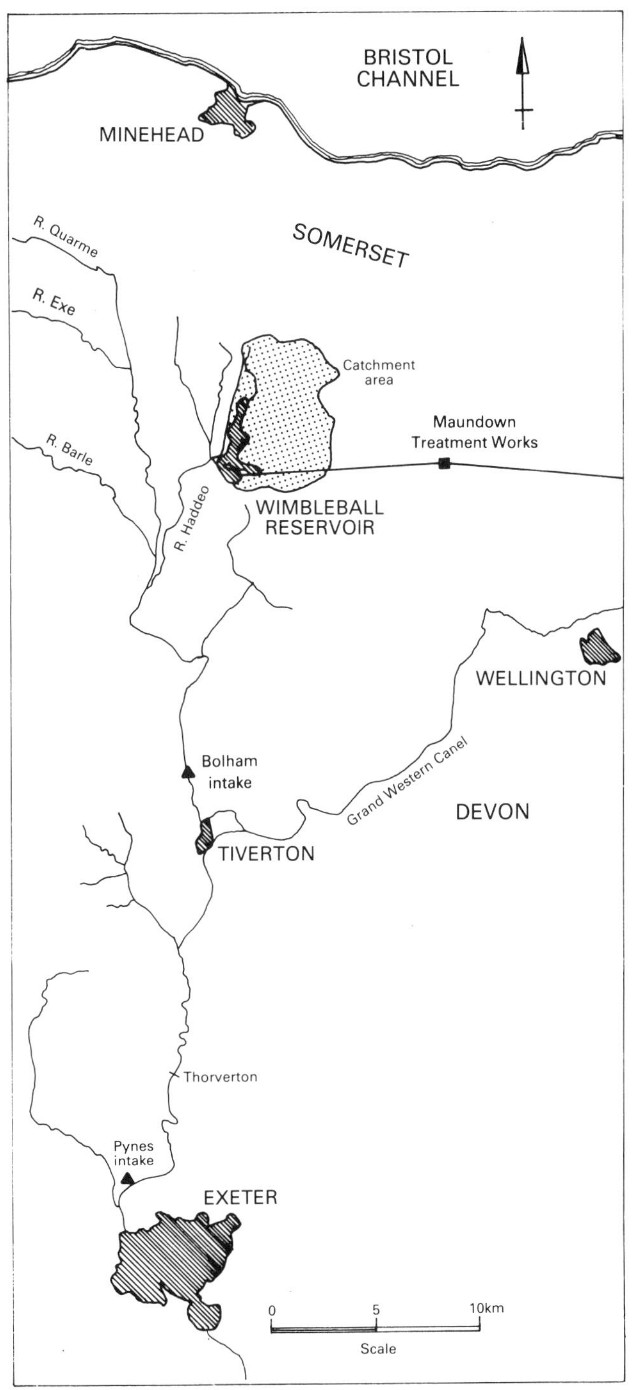

Fig. 6.5 — Wimbleball site and catchment.

relatively dense network of rainfall stations in the UK with records extending over three centuries had resulted in a valuable assemblage of information on extreme rainfall events.

Computations

As outlined in the introduction to this chapter, comprehensive guidelines have been compiled by practising engineers and the Institution of Civil Engineers' Interim Report of 1933 (ICE, 1960), takes into account all extreme events in upland areas where impounding reservoirs are constructed. The enveloping curve of plotted events, runoff intensity in discharge units per unit area against catchment area, which defines the NMF, is the foundation for the engineers' first estimation of the design flood. Using all the available information including the most recent recommendations resulting from The Flood Studies Report, a sequence of design flood estimates for the Wimbleball Dam proceeded as follows.

(1) From the Interim Report of 1933 (ICE, 1960),

$$\text{catchment area} = 29 \text{ km}^2 = 7166 \text{ acres.}$$

Following the recommendations of Appendix III it was decided to take 2.5 times the NMF. From Fig. 3 of the Report, the NMF for the catchment area is 600 cusecs per 1000 acres. Hence the extreme runoff from the catchment is

$$\begin{aligned} 2.5 \times 600 \text{ cusecs per 1000 acres} &= 1500 \text{ cusecs per 1000 acres} \\ &= 10\,750 \text{ cusecs} \\ &= 304 \text{ m}^3\,\text{s}^{-1}. \end{aligned}$$

This was the figure quoted at the feasibility stage.

(2) The design engineer used an empirical formula to obtain the 150 year flood and from experience of serious flooding elsewhere decided on 3.5 times this value to give a design flood for Wimbleball of 208 $\text{m}^3\,\text{s}^{-1}$. Guided by these calculations and noting the general location of the proposed reservoir, the experienced design engineers considered that it would not be prudent to adopt a value much below that previously quoted and took a value of 300 $\text{m}^3\,\text{s}^{-1}$ as the design flood inflow into the reservoir. After routing the flood through the reservoir a spillway design flood (SDF) of 210 $\text{m}^3\,\text{s}^{-1}$ was obtained which would give a discharge level of 1.39 m above the spillway crest.

(3) In 1973 when the design stage was well advanced, a preliminary estimate by the Institute of Hydrology gave a return period of 100 000 years for the 300 $\text{m}^3\,\text{s}^{-1}$ flood which may be compared with the following computation made after completion of the dam.

From *Floods and Reservoir Safety* (ICE, 1978) with the design data given in SI units, Fig. 8 gives PMF peak flows ($\text{m}^3\,\text{s}^{-1}\,\text{km}^2$) according to catchment area (km^2) and a variable RSMD which is the net rainfall in 1 day expected once in 5 years. A mean soil moisture deficit has been deducted from the 1 day, 5 year rainfall. The RSMD is taken from Fig. 5 to be 50 mm. From Fig. 8, the PMF peak flow per square kilometre is 9.4 $\text{m}^3\,\text{s}^{-1}\,\text{km}^{-2}$. For the catchment area of 29 km^2 the PMF = 9.4×29

= 272.6 $m^3 s^{-1}$. Following the noted recommendations of Fig. 8, 15% is to be added for a mountain area. Thus the PMF = 272.6 + 40.89 = 313 $m^3 s^{-1}$.

6.4 ITAIPU DAM

LOCATION Itaipu Dam is on the Rio Parana near the city of Foz do Iguacu, Parana State, Brazil.

SOURCES Brazilian Committee on Large Dams (1982). *Dams in Brazil.* Novo Grupo, Editoria tecnica Itda, Sao Paulo, Brazil.

De Moraes, J., Rodriguez Villalba, J., De Mello, W. F., Poty, L. C., Berny, O., Acosta, A. G., & Sarkaria, G. S. (1979) Selection of basic design of Itaipu spillway. *Transactions of the 13th ICOLD Congress, New Delhi, Vol. III, pp. 249–259.*

PROBLEM The site of the Itaipu Dam marks the downstream control point of the Parana River before it flows into neighbouring Argentina. The drainage area of 820 000 km^2 is being developed primarily for hydropower and Itaipu is the thirty-sixth in the ultimate plan for 58 projects. The dam constructed between 1975 and 1983 to incorporate a navigation lock as well as the power plant has a crest length of 7900 m and a maximum height of 185 m. It is a combined concrete gravity and earth and rockfill dam, with the concrete control works in the centre of the valley and earth embankments on both sides, a common design for the large Brazilian rivers. The impounded reservoir with an area of 1460 km^2 has a total storage of 29 000 Mm^3. In estimating an SDF, the flood flows for six major tributaries had to be merged with the main stream. An additional complication was provided by the regulation of the 35 reservoirs already existing upstream and the likely effects of the reservoirs to be built under future development.

Investigations

In the large drainage basin above Itaipu, there are hydrological records from 120 meteorological stations and 62 river gauging stations, many with good data back to 1932. Natural flows were evaluated for the regulated rivers at the existing dam sites and short or deficient records were extended or infilled by regression with the data from long term gauging stations. The stations used in the flood study are shown in Fig 6.6, a map of the catchment area also indicating the numerous sub-basins. The main river rises in the northeastern mountains, 1000 m in height, and flows down to a level of about 200 m near the dam. The catchment area spans two climatic regimes: tropical with summer rains and dry winters in the north and temperate but with hot summers and rain all the year in the south. The average annual areal rainfall is 1400 mm above Itaipu but the tropical regime dominates the seasonal pattern with the wettest months from November to February. The mean annual natural discharge of the Parana at Itaipu is 9070 $m^3 s^{-1}$. Of the several historic floods, a maximum daily discharge of 45 000 $m^3 s^{-1}$ was recorded downstream of the dam site in 1905.

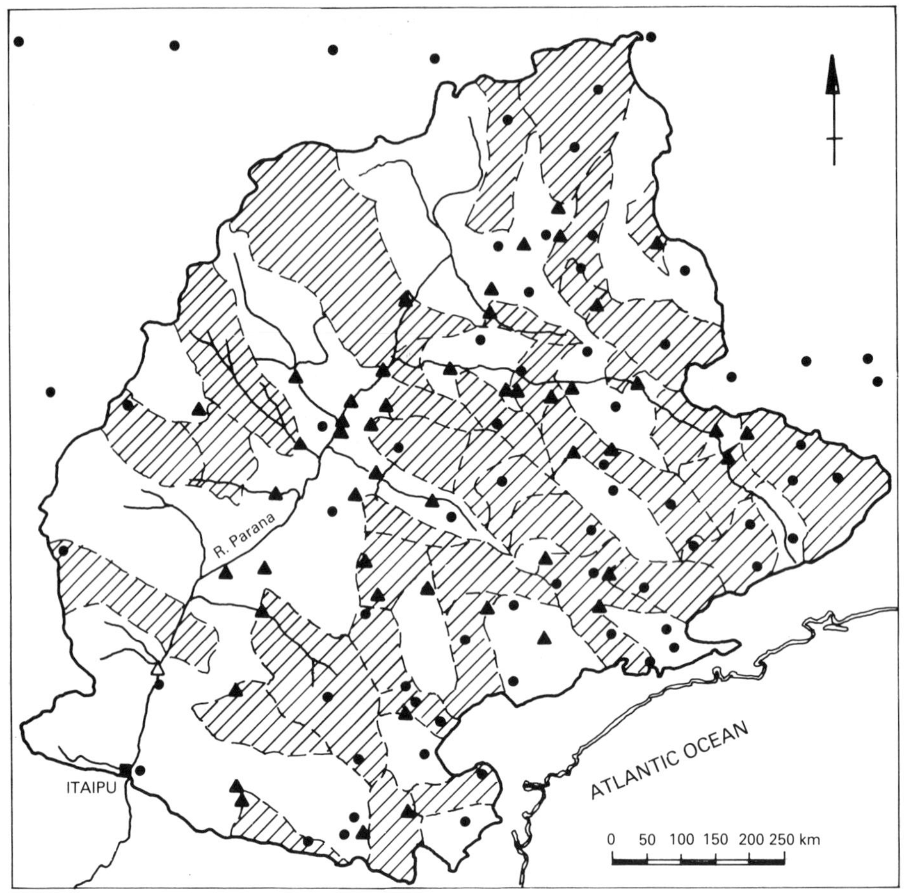

Fig. 6.6 — Rio Parana Basin and Itaipu Dam: ▭, sub-basins with computations by proportioning; ▨, sub-basins with unit hydrograph computations; ●, rain gauges; ▲, stream gauges.

Probable maximum precipitation

The vast size of the catchment area precluded the determination of the PMP from a design storm in the meteorological sense. Therefore a climatological PMP was synthesized by taking the maximum monthly catchment rainfalls for the wet-season months October– April and placing them in series. Thus the ensuing soil moisture conditions in the basin and the rainfall in each following month will exceed the worst historic conditions and produce a flood discharge to exceed the historic discharges. It was noted that the critical condition for a major flood at Itaipu, expected in March or April, would result from extensive storms over the whole basin for the first few months producing high baseflows and peaks from the upper catchment in February combining with high flows from severe storms over the lower reaches in March or April. The best PMP design sequence consisted of the following maximum rainfall months: October 1930, November 1939, December 1926, January 1951, February 1964, March 1928 and April 1956 with the March and April values coming from the lower contributing areas of the catchment.

Probable maximum flood

To derive runoff from the maximum rainfall, a deterministic catchment model similar to the Stanford watershed model was used and the computations done by computer. The model evaluated, on a daily basis, excess rainfall and net contribution to baseflow, converted excess rainfall to discharge by the unit hydrograph method, computed the groundwater component and then produced the total discharge at the catchment outlet. The model was applied to the sub-basins shown in Fig. 6.6. The parameters of the model were determined for each gauged sub-basin using records for the period October 1958–April 1959 when there was good rainfall coverage and little reservoir regulation. The unit hydrographs (1 day unit hydrograph and rainfall unit 10 mm) were derived from selected storm data. For ungauged sub-basins the unit hydrographs were obtained by scaling unit hydrographs from neighbouring catchments according to basin length and drainage area ratios. When each sub-basin had its own satisfactorily calibrated model, the maximum rainfall sequence was fed into the computer programs and then each discharge output was routed down to its confluence with the next river channel using the Muskingum method. Further computer programs were used to combine the discharges and, when proportionate values had been computed for unmodelled areas, the total tributary discharges were combined in sequence downstream to Itaipu. The PMF from the natural flood flows resulting from the 7-month PMP was 53 800 $m^3 s^{-1}$.

However, it was necessary to take into account the effects of the reservoirs in the catchment area. Reservoirs would reduce the flood travel time in a river valley and, if a reservoir was full, the peak discharge would occur downstream sooner than if it were passing down a natural channel. Thus, after the full development of the whole Parana scheme when all 58 reservoirs have been completed, there is going to be greater coincidence of the flood peaks from the various tributaries and therefore an enhanced total flood discharge at Itaipu. The basin models were modified to account for the reservoir effects and two further flood peaks obtained: 61 370 $m^3 s^{-1}$ for the present intermediate stage in development and 72 000 $m^3 s^{-1}$ for the final completed stage. The three SDFs are shown in Fig. 6.7.

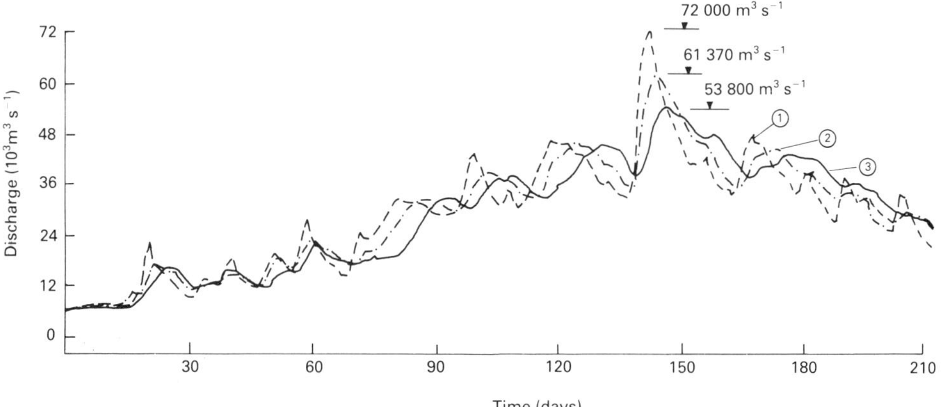

Fig. 6.7 — Design floods: ①, final development; ②, intermediate stage; ③, natural flood flow.

Conclusion

The design discharge choice for the spillway, 72 000 $m^3 s^{-1}$, was routed through the reservoir making a series of assumptions of initial reservoir state and spillway gate operations. A maximum design discharge capacity for the spillway was 62 200 $m^3 s^{-1}$ with an increase of 3.10 m above full supply level in the reservoir.

6.5 GARDINER DAM

LOCATION Gardiner Dam is on the South Saskatchewan River in Saskatchewan, Canada.

SOURCES Prairie Farm Rehabilitation Administration (1980) *The design and construction of Gardiner Dam and associated works*. Ministry of Supply & Services, Canada.

Berry, W. M., Durrant, E. F., & Booy, C. (1961) Hydrologic Investigations for the South Saskatchewan River Project. *Eng. J.* 44 No. 4, 61–68.

PROBLEM The Prairie Farm Rehabilitation Administration, a branch of the Canadian Department of Agriculture, was established to initiate studies into water supply for irrigation in the prairies. In 1947, a site for an earth dam on the South Saskatchewan River upstream from Saskatoon was chosen to store water for irrigation and power generation. A second dam near the headwaters of the Qu'Appelle River was needed to contain the reservoir across a low col separating the drainage basins. Authority for the scheme was granted in 1958 and the major works completed in 1967. The Gardiner Dam is a compacted-earth embankment 4968 m long with a maximum height of 64 m. The retained reservoir, Lake Diefenbaker, 225 km long has an area of 429 km^2 and a full capacity of 9363 Mm^2.

It was decided that the spillway and related features should be designed to pass the SDF, the discharge with the chance of occurrence of less than one in a thousand in any one year, and the PMF, the largest flood that could be expected to occur at the dam site. The design differences were in the allowances for freeboard on the dam and the hydraulic performance of the spillway structures. The hydrologists were to establish discharge values for the two required floods.

Investigations

The catchment area, 126 392 km^2, above the dam site is shown in Fig. 6.8. The numerous headwaters of the South Saskatchewan River rise in the Rocky Mountains (over 3000 m), a region of glaciers and heavy winter snowfalls. The annual pattern of the river flow results from the runoff characteristics of three distinct subdivisions of the catchment. The eastern slopes of the Rocky Mountains provide a large volume of runoff from summer melting snow but the foothills give the major contribution from spring snow melt together with spring and summer rainfalls. These upper regions produce about 92% of the average annual flow at the dam site. The largest proportion of the catchment area consists of the dry prairies with low precipitation

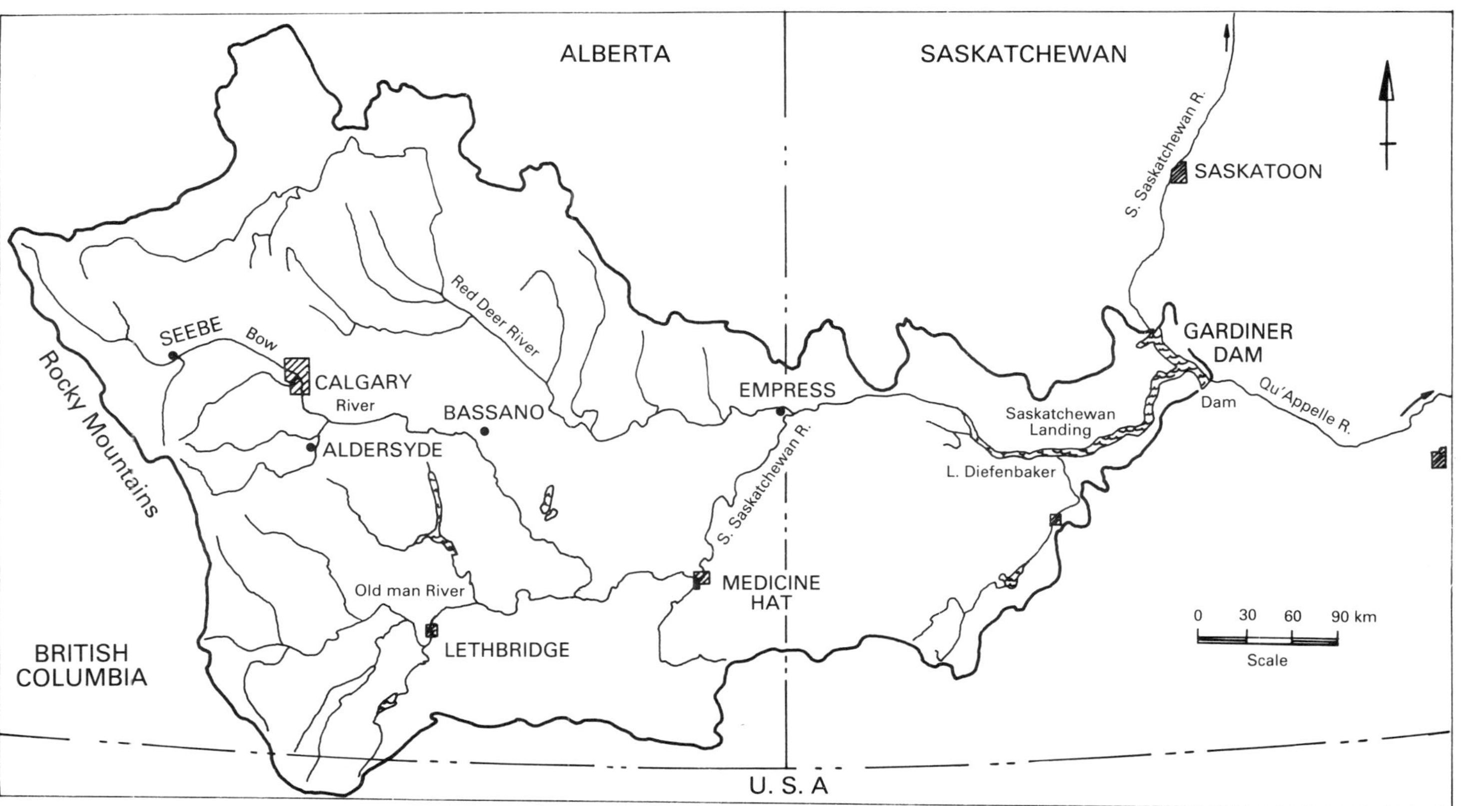

Fig. 6.8 — South Saskatchewan River drainage basin.

and usually low average runoff. There are great annual fluctuations in peak flow and volume of runoff due to variations in the amount and extent of the winter snowfall, the rate of rise of the spring melt temperatures and the timing and intensity of the summer rainstorms. From historic records at Saskatoon, downstream of the dam site, the highest discharge recorded was 4177 $m^3 s^{-1}$ in June 1953 but there was an estimated 5040 $m^3 s^{-1}$ in 1897 before records began. Major floods resulted from the coincidence of rapid melting of an extensive thick snow cover with a heavy summer rainstorm.

Probable maximum flood

Since detailed meteorological information was not available, an extreme design rainstorm pattern was devised from consideration of all recorded 48 h storms that had occurred on or near the catchment area. The most severe storm was transposed to the centre of the drainage basin. The resultant isohyetal pattern of the design storm is given in Fig. 6.9. The recorded storm hydrographs from the major tributaries were analysed to assess retention and infiltration losses and runoff from snow melt and baseflow. For loss values, the lowest values occurring in the major recorded rainstorms were taken. For snow melt runoff, an investigation into the relationship of accumulated runoff and accumulated degree days above 32°F was made for two sample tributaries. A major snow melt runoff was assumed to result from the maximum recorded rate of discharge per degree day over a very warm period of several days. Since the resultant snow melt runoff and baseflow were found to be of lesser importance than the rainfall effects and further data were lacking, the major snow melt runoff was estimated for the other tributaries. The 24 h unit hydrographs were derived for each of the tributaries. It was observed that the unit hydrograph peaks were too small when derived from ordinary floods and so it was decided to increase arbitrarily the peak ordinates. The chosen unit hydrograph for each tributary thus had its peak obtained from a 15% increase of the average unit hydrograph peak or from a 7.5% increase of the highest peak. The tributary floods were then obtained by applying the storm rainfall excesses to the unit hydrographs and adding the snow melt and baseflow.

The combining and routing of the tributary floods down to the dam site then provided a major problem. The northern group of headwaters were combined and routed to Bassano on the Bow River and the southern group combined at Lethbridge on the Oldman River. These were then routed to Medicine Hat on the South Saskatchewan River. Recorded flood flows were studied to assess translation and attenuation of the peaks and how the two streams combined. An assumed discharge value for ungauged areas and lateral inflow between the gauging stations was added to the flood hydrograph. The worst-flood conditions would result from the coincidence of all the tributary peaks and studies were made of the effects of storm movements over the catchment. The effect of storm movement was found to be useful only in the coincidence of the peak from the long northern tributary, Red Deer River, with the peak from South Saskatchewan River at Empress. The flood hydrograph from Medicine Hat was routed downstream, joined by the Red Deer River flood flow at Empress and then continued down to Saskatchewan Landing. The resulting estimated PMF had a peak flow of about 18 400 $m^3 s^{-1}$ and a total flood volume of about 8634 Mm^3.

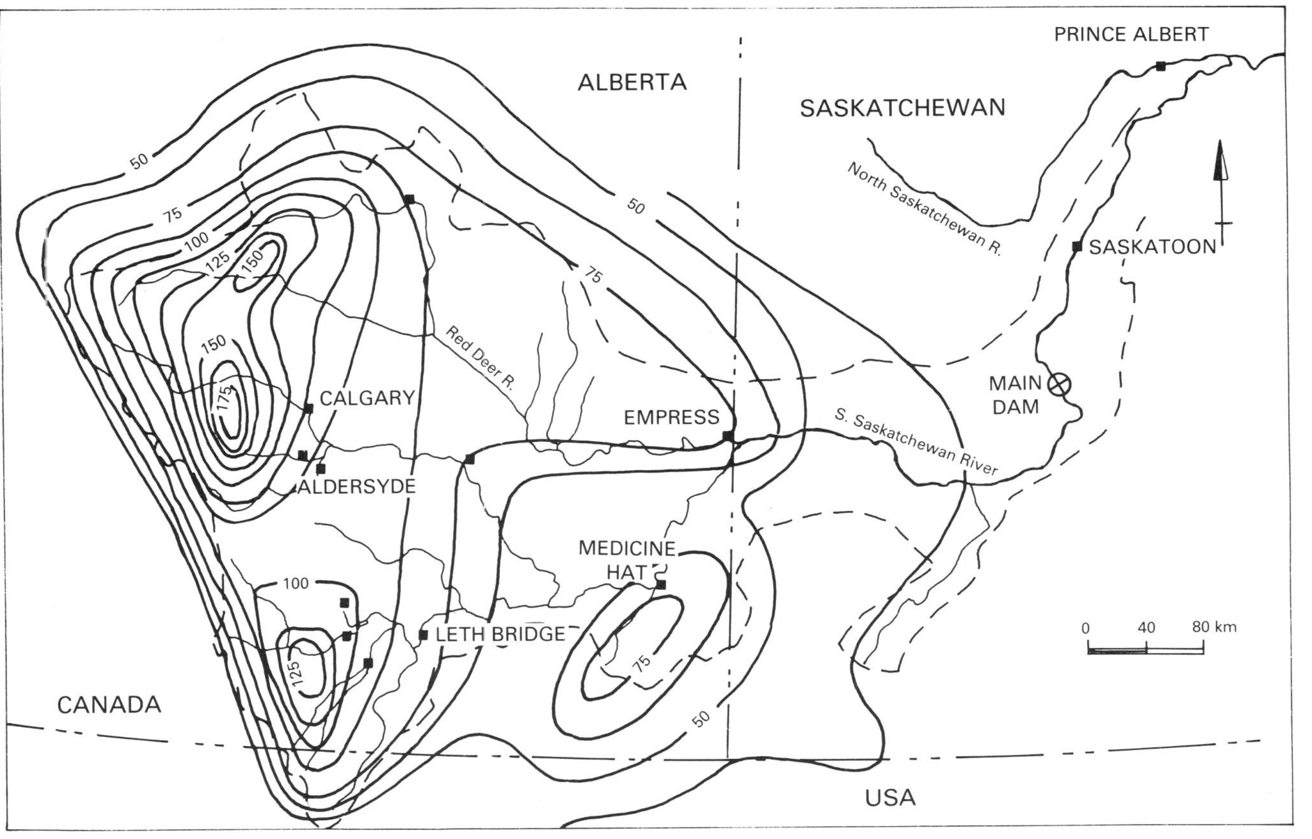

Fig. 6.9 — Design storm. The isohyets are in millimetres.

Spillway design flood

The derivation of the SDF used the same analytical method but made different assumptions of the magnitude and distribution of the rainstorm. The worst-loss conditions used for the PMF were also modified. The resultant SDF produced a peak flow of about 11 330 $m^3 s^{-1}$ and a total flow volume of about 4934 Mm^3.

A comparison was made with information from other rivers rising on the eastern slopes of the Rocky Mountains having the same climatic regime. By plotting the maximum peak discharge per unit area against catchment area, it was shown that the SDF fell on the enveloping curve of the largest recorded flood on each river and the PMF was placed well beyond it. This together with favourable points on a Gumbel plot of the annual peak flows at Saskatoon (1908–1953) supported the results of the hydrological study for the required flood flows.

Conclusion

To provide further guidance for the design engineers, the hydrologists routed the two discharge hydrographs through the reservoir, making assumptions on the shape of the inflow hydrographs and on the operation of the spillway gates. They obtained maximum water levels during the simulated passage of both floods and the resulting minimum freeboards below the top of the dam were 4 m for the SDF and 1.2 m for the PMF.

6.6 KEKRETI DAM

LOCATION Kekreti Dam is on the Gambia River, West Africa.

SOURCE Agrar- und Hydrotechnik Gmbh and Howard Humphreys & Partners (1986) *Kekreti Reservoir Project, feasibility study*, Report for Gambia River Basin Development Organization and the Federal Republic of Germany.

Howard Humphreys (1985) *Kekreti Dam*, Fact Sheet.

PROBLEM Plans for a dam 37 m high on the Gambia River at Kekreti in Senegal have been prepared for the Gambia River Basin Development Organization. The dam construction is to be part embankment and part concrete gravity and the impounded reservoir of 3500 Mm^3 capacity is to have a multipurpose function. It will regulate the river for irrigation, provide hydroelectric power and help to limit saline intrusion up the river estuary. One of the many requirements of the consultant engineers was the evaluation of the flood flows in the Gambia River and the determination of a recommended maximum discharge for the dam spillway design.

Kekreti catchment

The location of the Kekreti Dam is shown on the general map for the Gambian River Basin development study (see section 9.3) and the catchment area to the dam site (13 890 km^2) is illustrated in Fig. 6.10. The Gambia River rises in the Fouta Djallon Mountains of Guinea and flows in a general northerly direction until it turns westerly in Senegal to flow through the Gambia into the Atlantic Ocean. The climate of the

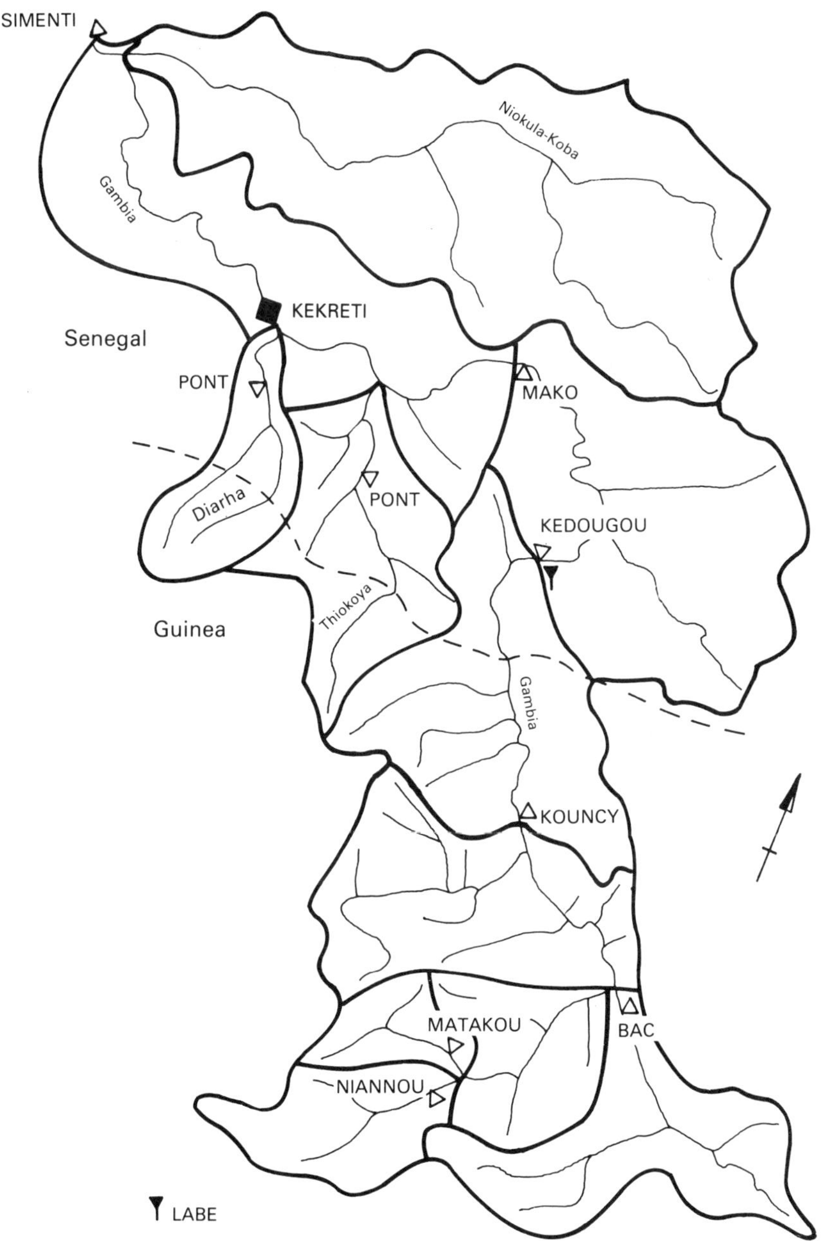

Fig. 6.10 — Upper Gambia River Basin and Kekreti Dam: ▽, river gauging stations; ▼, climate stations.

region is tropical with a long hot dry season from November to May. In the shorter wet season, the heavy thundery rainfalls reach a maximum from July to September. The average annual rainfall at the dam site is about 1300 mm with the average annual evaporation estimated at around 2000 mm.

As part of previous studies in the Gambian River Basin, river gauging stations had been established at key points on the river and on some of its tributaries. The record at Kedougou from 1970 was particularly valuable for the analysis of flood flows at the dam site. Rainfall data were obtained from Kedougou Meteorological Station where autographic rainfall charts were available from 1968 and from Labe in Guinea to the south of the catchment.

Flood studies

Two methods were adopted: the probability analysis of the annual maximum flow series at Kedougou and the application of storm rainfall profiles to a unit hydrograph derived for Kekreti.

Peak flow frequencies

The 12 annual maximum flows at Kedougou (1970–1981) range from a maximum 1110 $m^3 s^{-1}$ in 1970 to a minimum 489 $m^3 s^{-1}$ in 1979. After ranking the flows in order of magnitude, each value was allocated an unbiased plotting position using the Gringorton formula

$$F_i = \frac{i - 0.44}{N + 0.12}$$

where i is the ranking and N the number of record years. The return periods T are given by $T = 1/F$ and the reduced variate y of the Gumbel EV1 distribution by

$$y = -\ln\left[-\ln\left(1 - \frac{1}{T}\right)\right].$$

The plotted values are shown in Fig. 6.11 together with the best-fit frequency curve extrapolated to the 1000 year return period. The 90% confidence intervals demonstrate the increasing uncertainty of the estimates for higher values of return period.

To derive peak flows for the Kekreti site, the estimated flows at Kedougou were scaled according to their catchment areas using the following relationship:

$$Q_{\text{kek}} = Q_{\text{Ked}}\left(\frac{A_{\text{Kek}}}{A_{\text{Ked}}}\right)^n = Q_{\text{Ked}}\left(\frac{13\,890}{7550}\right)^n$$

where n is taken to be 0.75. The derived peak flows at Kekreti dam site for a selection of return periods are given in Table 6.3. In view of the short length of record, the results from this method must be treated with reserve.

Unit hydrograph method

Twelve flood events at Kedougou gauging station between 1975 and 1980 were selected for analysis. For each event, flow ordinates at 3 h intervals were listed with a baseflow component deducted and the corresponding 3 h rainfall totals at the Meteorological station abstracted from the autographic charts. 3 h unit hydrographs were derived from the rainfalls and direct storm runoffs by the matrix inversion technique incorporating the least-squares fitting and smoothing technique. Only five

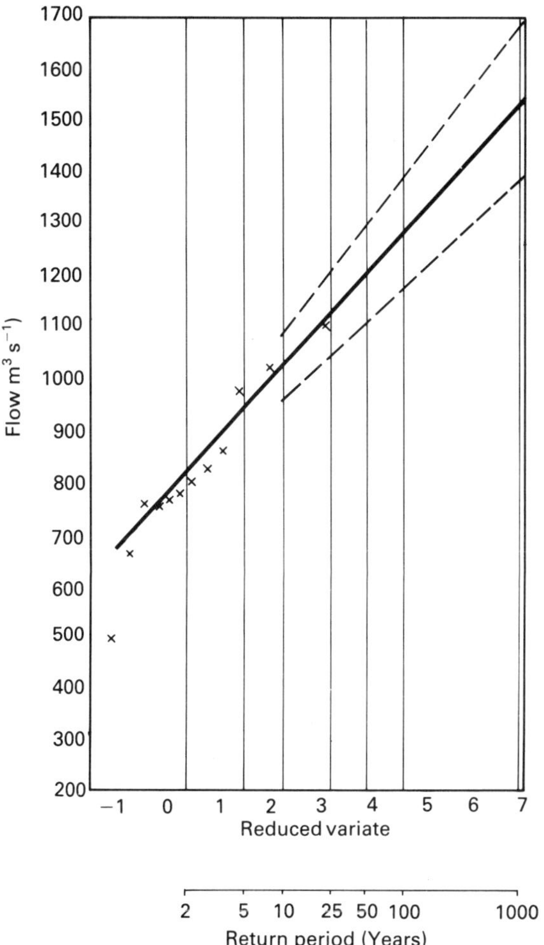

Fig. 6.11 — Annual maximum peak flows at Kedougou: ———, mean values; - - - -, upper and lower 90% confidence limits.

Table 6.3 — Flood peaks at Kekreti dam site

Return period (years)	Peak flow (m^3 s^{-1})	Confidence limit (m^3 s^{-1})
10	1611	± 96
50	1895	+153
100	2020	±175
1000	2417	±232

of the events were single-peak floods and the range in percentage runoff values from 7 to 116% demonstrated the inadequacy of the single point rainfalls at the outfall of the catchment at Kedougou to represent the variable areal rainfall over the 7550 km^2 catchment. (The records from Labe were not available at the time of this analysis.) However, a realistic form of unit hydrograph was computed.

In view of the uncertainties involved in translating this unit hydrograph downstream to Kekreti, a synthetic unit hydrograph was derived for Kedougou using the US Department of Agriculture method (USDA, 1972). The synthetic unit hydrograph is a triangle defined by the following.

(1) Time to peak

$$T_p = \frac{T_r}{2} + \frac{T_c}{0.6}$$

where T_r (h) is the rainfall time increment and T_c (h) is the time of concentration. T_c is obtained from the Kirpich equation.

$$T_c = \frac{0.0195(L^3/H)^{0.385}}{60} \qquad \text{(h)}$$

where L (m) is the main stream and H is the catchment maximum–minimum elevation (m).

(2) Total base time of the hydrograph

$$T_B = 2.67T_p \qquad \text{(h).}$$

(3) Peak runoff rate

$$Q_p = \frac{0.2081AP}{T_p} \qquad (\text{m}^3\,\text{s}^{-1}).$$

The completed 3 h unit hydrograph for Kedougou closely resembled the shape of the best unit hydrographs derived from the storm rainfalls and runoffs and thus it was considered appropriate to use the US Department of Agriculture method to calculate the triangle parameters for the dam site at Kekreti. The variables for the two catchments are given in Table 6.4.

Table 6.4 — Variables for the synthetic 3 h unit hydrographs (1 mm)

	Kedougou	Kekreti
L (m)	243 000	393 000
H (m)	1020	1075
A (km^2)	7550	13 890
T_c (h)	34.8	64

Storm rainfall

Rainfall depth–duration–frequency curves were derived from the autographic rainfall records at Kedougou for durations up to 12 h and for durations 24–120 h from the available records at Labe. By extrapolation, rainfalls for different durations for return periods of up to 1000 years were obtained. A statistical estimate of the PMP was calculated for different durations using the Hershfield formula

$$\text{PMP} = P_n + 15S_n$$

where P_n and S_n and are the mean and standard deviation of the annual maximum series for n years. Here again the value of the estimate is limited by the shortness of the records.

A number of storm profiles for different durations were tested and a bell-shaped profile of duration 48 h gave the worst floods and therefore was adopted for convolution with the synthetic unit hydrograph to give the required design hydrographs. The profiles for the 50 year, 1000 year and PMP rainfalls derived from the Labe records were used.

The lack of information on storm rainfall losses meant that estimates of a runoff coefficient had to be made. Indications from the individual storm events suggested a realistic figure of 10%. Therefore the storm profiles reduced by 90% were convoluted with the synthetic unit hydrograph for Kekreti to give the assumed corresponding return period flows (Table 6.5). Although the 50 and 1000 year peak flows are

Table 6.5 — Flood hydrograph features

	50 year	1000 year	PMF
Direct runoff over 5 days (Mm^3)	241	346	621
Peak flow ($m^3\,s^{-1}$)	1055	1462	2888
Baseflow ($m^3\,s^{-1}$)	550	550	550
Total peak flow ($m^3\,s^{-1}$)	1605	2012	3438

lower than those obtained by frequency analysis, these are considered more reliable by nature of their derivation.

Conclusion

The total 1000 year and PMF hydrographs with a base length of 5 days were routed through the reservoir to investigate the necessary spillway capacity to pass the attenuated floods. However, the volumes of these events were small relative to the reservoir storage and thus a maximum inflow volume over a longer period of approximately 30 days (the reservoir lag time) was required to specify enhanced spillway capacities.

From a statistical analysis of the estimated mean annual flows at Kekreti, the 1000 year mean annual flow of 244 $m^3\,s^{-1}$ was distributed in 10 day increments through the wet season based on the worst-known conditions of 1975–76. With the first part of the seasons flow replenishing depleted storage by mid-September, the residual volume, with deductions for evaporation loss and power demand, would represent the net 1000 year flood runoff. This net volume converted to a daily time interval hydrograph with a peak of 2100 $m^3 s^{-1}$ and total duration of 44 days, was taken to represent the baseflow component of the PMF, with the direct runoff ordinates of the 5 day PMF superimposed on the peak period of the baseflow (Fig 6.12).

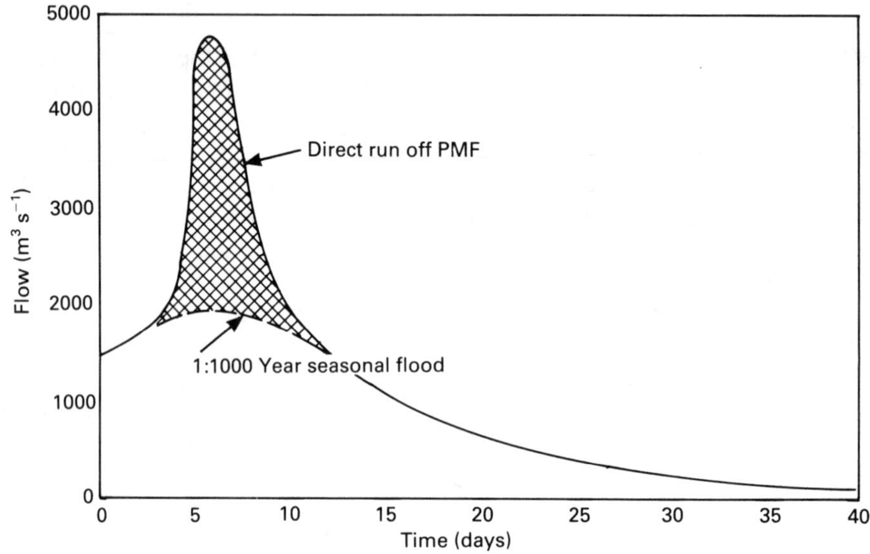

Fig. 6.12 — PMF inflow hydrograph.

The resultant hydrograph with a short-duration peak discharge of 48 000 $m^3 s^{-1}$ was considered to represent the severest runoff conditions likely to occur during any year in the lifetime of the project. When routed through the reservoir the peak outflow would be 1570 $m^3 s^{-1}$ with a related rise of 3.07 m at the service spillway.

7

Water resources

INTRODUCTION

The evaluation of water resources to ensure a continuous supply of water to satisfy demands usually entails the study of periods of water shortage and the means of storing supplies to cover those periods. Deficiencies may result from short dry spells or long-term droughts according to climate and/or requirements. The many definitions of drought relate to water use and availability. A period of water shortage for domestic consumers in a developed country may be termed a drought, which has quite a different meaning in a semiarid region when the seasonal rains desperately needed for subsistence agriculture have failed for several years.

The resources of an area or of a catchment are assessed by the analyses of both surface water and groundwater. For surface water supplies, river flow records are desirable but very often estimates of river flows have to be obtained by water balance methods involving precipitation inputs and evapotranspiration losses. Such hydrometric records required are analysed on a monthly basis, a convenient time period for water resources evaluation. However, long-time series of records are necessary for identifying the dry periods in order to define reliable yields and to estimate the probability of shortages. In investigating groundwater supplies, records of piezometric levels are required and pump tests may be carried out to assess potential well yields and aquifer characteristics may be determined to verify models of groundwater movement.

Under conditions of regular ample rainfall, the need for storage may not arise. Adequate volumes to satisfy demands are pumped directly from a river offtake to purification works or other demand centres. The main concern with such an arrangement is to ensure that sufficient water is left in the river at all times to maintain an acceptable flow and to supply other users downstream. Natural storage is provided in the ground. Groundwater resources are replenished by infiltration with percolation through the unsaturated zones and seepage from river beds to the water table. A thorough knowledge of the characteristics of the natural aquifers is needed to assess the regularly available quantities and to avoid over-exploitation.

In this chapter, schemes involving natural river flow and groundwater resources are described and man-made storages are considered in the following chapter.

7.1 WATER RESOURCES FOR KUANTAN

LOCATION Kuantan is an area of Pahang on the east coast of Peninsular Malaysia.

SOURCE Binnie Dan Rakan M. (1977) *Water resource development in the Kuantan region*, Report to the Government of Malaysia.

PROBLEM The area round Kuantan Town at the mouth of the Sungai Kuantan had been chosen as a major development area with rapid growth of industry and urban population (Fig. 7.1). Water demand had been estimated to rise from 23 Ml day^{-1} in 1977 to a total of 227 Ml day^{-1} in 1990 to meet natural growth and heavy industry needs. For increased storage, proposed reservoirs in the upland reaches of the Sungai Kuantan presented many difficulties among which the dam sites would be endangered by frequent landslides. In particular, excessive volumes of water would be required to provide sufficient releases to limit salt water intrusion in the Kuantan estuary. Thus the development of further water supplies was to be based on lowland works. The hydrological studies were concerned with the provision of information on overall water resources, agricultural needs and flood mitigation. In this study attention is focussed on resources for domestic and industrial supplies.

Water resources

The Sungai Kuantan draining an area of 1658 km^2 flows into the South China Sea at Kuantan Town. The port and industrial districts of Gebang in the development area could also be economically served by the resources of the Sungai Kemaman draining an area of 2253 km^2 of the State of Trengganu to the north.

The Sungai Kuantan catchment ranges from rugged mountainous areas with peaks of over 1220 m to low-lying swamp lands near the coast. A large proportion of the basin composed mainly of granite and volcanic rocks is covered by tropical rain forest. Some areas bordering main roads have been cleared for rubber and oil palm plantations.

Rainfall

The average annual areal rainfall (AAAR) over the whole Kuantan catchment is about 3124 mm, ranging from 2794 mm near the mouth to more than 3556 mm over the mountain headwaters. The monthly rainfall pattern is given for two stations in Table 7.1. Although rainfall occurs throughout the year, the heaviest falls are brought by the northeast monsoon from October to January.

Evapotranspiration losses are estimated to average 1448 mm year^{-1}.

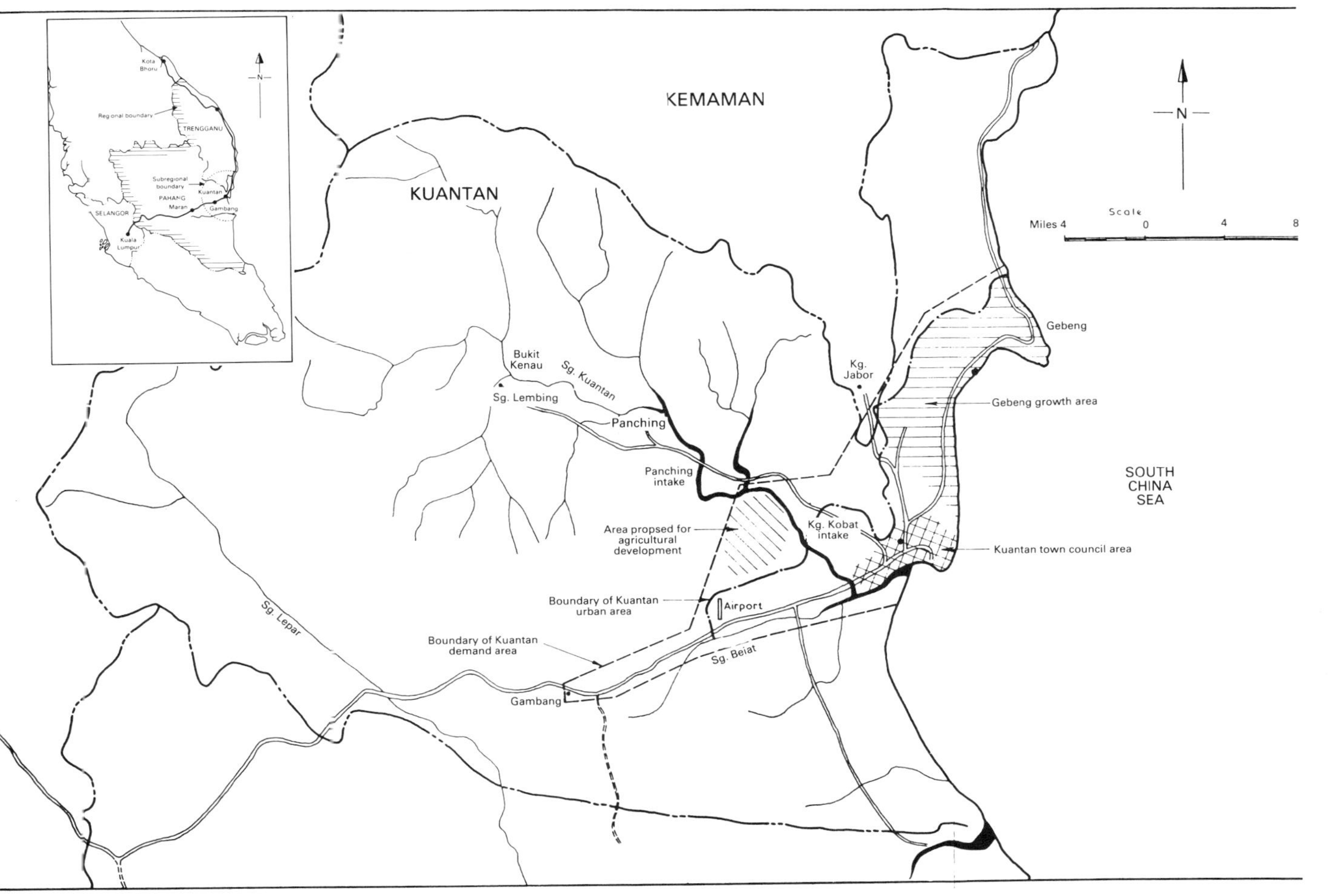

Fig. 7.1 — Kuantan development area: ▲, gauging stations; ○, rain gauges.

Table 7.1 — Monthly mean rainfalls 1947–1975

	Mean rainfall (mm)											
	January	February	March	April	May	June	July	August	September	October	November	December
Sungai Lembing	350	236	218	218	244	183	170	224	259	340	396	561
Kuantan	348	196	178	180	193	157	155	173	231	282	338	610

Table 7.2 — Average annual runoff in the Sungai Kuantan Basin

Station	Catchment area (km^2)	AAAR (mm)	Losses (mm)	AARO (mm)	ADF ($m^3 s^{-1}$)
Bukit Kenau	583	3200	1422	1778	33
Panching intake	932	3251	1448	1803	53
Kg. Kobat intake	1243	3226	1448	1778	70

Runoff

No long river flow records were available for the Sungai Kuantan and so flow estimates were first made from catchment rainfall minus estimated catchment losses to give average annual runoff (AARO). The average daily flows (ADFs) derived for three points on the river are shown in Table 7.2. The short records at the few river gauging stations were insufficient to provide runoff–frequency relationships on a monthly or even shorter time period basis. Twenty-nine years of record were available for a gauging station on the Sungai Trengganu to the north with a catchment similar to that above Bukit Kenau. The monthly flows were analysed to determine the minimum runoff volumes for periods from 1 to 24 months for selected return periods. These were then expressed as percentages of the AARO and transposed to the Sungai Kuantan via the estimated AARO on the Kuantan. Minimum flows for 1 and 7 day periods were calculated for drought periods assuming no rain and applying a daily recession constant of 0.98. These estimates were valid only for undeveloped catchments and it was found from spot flow measurements from tributaries and points in the lower catchment that baseflows there were not so well sustained. A shortage of local information prevented precise yield return period estimates for the more developed parts of the basin.

A variety of techniques were used to derive minimum runoff estimates including the transposition of runoff rates per unit area. From the range of results, the lowest values have been abstracted for Sungai Kuantan at the Panching Intake in Table 7.3.

Table 7.3 — Best minimum flow estimates at Panching intake

Return period (years	Minimum flow estimate ($m^3\ s^{-1}$) for the following periods			
	60 days	30 days	7 days	1 day
50	4.4	3.5	1.1	0.9
30		4.0	1.5	1.3
20		4.6	2.0	1.8
10		5.5	3.5	2.9

The lack of data for the Kuantan Basin and the need to transpose records from other areas resulted in considerable uncertainty in yield estimates.

In conjunction with the investigations into water quantities, research into water quality was also undertaken. Sediment yield of the catchment was estimated and samples near the river mouth analysed. Analysis of water quality samples showed soft, mildly acidic water, markedly turbid with high iron content. Direct abstractions would cause operational difficulties and pre-treatment storage was recommended.

Resource development

The estimated yields indicated that for 30–50 year return periods, the Sungai Kuantan could meet all demands up to 227 Ml day^{-1} (2.63 $m^3\ s^{-1}$) and from the

Sungai Kemaman a further 409 Ml day^{-1} (4.74 m^3 s^{-1}) could be available. Existing supplies are pumped from the Kg. Kobat intake on the Sungai Kuantan up to a treatment works whose capacity could be increased from 27 to 36 Ml day^{-1}. A further 9 Ml day^{-1} is to be available from the new intake and treatment works at Panching. Both of these intakes are within the tidal reach of the river which extends about 40 km inland. A storage for about 7 days supply of raw water could be provided to meet demand when salinity levels prevent abstraction but this was later rejected.

The 227 Ml day^{-1} could be supplied without storage if sea water intrusion were to be prevented. Detailed salinity intrusion studies heavily weighted the choice in favour of a tidal barrage across the river downstream of the lower intake with control gates to exclude the sea water at times of low river flows. It would, however, require a lock for barge traffic on the river.

Twelve different water supply systems were considered taking into account all possible practical means of utilizing the main resources of the Kuantan and Kemaman rivers and the minor yields of the lowland tributaries. Capital expenditure for the construction of intakes, pumping stations, barrage, storage reservoirs and new treatment works and the operating costs for electric power for pumping, treatment chemicals and labour for operation and maintenance were assessed for each scheme, providing for four demand options. A selection of the most economical schemes were presented to the clients and a recommended development programme of works up to 1990 was set out.

7.2 WATER RESOURCES OF THE WORFE CATCHMENT

LOCATION The River Worfe is a left-bank tributary of the River Severn, northwest of Birmingham.

SOURCES Miles, J. C., & Rushton, K. R. (1983) A coupled surface water and groundwater catchment model. *J. Hydrol.* **62**, 159–177.

Rushton, K. R., & Ward, C. (1979) The estimation of groundwater recharge. *J. Hydrol.* **41**, 345–361.

PROBLEM The evaluation of the water resources of a catchment requires a thorough understanding of the interchange of water between the river and the underlying strata. A practical way to achieve this is to model the spatial and temporal variations in surface and subsurface flows for the catchment. In solving this problem for the Worfe catchment (260 km^2) a coupled mathematical model of surface and groundwater was devised using 20 years of monthly data for calibration. The model was constructed to represent the spatial variation of surface and subsurface water flows calculated for 122 different locations. The applications of the model to regulate aquifer abstractions and to determine long-term effects of climatic variations on the resources were prime objects of the study, carried out with the assistance of the Severn–Trent Water Authority.

Catchment model

The relatively small Worfe catchment of moderate relief ranging in height from 160 to 32 m is 70% underlain by the outcrop of the Bunter Sandstone aquifer dipping to the east and with an average thickness of 350 m. The northwestern and eastern parts are composed of low-permeability Carboniferous and Triassic strata respectively forming aquitards. These solid geological strata are covered in places with impermeable drift deposits (Fig. 7.2). The catchment boundaries are well defined, in some

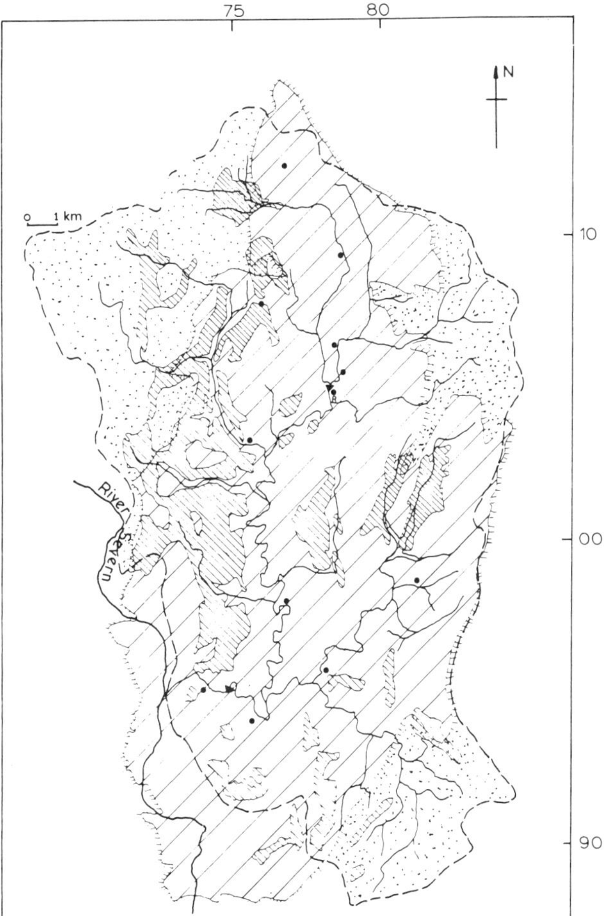

Fig. 7.2 — Worfe catchment plan: , bunter sandstone; , impermeable drift; low permeability strata; , catchment boundary;; , edge of aquifer; ●, abstraction borehole; R, river abstraction; ▲, gauging station.

places by groundwater divides. From this physical structure, three hydrologically significant land types have been identified.

(1) Aquifer covered by permeable deposits.

(2) Aquitard (low permeability) covered by permeable deposits.
(3) Areas covered by impermeable drift.

The three subdivisions form the basis for the catchment model (Fig. 7.3).

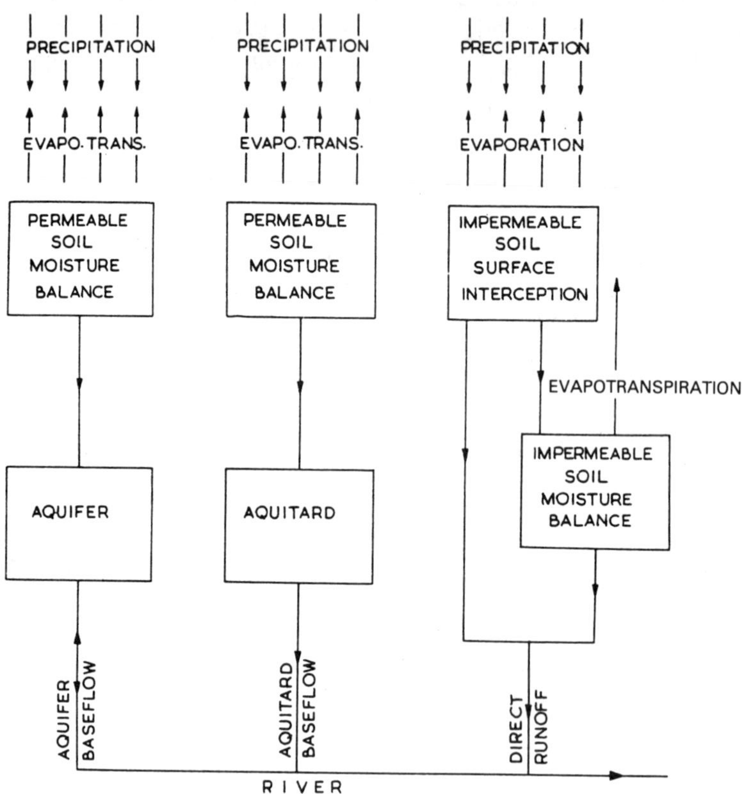

Fig. 7.3 — Conceptual arrangement of catchment model.

For area types (1) and (2), the water recharge to groundwater is calculated by the soil moisture balance method (Grindley, 1967; Rushton & Ward, 1979) with the land use divisions, permanent grassland 32%, annual crops 57%, woodlands 8% and the remaining 3% urban and riparian areas. For area type (3), with impermeable drift, an estimated 50% of the effective rainfall bypasses the soil moisture zone and flows directly into a watercourse.

The river flows are obtained from the effective rainfall according to the land type.

(1) In the aquifer areas, the surplus water flow may be considered in three phases.
   (a) Flow within the aquifer is represented by the differential equation

$$\frac{\partial}{\partial x}\left(T_x\frac{\partial h}{\partial x}\right)+\frac{\partial}{\partial y}\left(T_y\frac{\partial h}{\partial y}\right) = S\frac{\partial h}{\delta t}-G(x, y, t) \tag{7.1}$$

where T_x (m^2 day^{-1}) and T_y (m^2 day^{-1}) are x and y direction transmissivities, h (m) is the groundwater potential, S is the specific yield and G (m day^{-1}) is the recharge with t in days. The equation is solved for h using finite-difference approximations.

(b) Flow Q from aquifer to river R for month m at river node n is

$$Q_{R,m,n} = KL_n(h_{m\text{-}1,n}-g_n) \ (\text{m}^3 \text{ day}^{-1}) \tag{7.2}$$

where L_n (m) is river length at node n, $h_{m-1,n}$ is the groundwater model head (m) at node n at the end of the previous month and g_n (m) is the river stage level at node n. K (m day^{-1}) is a flow resistance coefficient.

(c) (c) River flows across the aquifer are calculated from water balances at the nodes of the groundwater model in contact with the river:

$$R_{n,m} = R_{n-1,m}+Q_{\text{Im}}+Q_{R,m,n} \ (\text{m}^3 \text{ month}^{-1}) \tag{7.3}$$

where $R_{n,m}$ and $R_{n-1,m}$ are the m month flows at river node n and upstream node $n-1$, Q_{Im} is the impermeability drift flow near n and $Q_{R,m,n}$ comes from equation (7.2), according to the number of days in the month.

(2) From the aquitard, the low-permeability strata, the flows are only required as input to the river–aquifer part of the model. The monthly flow is calculated from an adaptation and summation of Horton's empirical equation for the normal flow depletion curve:

$$Q_{\text{A}} = A\sum_{m=1}^{m=mt} I_m \exp(-\beta m^{\alpha}) \ (\text{m}^3 \text{ month}^{-1}) \tag{7.4}$$

where A (m^2) is the aquitard area, I_m (m) is the recharge water in month m, and β and α are constants.

(3) Flows from the impermeable drift are also obtained simply by apportioning the contribution from 2 months:

$$Q_{\text{Im}} = A(0.9I_m+0.1I_{m-1}) \ (\text{m}^3 \text{ month}^{-1}) \tag{7.5}$$

where A is the area, and I_m and I_{m-1} (m) are recharges for the appropriate months. These flows are added to the total river flows at the aquitard boundary or to the nearest relevant node of the river–aquifer model.

The calibration of the model, i.e. the determination of parameter values was effected by comparison of model runs with continuous river flow data from two gauging stations (Fig. 7.2) and regular gaugings at a further 43 sites in the catchment. The sensitivity of the model parameter values was assessed by the objective function

$$\sum (\ln Q - \ln \hat{Q})^2$$

where Q is the measured flow and $\hat{Q}$ is the equivalent model flow and also by calculating the correlation coefficient between the two sequences of monthly values. Pumping tests on boreholes in the aquifer gave a transmissivity of about 300 m^2 day^{-1} in the centre of the catchment with a specific yield between 0.07 and 0.16, indicating high storage effects.

From several trial values, a K of 1300 m day^{-1} for equation (7.2) was obtained and the g_n values were also determined from the simulation. In equation (7.4) the model is not particularly sensitive to the values of α and β and 0.5 was found suitable for both parameters. It was also found that the number of months mt was also insensitive for the flows from the aquitard.

Results

The starting year for the simulation needed careful consideration in order to have a dynamic balance in the model. This was difficult owing to the number of groundwater abstractions in recent years. By starting the model run in 1960, comparison of calculated heads with actual water levels at selected nodes from 1975–1979 showed good agreement. Differences in absolute values were due to model areal averages over a month being compared with spot locational values.

When the total monthly flows for a sequence of years were plotted against the measured flows at the lower gauging station, there was generally good agreement between the recorded data and the model. The results for two years given in Table 7.4 indicate that the model tends to overestimate flows in the winter months and to underestimate in the summer. Plots of the three model component flows showed that the runoff from the impermeable drift was highly variable, that the contribution from the aquitard was more regular with appreciable volumes of flow in the summer months and that the aquifer baseflow showed a general net flow of water from the aquifer but after droughts significant amounts of water passed from the river to the aquifer. The variability in the contributions from the three hydrologically different land types over the catchment was demonstrated by results for the drought month of August 1976 and the following wet September.

This catchment model provides a means of estimating the impact of future abstractions on the river flows and thereby is a valuable tool in the management of the resources of the Worfe catchment.

7.3 WATER RESOURCES FOR TARAWA

LOCATION Tarawa atoll is in the Gilbert Islands, western Pacific Ocean.

SOURCE Lloyd, J. W., Miles, J. C., Chessman, G. R., & Bugg, S. F.

Table 7.4 — Comparative monthly flows

Month	Monthly flow (Ml day^{-1})			
	1980		1981	
	Measured	Model	Measured	Model
January	158.2	220.7	154. 8	177.2
February	239.0	297.8	140.1	169.8
March	221.8	259.6	274.0	275.2
April	133.0	117.6	138.0	120.3
May	71.0	65.9	141.9	125.0
June	97.7	85.4	115.8	85.2
July	66.9	48.0	53.0	49.7
August	66.4	54.4	53.3	38.7
September	73.7	55.9	76.6	86.0
October	111.8	109.1	103.0	101.0
November	131.3	134.4	83.4	81.8
December	154.9	172.2	174.5	196.3

(1980). A groundwater resources study of a Pacific Ocean atoll — Tarawa, Gilbert Islands. *Water Res. Bull.* **16**, No. 4, 646–653.

PROBLEM The provision of reliable water supplies to the developing communities of remote Pacific islands requires larger storages than provided now by roof catchment tanks. The advantages of using groundwater supplies are well recognized in tropical countries but the limitations of the geological composition and extent of water-bearing material of atoll islands make for the real danger of salt water intrusion. The problem of adequate water supplies for Tarawa has been investigated by a team of research hydrologists and hydrogeologists for the UK Ministry of Overseas Development.

Investigation

Tarawa atoll consists of an angled chain of elongated islands of unconsolidated calcareous debris accumulated upon a stable coral limestone reef platform (Figs 7.4 and 7.5). The island material results from the erosion of the coral limestone; the highest parts, up to 5 m, of coarse boulder deposits occur on the ocean side of the islands while the lower leeward sides bordering the 'lagoon' are composed of deposits of calcareous sand and silt, much reworked by wave action in the lagoon. The combined wind and wave actions have resulted in a central depression, developed for crop growing.

The determination of the geometry of the freshwater lens (Fig. 7.5) was the first major problem. The WHO recommended minimum standard of salt content for potable water resources is 600 mg of chloride (Cl^-) per litre and so, giving a margin of safety, a value of 500 mg of chloride per litre was adopted to define the limit

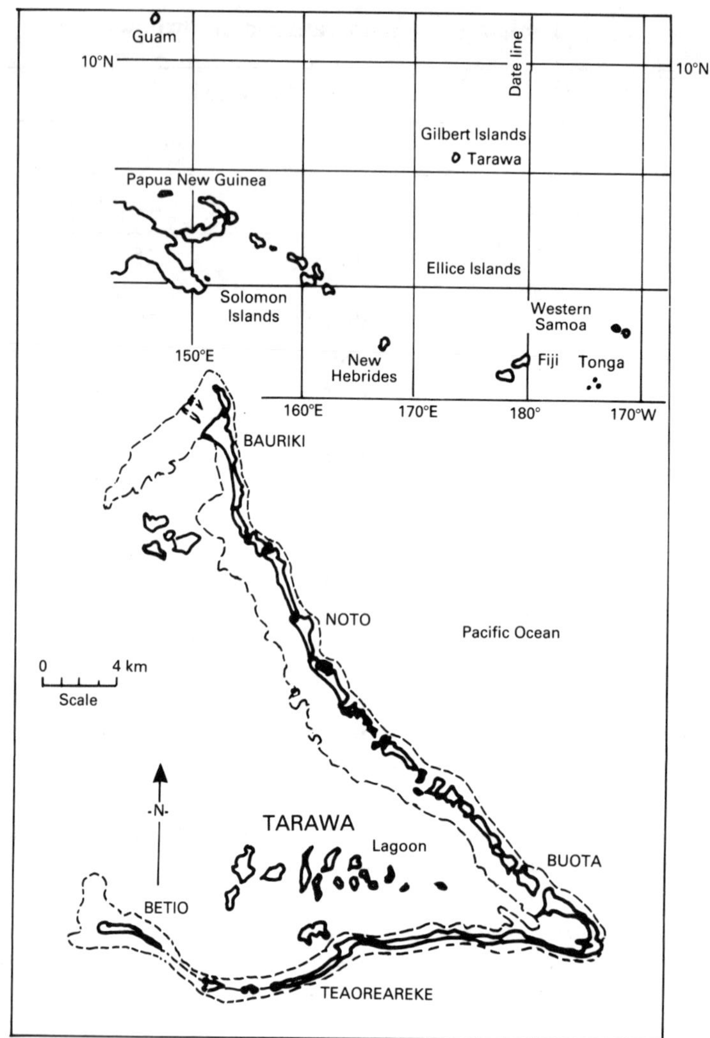

Fig. 7.4 — Geographical location of Tarawa and local detail.

between freshwater and saline water. Owing to the lack of drilling equipment and hence the lack of boreholes, the thickness of the freshwater lens was obtained by defining the lens base by resistivity measurements and calculating the water table height on a freshwater density–saline water density relationship. Controls and checks on the calculations were effected where the lens base approached the surface. The density relationship for a lens using the limit of 500 mg of chloride per litre was of the order of 1:20 to 1:30. The maximum depths of the lenses of freshwater were found to range from 20 to 30 m below sea level on the four larger islands in the main chain studied in detail.

Water resources

A monthly groundwater balance was calculated using the simple relationship

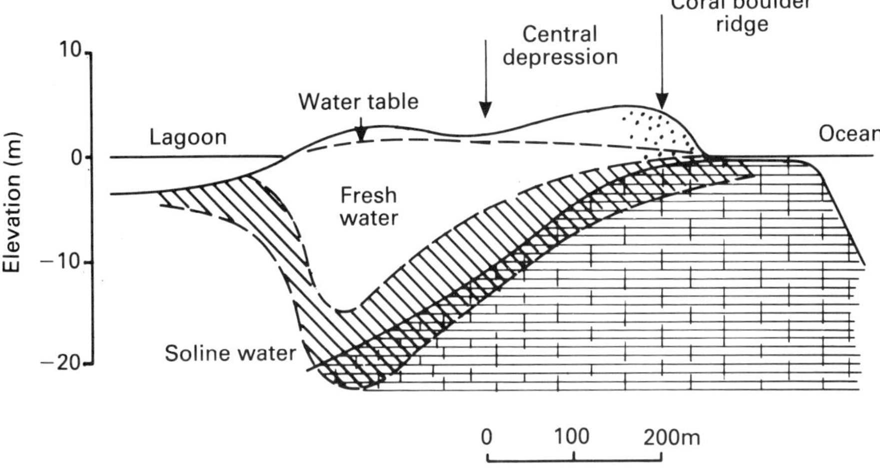

Fig. 7.5 — General geological cross-section, and freshwater lens location: [sand pattern], sand and silt; [boulder pattern], boulder beds; [limestone pattern], coral limestone.

$$P = I + \mathrm{ET} + \mathrm{Re} + Q_{ab} + Q_s + \mathrm{Si}$$

where P is the precipitation, I the intercepted rainfall, ET the evapotranspiration, Re the recharge, Q_{ab} the abstraction, Q_s the groundwater flow to the sea and Si the change in groundwater storage. Surface water flow is very small and omitted.

With Tarawa lying in the Intertropical Convergence Zone, the rainfall occurs in short heavy localized storms but, over a month, the distribution may be considered less variable and the data from the meteorological station on Betio, in the west of the island group, were used in the studies. A long-term rainfall record of 51 years (average annual rainfall, 1808 mm) enabled significant drought periods to be identified. The 2% drought events for 1 and 2 year sequence were 330 and 1400 mm respectively well below the annual mean. From previous studies on Pacific islands, about 15% of the total rainfall is intercepted by the vegetation and is thereby lost but with the experience of the model on Tarawa this had to be halved. There were insufficient data for Penman ET calculations; therefore, estimates of this loss were made using the Thornthwaite formula. In assessing the recharge from the soil moisture balance, from field measurements the soil moisture level before reduced evapotranspiration occurs was taken as 50 mm and the wilting point as 120 mm. Between these two values, evapotranspiration was taken as 10% of ET. An example of the recharge calculations for the months of 1977 used for all the groundwater lenses is given in Table 7.5. A comparison between the calculated recharges and resulting decrease in groundwater salinity showed good agreement on Betio.

To quantify the groundwater resources, the irregularity of the recharge events meant that a non-steady digital modelling method had to be used. A time-variant drawdown equation neglecting the small vertical flow components was used:

Table 7.5 — Rainfall–recharge calculations, 1977

Month	*P–I*	ET	Effective rainfall	Soil moisture deficit	Recharge (mm)
January	587	146	441	0	441
February	274	135	139	0	139
March	372	149	223	0	223
April	189	147	42	0	42
May	157	152	5	0	5
June	80	149	– 69	– 52	0
July	248	155	93	0	41
August	151	156	– 5	– 5	0
September	283	149	134	0	129
October	123	154	– 31	– 31	0
November	440	150	290	0	259
December	179	156	23	0	23

$$\frac{\partial}{\partial x}\left(T_x\frac{\partial s}{\partial x}\right)+\frac{\partial}{\partial y}\left(T_y\frac{\partial s}{\partial y}\right) = S\frac{\delta s}{\delta t}+Q$$

where T_x and T_y are the transmissivities in the x and y directions, s is the groundwater potential, S is the specific yield and Q is water entering or leaving the aquifer per unit area per unit time. However, applying the Ghyben–Herzberg approximation for freshwater–saline water, the transmissivities varying with the water table height were modelled by

$$T = Kh(1+\alpha)$$

where α is the lens depth or density ratio, taken to be 1/20, h is the height of the water table above mean sea level and K is the permeability (Chidley & Lloyd, 1977).

In applying the groundwater simulation model to each of the four island lenses, rectangular grids of 50 m × 150 m or 100 m × 100 m were used; permeability and specific yield were assumed constant across a lens with the recharge evenly distributed. Calibration of the models was difficult owing to lack of historical water level data; a range of permeabilities 3.0–7.0 m day^{-1} and specific yields of 0.10–0.15 were tried. Eventually a permeability of 4.5 m day^{-1} was found to be the most appropriate and a specific yield of 0.15 was chosen since with lower values water levels rose to too high a level during recharge events.

Conclusion

The paucity of data resulted in considerable subjectivity in modelling the groundwater resources and it was decided to concentrate on critical drought conditions. For each lens the models were run with the recharge from the 2% rainfall data following 3 years of typical data run to allow the lens to achieve an initial dynamic balance. For the largest Bauriki lens, with small abstraction rates, the depth of the lens remained similar to that for no abstractions during the initial period. For the drought period with no abstractions the lens reduced to about 4 m in thickness and an abstraction of only 20 $m^3 day^{-1}$ reduced the minimum lens thickness to about 2 m. It was concluded that the lenses could only be safely used for abstraction during a 2% drought of 1 year duration. Sustained abstraction during a second year of drought would result in saline intrusion and disruption of the lens of freshwater. However, during years of consistent recharge, the four lenses are each capable of sustaining yields of about 50 $m^3 day^{-1}$.

7.4 GROUNDWATER RESOURCES FOR LIMA

LOCATION Lima, the capital of Peru, is situated on the coastal plain between the Andes and the Pacific Ocean.

SOURCES Lerner, D. N., Mansell-Moullin, M., Dellow, D. J., & Lloyd, J. W. (1982) Groundwater studies for Lima, Peru. *Optimal Allocation of Water Resources, Exeter Symposium*, Publication No. 135. International Association of Hydrological Sciences, 17–30.

Wild, J., & Ruiz, J. C. (1987) Lima groundwater modelling revisited. *National Hydrology Symposium, Hull.* British Hydrological Society, 14.

PROBLEM The population of the City of Lima, 5.5 million in 1985, is expected to rise to 8.3 million by the end of the century and the resultant water demand to increase from the estimated average value of 33 $m^3 s^{-1}$ in 1978 to over 50 $m^3 s^{-1}$. At present, 64% of the demand for domestic, industrial and irrigation purposes is met from two rivers and the rest from groundwater. A major scheme to transfer water across the Andes from the Rio Mantaro, one of the Amazon headwaters, has been studied but construction has been delayed since the scheme is very costly. Meanwhile further supplies must be found within the Lima region. Detailed investigations of the groundwater availability were made by Binnie and Partners, consultant engineers, in 1979–1980. As a result of a decrease in the estimated demand growth rate and the consequent deferment of the Mantaro transfer project, a further groundwater study was commissioned in 1984 to find the required increase in supplies.

Lima region

The location of Lima is shown in Fig. 7.6. The two rivers, the Rimac and Chillon, draining 3490 and 2310 km^2 respectively, rise in the Andes at heights of over 4000 m. The mean annual precipitation ranges from 700–800 mm in the mountains to only 10 mm near the coast. On the coastal plain although only 12°S of the Equator, the mean annual temperature is moderated to around 18.5°C owing to cool ocean currents offshore. With virtually no rainfall, vegetation is dependent on irrigation.

The larger Rio Rimac has monthly discharges ranging from 20 to 60 $m^3 s^{-1}$ with

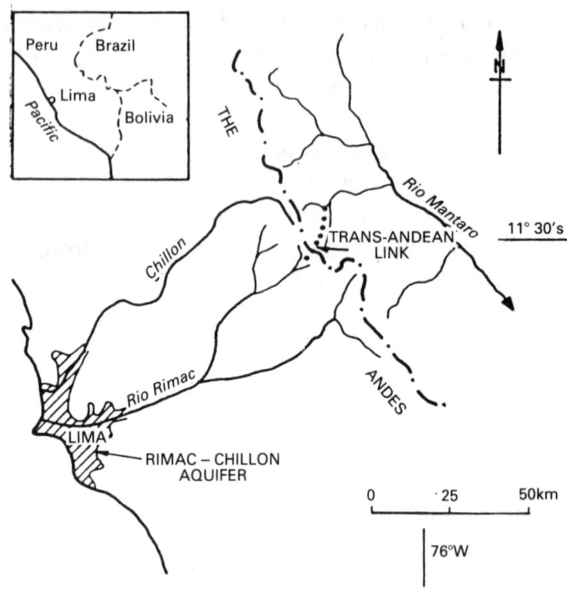

Fig. 7.6 — Location of Lima.

the main runoff season January–April. In the dry season the smaller Rio Chillon reduces to 1–2 $m^3 s^{-1}$ and, during this period June–November, the total flows in the two rivers are used for water supply.

The Lima aquifer, extending over 400 km^2 and reaching a maximum thickness of over 500 m near the coast, is unconfined and consists of thick sequences of varied alluvial fan sediments. The basement rocks, limestones, sandstones and tertiary volcanics contribute little to the groundwater resources. The aquifer supplies about 9.5 $m^3 s^{-1}$ from 330 municipal wells and over 750 private industrial, domestic and agricultural wells. The decline in groundwater levels from 1969 to 1985 ranged up to 40 m owing to increasing abstractions and reduction of recharge on irrigated land replaced by urban developments. Nevertheless, the aquifer contains the only available extra resources to meet increases in demand in the short term.

Groundwater model

For these additional studies, the two-dimensional finite-difference computer model originally used, was developed to provide three options:

(1) Constant transmissivity.
(2) Transmissivity varying linearly with saturated thickness.
(3) Transmissivity varying non-linearly with saturated thickness (permeability varying linearly with depth).

The aquifer area was modelled (with a final calibration using the second option) within a 30 km × 36 km regular grid of 1080 nodes orientated approximately to accord with the aquifer boundaries. Only 444 of the nodes were active and 116 had fixed heads related to sea level (Fig. 7.7). A monthly time step was used for the flow calculations on the uniform 1 km grid.

To provide initial conditions for the model, an estimate of the aquifer water balance was made (Fig. 7.8). Recharge from rainfall was discounted since it is negligible. Monthly estimates of flows at each node were required and, for the first run, only outflows to the sea were unknown. The early calibration of the model was improved by incorporating results of later fieldwork, in particular from a map of the varying transmissivity values prepared from all available data as a check on the model performance. The calibration runs allowed variations in both flows and aquifer properties so that the transmissivity values were much closer to the field results. A realistic set of starting conditions for simulation runs of the calibration period (1969–1985) was obtained in two stages.

(1) By derivation of steady-state levels using aquifer inflows and outflows for the average year 1970 (1969 was an untypically wet year).
(2) By derivation of a dynamic balance by running five annual cycles with flow data for the same year.

The model was calibrated against piezometric levels since there was more information on well levels than flows. The final calibration was achieved after 86 trial runs; 28 of these were over the full 17 year period. The sensitivity of the model was assessed to changes in model transmissivities, the magnitude of river valley groundwater inflows and outflows, river bed infiltration and leakage from the water distribution system.

The results of the calibration runs identified the reduction in leakage losses during the later years of the period and several locational anomalies, one of which was a ridge of basement rocks at an unknown higher level. The improved model was found to provide a closer match with observed well hydrographs. The maximum falls in water levels around 1980 were matched in most areas and the recovery of levels in some zones was also adequately modelled. The comparison of the performance of the two models is shown in the average 1969–1978 water balances in Table 7.6.

Future supplies

The improved calibrated groundwater model was used to investigate the availability of further groundwater supplies over the next 15 years, 1986 to the year 2000. Three main operational methods were studied.

Option 1 Maintaining the existing distribution and rate of abstraction.
Option 2 Increasing groundwater abstractions to meet increasing demand.
Option 3 Introducing conjunctive use and artificial recharge schemes.

Options 1 and 2 would certainly need to consider the increased lowering of water levels and in operating the model practical limitations were imposed — the rate of abstraction from 1 km^2 nodal area to be 10 000 m^3 day^{-1} or the minimum saturated thickness to be 20 m before wells considered dry. For each option the model predicted the following.

(1) Aquifer areas which became dry.
(2) Areas at risk of saline intrusion.

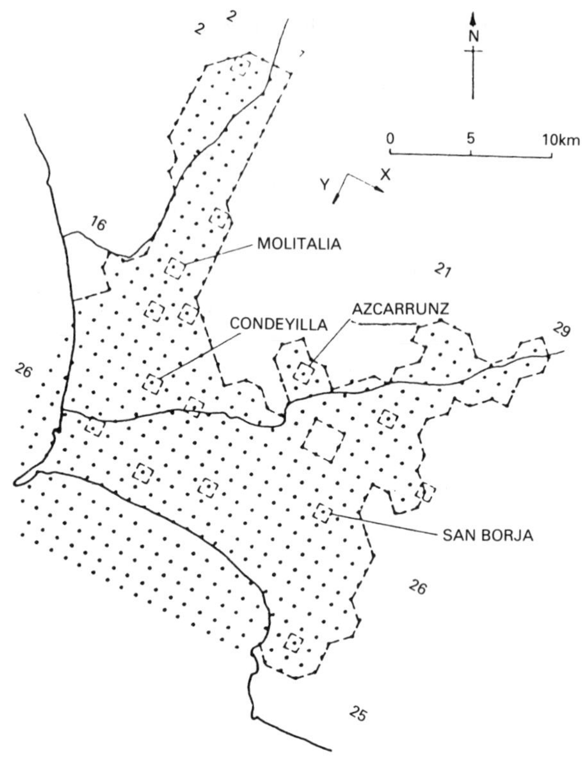

Fig. 7.7 — Location of observation wells: ⁰∘, model nodes; ⌜·⌟, nodal areas with well hydrograph used for calibration.

(3) Drawdowns below 1986 water levels.
(4) Yield reduction of existing wells caused by falling water table.

The introduction of conjunctive use and artificial recharge schemes combined with the redistribution and increase of groundwater abstraction was predicted to succeed in delaying the need for the Mantaro transfer scheme until 1999.

The recommended plan assumed that a new reservoir to be operational in 1990 would increase the reliable low flow of the Rio Rimac from 11.5 to 13.5 $m^3 s^{-1}$ and that treatment works capacity should be raised to 20 $m^3 s^{-1}$ with an enlarged distribution network. Recharge by lagoons from January to April could provide increased groundwater resources in the two river valleys and induced river bed recharge in the Upper Rimac might produce further supplies. However, the model simulations showed that, from 1990 to 1999, about 30 new wells per year would be required and about 25 wells per year would need to be deepened to meet the increasing demand and to keep pace with the falling-water table.

The groundwater model is installed on microcomputers in Lima and can be used as a planning and management tool in the difficult period ahead to 2000.

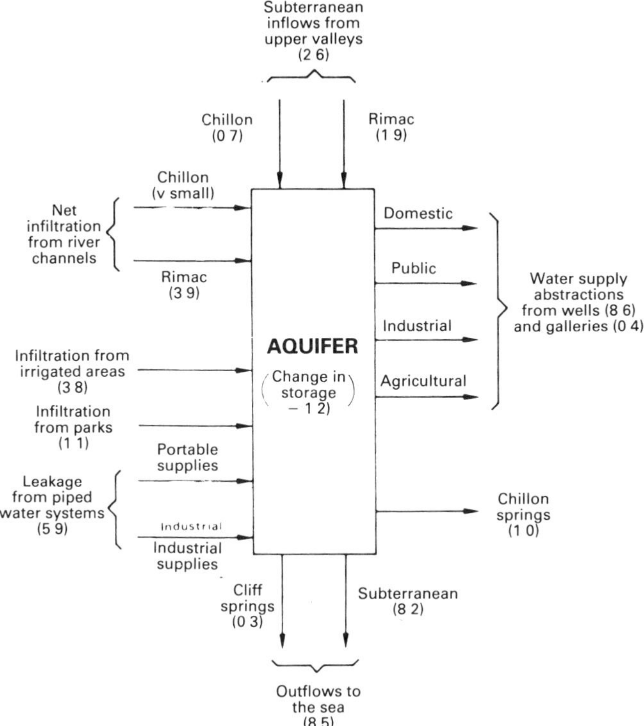

Fig. 7.8 — Main groundwater inflows and outflows. The numbers in parentheses are the main flows ($m^3 s^{-1}$) (1967–1978). Recharge from precipitation is negligible.

Table 7.6 — Average annual groundwater balances

	1969–1978		1969–1985
	Early study	Later model	Later model
Inflows ($m^3 s^{-1}$)			
Ground water	2.65	2.07	2.05
River bed recharge	4.31	2.95	3.21
Irrigation for agriculture and parks	4.91	5.44	4.56
Leakage from water supply network	5.84	4.07	3.99
Total inflow	17.71	14.53	13.81
Outflows ($m^3 s^{-1}$)			
Springs	1.31	1.07	0.87
Abstractions	9.00	8.96	0.43
River bed exfiltration	0.40	0.57	0.43
Ground water	8.22	4.58	4.17
Total outflow	18.53	15.18	14.80
Inflow − outflow	−0.82	−0.65	−0.99
Storage change (M m^3)			−31.5

7.5 QATAR WATER RESOURCES AND AGRICULTURAL PLAN

LOCATION The State of Qatar occupies a peninsula projecting northwards from Arabia into the Persian Gulf.

SOURCES Hall, M. J., & Hill, N. A. (1983) A master water resources and agricultural plan for the State of Qatar Part I, Physical setting and resources. *Water Resour. Dev.*, **1**, No. 1, 15–30.

Annesley, T. J., Hall, M. J., & Hill, N. A. (1983) A master water resources and agricultural plan for the State of Qatar, Part II, The systems model and its application. *Water Resour. Dev.*, **1**, No. 1, 31–49.

Pike, J. G. (1983) The planning of water resources development in an Arabian Gulf State: Qatar, a case study. *J. Inst. Water Eng. Sci.*, **37**, 75–88.

PROBLEM The rapid development and modernization of Qatar stemming from the exploitation of oil and natural gas during the last two decades has resulted in the increasing depletion of the groundwater resources for the extra demands for agricultural food products. In addition, the rising population served by cheap desalinated water and the growing availability of sewage effluent for agricultural purposes stimulated a special study of the optimum allocation of the three major water resources to satisfy the expanding domestic, industrial and agricultural needs. The consultants, Halcrow Balfour Ltd, were retained to undertake a master planning study which involved the setting up of a systems model incorporating the sources of water supply and the changes in demand likely to occur up to 2000 AD with the expansion of the state economy.

Water resources

The State of Qatar covers an area of 11 600 km^2 (Fig.7.9). The landscape is generally flat except for a line of hills parallel to the west coast and a belt of large barchan sand dunes in the south. The arid desert climate experiences a slight amelioration from the surrounding seas but the average annual rainfall is only 75 mm and high temperatures are usual from June to September (mean daily maxima, over 40°C). Most of the rainfall occurs in high-intensity storms between November and April, the cooler season when mean daily maxima are still 21°C. Open-water evaporation can reach 20 mm day^{-1} in summer, a factor to be considered in crop irrigation. Diurnal variations in temperature and humidity also affect horticultural practices and high winds are known to damage crops in open locations.

The three distinct sources of water are groundwater, desalinated sea water and treated sewage effluent (TSE).

(1) *Groundwater* The main structural feature of the Qatar geology is a broad-

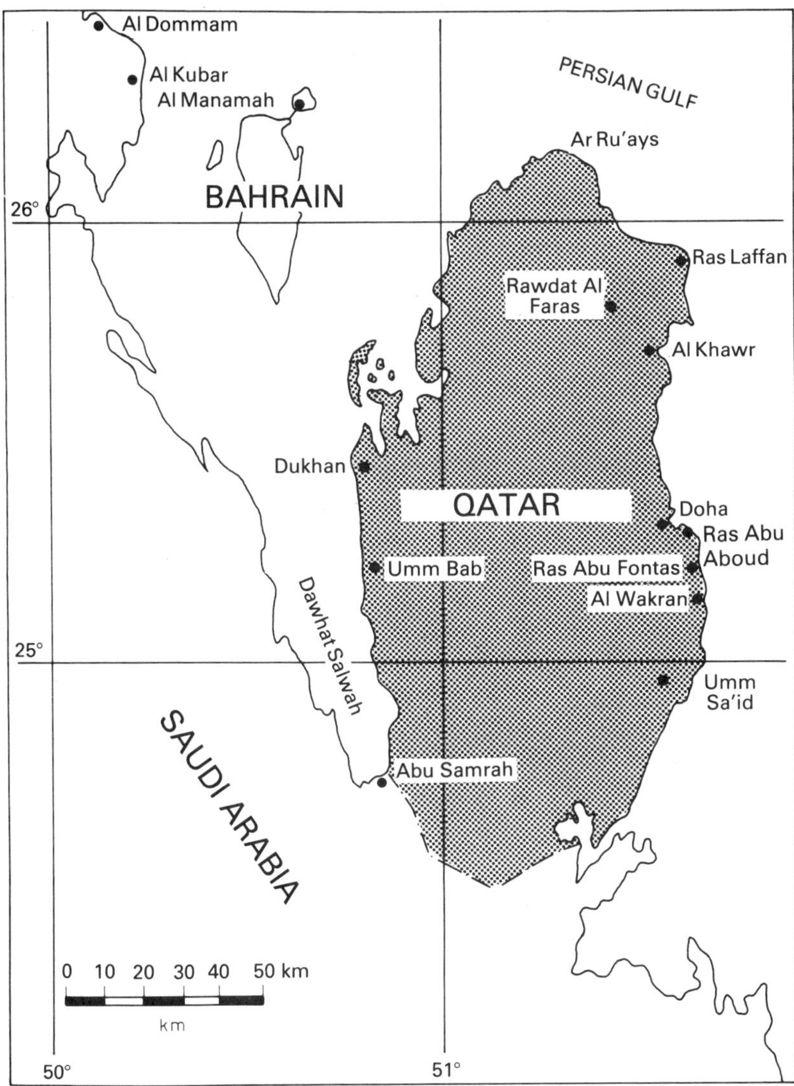

Fig. 7.9 — Location of Qatar.

layered anticlinal arch, axis north to south, with many of the formations composed of permeable limestones and dolomites. Nevertheless the hydrogeology is extremely complicated in detail and special digital groundwater modelling studies have been made to assess yields in the strategically important areas. A finite difference model was used to predict temporal and spatial effects of different abstraction policies. This model demonstrated that some 13–15 Mm^3 $year^{-1}$ could be extracted from the major upper aquifer and the lower aquifer could provide 20 Mm^3 $year^{-1}$ for about 50 years. The safe yield of the groundwater system was estimated to be about 33 Mm^3 $year^{-1}$.

(2) *Desalinated sea water* Domestic potable supplies are mainly served by desalination. Two major plants are at Ras Abu Aboud and Ras Abu Fontas power stations with capacities of 52 000 m^3 day^{-1} and 180 000 m^3 day^{-1} respectively. A reliable yield of 242 000 m^3 day^{-1} in 1983 could be enhanced by further supplies from a new power station at Ras Laffan. In addition, large industrial undertakings have their own desalination plants.

(3) *Treated sewage effluent* The capital Doha has the most extensive sewerage system but at present most of the effluent is only partially treated. The quantities available are dependent on sewage works capacities but with the expected population increases and comparable works expansions it is estimated that the available volume of TSE will be approximately the same as the groundwater safe yield and will be a very significant resource for agriculture.

Water demand

The potable water demand up to 2000 AD based on the present population structure proved very difficult to estimate. Qatar has a high transitory population of immigrant workers such that the expatriate-to-Qatari ratio at present is 2.14 to 1 and the number of expatriates is closely related to the expansionist programme and government policy. The varied demand per capita also added to the complexities and the problem was resolved by allocating households into four income groups in which the per capita consumptions were assumed to be constant up to 2000 AD. The projected requirements are shown in Fig.7.10 in which the four curves represent different combinations of population growth rates and expatriate to Qatari ratios.

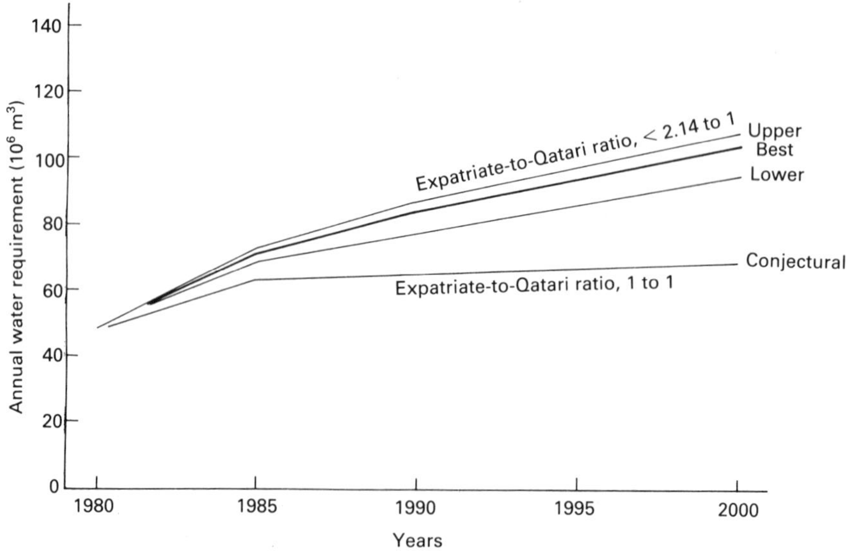

Fig. 7.10 — Projected annual domestic, commercial and industrial water requirements.

The changes in water demand for agriculture proved even more difficult to assess. Information on suitable soils, farm sizes and crops grown was gleaned from sample surveys and crop trials. The most favourable areas are in the north and east with ready well access to groundwater. However, only 248 out of the 377 farms operating in 1980 were considered to produce crops in commercial quantities. The remainder were operated for amenity value and for small quantities of fruit and vegetables for private consumption. All suitable crops are already produced in Qatar and growth in water use would be by increased production or the introduction of new large-scale farms. Fifteen farming type options were identified and 15 model farms were assessed according to water quantity and quality requirements, labour requirements, irrigated area, minimum farm size, capital requirements and operating costs. Financial budgets for existing and proposed farms showed that, in the absence of subsidies, only vegetable farms are profitable. Thus there is little incentive for private investment in agriculture.

Master plan model

The analysis of the water resource and water demand systems of Qatar were carried out using an optimization model in which the system relationships are expressed in linear form. Then a standard linear programming computer package was used to derive the optimal strategy. The objective was the allocation of the available water resources so that sufficient food could be produced for the inhabitants either by domestic agriculture or by means of imports, at a minimum cost under any given policy. An assumption was made that domestic, commercial and industrial water demands should be met first; so these were subtracted from available supplies before optimizing the agricultural system. Regional variations were considered by dividing the country into six areas within which the variables and constraints on the model needed definition. The variables occur in three groups: the food imports within nine selected product categories, the number of farm units in each of 15 types of model farm and the sources of available water to serve the different farm types. The main constraints were the following.

(1) Total annual water demand in each area should not exceed available water supply and water quality to be appropriate to the need.
(2) Volume of water supplied from any source should not exceed amount available.
(3) Total land requirements for farms and large plots should not exceed amount available.
(4) Labour requirements of all farms should not exceed amount available.

The requisite data were assembled in two- or three-dimensional matrices according to the variable. The 13 data files for input into the allocation model were as follows.

(1) Water quality requirements of each farm type.
(2) Total water requirements of each farm type.
(3) Supply of water available from each source in each year 1980–2000.
(4) Area of land in large plots—requirements of each farm type.
(5) Total land requirements of each farm type.
(6) Land availability in large and small plots in each area.
(7) Food production in each food category in each farm type.
(8) Maximum labour requirements in each farm type.

(9) Qatar's food requirements in each food category in each year, 1980–2000.
(10) Import costs of food in each food category.
(11) Total operating costs of each farm type.
(12) Maximum number of each group of farms in each year.
(13) Maximum volume of imports in each food category in each year.

The running of the model on an Apple II microcomputer with the problem having 50 variables and 40 constraints required 3–6 h for the first year 1980. For the later years the run took 10 min to 4 h depending on changes to the constraints affecting the linear programming solution.

To evaluate operational policies, the master plan model was applied making a range of basic assumptions on the population estimates, decreases in water supply distribution losses, unlimited desalination water, energy costs, etc. Eight trial policies were examined and sensitivity of changes in constraints and assumptions was tested.

Conclusions

The most attractive agricultural development plan resulting from the model was as follows.

> 'Increased production by the provision of desalinated water to enable Qatar to achieve the maximum degree of self-reliance in all crops except cereals and sheep-meat but using TSE only in central Qatar.'

This plan would entail the expansion of the traditional vegetable and fruit farms for which financial incentives to the tenant farmers would be needed to encourage increases in crop yields by more modern farming practice. The expansion of milk and beef production would be encouraged with the establishment of new modern farms in favourable areas. A subsidy system would be necessary to ensure proper developments. Assuming that the government adopted this strategy, the water resources development plan indicated that the water demand for commercial agriculture would rise from 53 Mm^3 $year^{-1}$ in 1980 to 69 Mm^3 $year^{-1}$ by 2000.

The overall water demand and supply for Qatar over the study period is given in Fig. 7.11. The analysis demonstrated that the water demands could be well met by existing supply sources to avoid the need for large costly storage reservoirs. Increased demands would be economically met by the cheap supply of desalinated water guaranteed by the abundant energy resources.

7.6 SUKHOTHAI GROUNDWATER DEVELOPMENT

LOCATION Sukhothai is in the River Yom Basin in northern Thailand.

SOURCE Howard Humphreys & Partners (1986) *Sukhothai Groundwater Development Project, aquifer modelling studies*, Main Report to the Royal Irrigation Department, Kingdom of Thailand.

PROBLEM From exploratory studies in the Yom Basin in the early 1970s, it was found that the Sukhothai Plain was composed of alluvial

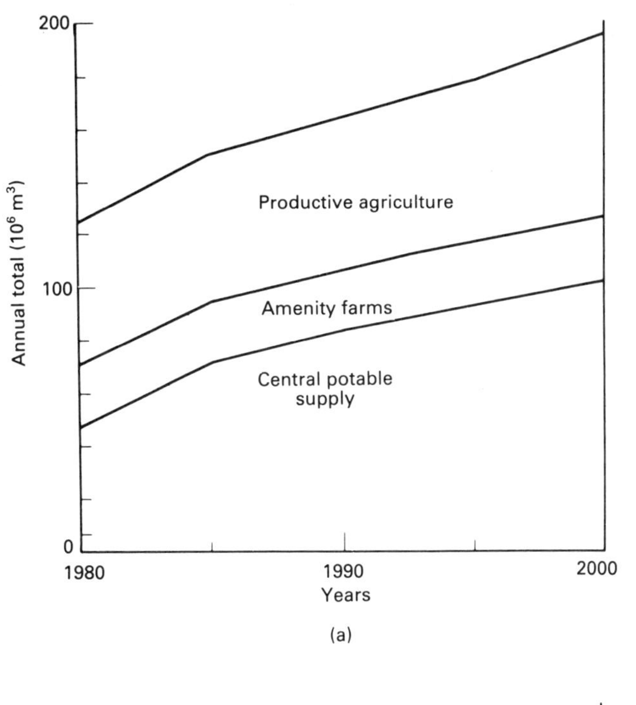

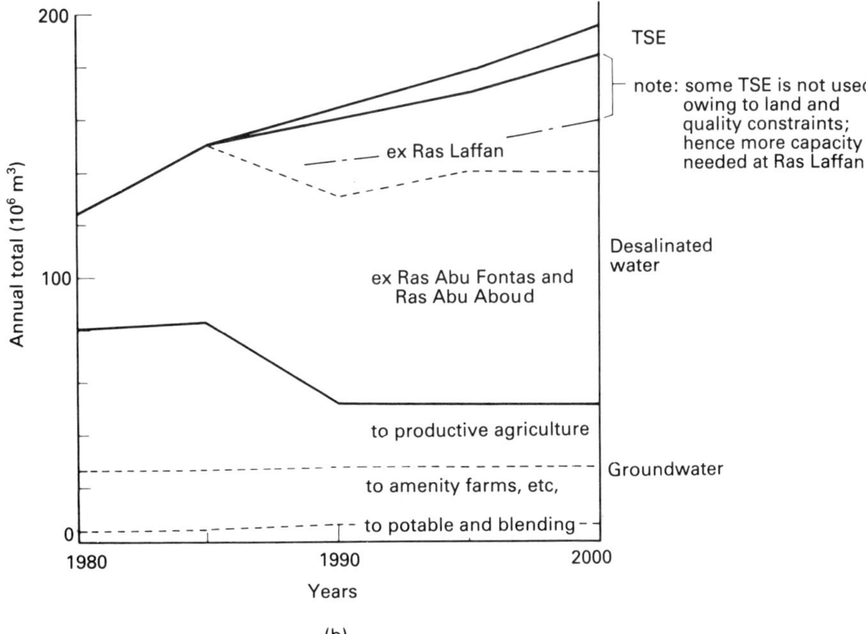

Fig. 7.11 — Main water balance for Qatar, 1980–2000: (a) demand; (b) supply.

aquifers with groundwater resources suitable for the development of irrigated agriculture (Fig. 7.12). Two major groundwater schemes are being implemented, each to increase abstraction gradually to a planned capacity of 40 Mm^3 $year^{-1}$ and in addition local farmers are actively exploiting groundwater by using shallow hand-dug wells. Since the early studies, more hydrogeological data have become available, and although the planned levels of abstraction are not expected to exceed the annual aquifer replenishments, it was considered expedient to make new more reliable assessments of groundwater resources by mathematical modelling techniques. The consultants, Howard Humphreys & Partners were commissioned to undertake the aquifer modelling studies by the Royal Irrigation Department of the Kingdom of Thailand.

Sukhothai Plain

A northern extension of the upper central plain of Thailand, the Sukhothai Plain, is bounded to the north, east and west by mountains rising to 500 m (Fig. 7.12). There are a few isolated hills near the edge of the plain but the topography is subdued, falling from 60–70 m in the north to around 50 m. Two rivers, the Yom and the Nan, follow parallel meandering courses across the plain from north to south and eventually join together beyond the study area. The plain is composed of a series of alluvial deposits laid down in the basin between the mountains on an undulating pre-Tertiary Basement rock surface. Major fault systems have resulted in sediment depths of nearly 4000 m but the alluvial succession has only been proved to a depth of 335 m. There is great lateral variation in the series of sands, gravels, cobbles, clays and silts. However, three main groupings have been identified.

(1) Surface clays and silts (from less than 5 m to more than 50 m).
(2) Upper alluvials, interbedded sands and gravels with minor clays and silts (down to 65 or even 100 m).
(3) Lower alluvials, silty clays and sands with cleaner sand and gravel horizons (below the upper alluvials to maximum borehole depth of 335 m).

The recent exploratory studies have shown that the study area is underlain by a single alluvial aquifer system whose character varies according to lithological changes. It is bounded at the top by the surface clays which act as an aquitard. In the west, the basement rocks often with an overlying clay form the base of the aquifer and this dips so steeply eastwards that it has not been identified at the eastern edge of the study area.

Hydrometric data

The area is well served with two climate stations, at Sukhothai and Si Samrong, and 18 rainfall stations mostly with records back to 1952. The average annual rainfall ranges from 942 to 1458 mm (Fig. 7.13). The seasonal pattern of the monsoon climate with an annual mean temperature of 27.5°C is demonstrated in Table 7.7. Owing to some unexplained exceptionally extreme annual rainfall totals at Sawankhalok, the data from Si Samrong have been used as representative of the modelled area.

River flow records are available for three stations on the Yom River, Kaeng

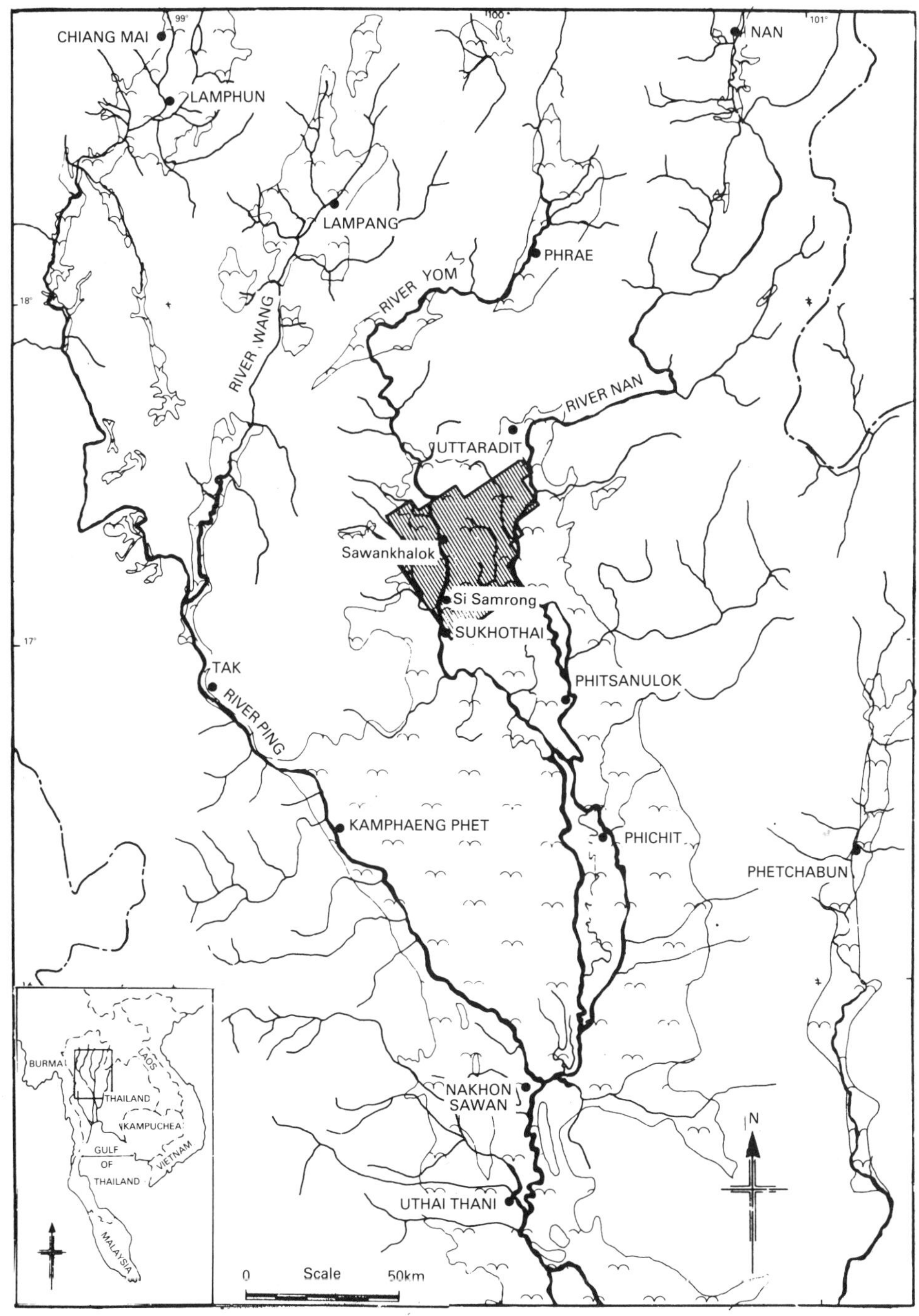

Fig. 7.12 — Sukhothai location map.

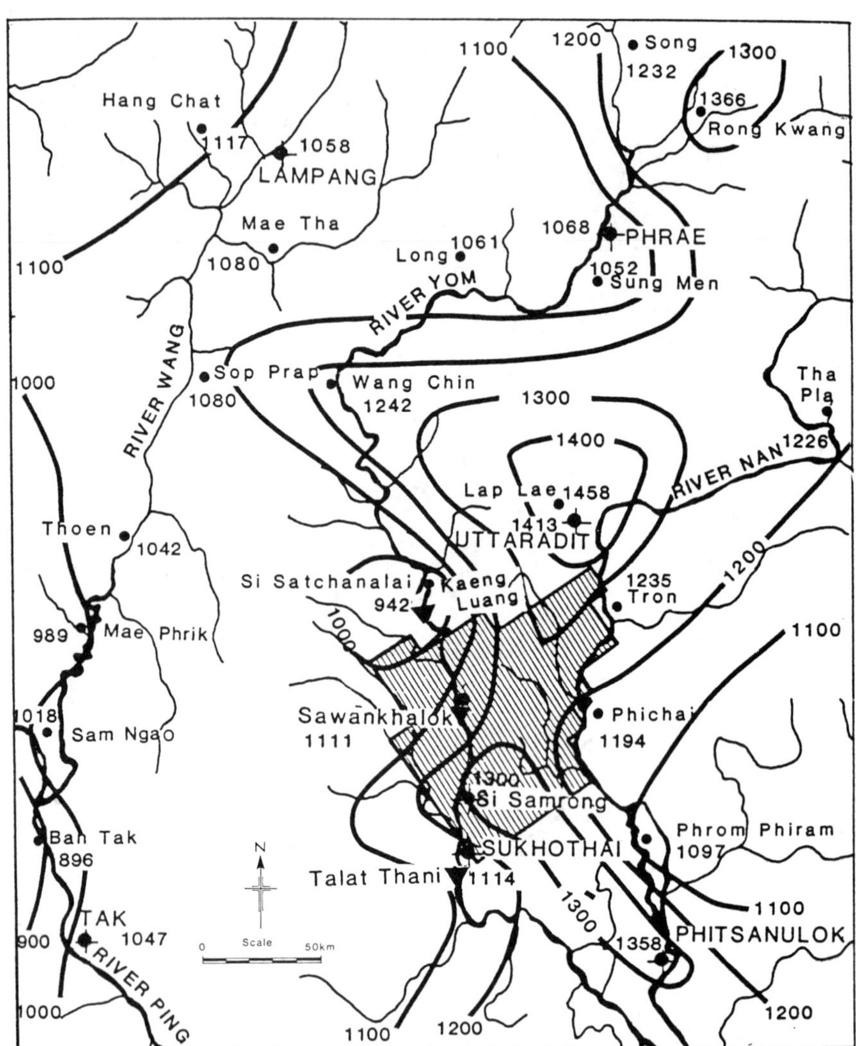

Fig. 7.13 — Mean annual rainfall map, 1952–1984: ●, rainfall stations, the numbers give mean annual rainfall (mm) (1952–1984); -✦-, climate stations, ———, rainfall isohyets, the numbers give rainfall per year (mm year^{-1}); ▼, gauging stations.

Table 7.7 — Monthly mean rainfall and evaporation

Month	January	February	March	April	May	June	July	August	September	October	November	December
Sawankahlok, rainfall, 1952–1984 (mm)	5.4	4.8	18.2	45.7	145	144	147	167	234	140	15.7	2.9
Si Samrong, rainfall, 1968–1984 (mm)	9.9	5.5	31.0	49.0	182	138	152	180	297	147	24.1	7.7
Pan evaporation (mm)	119	142	200	224	209	167	167	162	140	131	120	115
PE grass, reference crop (mm)	96	111	150	172	158	132	122	117	114	121	1101	90

Table 7.8 — Monthly mean river flows

	Mean river flow (Mm^3)											
	January	February	March	April	May	June	July	August	September	October	November	December
Yom River												
Kaeng Luang, 1952–1983	24.9	12.7	8.6	12.6	70.2	121	207	627	987	442	140	52.0
Sawankhalok, 1967–1983	30.2	13.8	10.0	17.4	81.9	129	264	707	919	438	146	57.2
Talat Thani, 1950–1981	29.0	14.2	10.2	11.2	61.2	115	151	375	534	341	144	60.4
Nan River												
Phichai, 1951–1970	138	96.7	71.7	73.2	177	352	765	1836	2424	1036	405	219
Phichai, 1975–1983	528	629	819	858	790	767	955	1166	1125	921	708	576

Luang (catchment area, 12 585 km^2, mean daily flow 86 $m^3 s^{-1}$), Sawankhalok (catchment area, 13 583 km^2; mean daily flow 94 $m^3 s^{-1}$) and Talat Thani (catchment area, 17 731 km^2, mean daily flow, 61 $m^3 s^{-1}$) (Fig.7.13). Of four stations on the Nan River, the mean monthly values at Phichai (catchment area, 19 383 km^2, mean daily flow, 316 $m^3 s^{-1}$) have been given in Table 7.8 to show the regulation of the flows in the later period following construction of the Sirikit Dam upstream.

Hydrogeological measurements are obtained from water levels in hand-dug wells, production wells and piezometers. The water levels vary between 5 and 10 m below ground level except in the exploitation areas where they are 15–18 m. There is a regular annual response to the summer monsoon rainfall with the maximum levels in the shallow wells occurring within 1 month of the peak rainfall and the deeper wells responding about a month later. The water level annual fluctuation ranges from 3–6 m. From pumping tests the estimates of transmissivity, permeability and storage coefficients have been derived over the study area. Abstractions from the two development zones together with private abstractions have grown to exceed 33 Mm^3 $year^{-1}$(1984).

Aquifer recharge

Three main sources of recharge to the modelled aquifer system have been identified: lateral inflow, river bed leakage and rainfall infiltration.

From the general slope of the piezometric surface, lateral subsurface inflow could only be expected across the northern boundary of the study area which is 50 km long and an estimate of 2500 $m^3 day^{-1}$ was made. Both rivers, the Nan and the Yom, are potential sources of recharge. From studying the piezometric surface in the surface alluvials and deeper alluvials, there could be a potential for discharge from the deeper alluvials towards the river and actual recharge from the river bed is small. However, in extreme drought conditions the river contribution could be assessed as 12 000 $m^3 day^{-1}$. Pumping tests near the Yom indicated that the river did not act as a source of recharge.

The high rainfall of the monsoon season is the main source of recharge to the groundwater aquifer with surplus irrigation water forming a significant contribution. The downward penetration of the net rainfall (evapotranspiration deducted) is inhibited by the surface aquitard and the recharge of the aquifer below is dependent on the leakage from water stored in the aquitard. Owing to the variability of these alluvial deposits, the infiltration rates vary over the area from less than 0.5 to over 10 mm day^{-1} and the permeability of the aquitard range between 1 and 23 mm day^{-1}. Several estimates of the total recharge over the 1575 km^2 of the study area were made making assumptions of the hydraulic gradient across the aquitard and of the possible effects of groundwater abstractions from the aquifer. From the latest calculations and measurements (1982–1984) a value of 213 Mm^3 $year^{-1}$ was obtained, corresponding to 12.7% of the rainfall.

Groundwater model

The modelled area shown shaded on Fig. 7.13 has been represented in the vertical by a single aquifer overlain by the aquitard, the surface clay layer, and underlain by an aquiclude. Lateral groundwater movement is calculated by the general Laplace equation

$$T_x \frac{d^2h}{dx^2} + T_y \frac{d^2h}{dy^2} = S \frac{dh}{dt} + W(x, y, t)$$

where T_x and T_y are the aquifer transmissivities in the x and y directions, S the storage coefficient per specific yield, h the water level or head, t the time and $W(x, y, t)$ the nodal recharge or discharge. The solutions for head h at nodes on a variable grid network allowing detailed results where necessary, are made by using the equation in finite-difference form. For the backward difference scheme, only the h from the previous time step is assumed and thus for each nodal calculation there are five unknown h values (with the neighbouring nodes). The resulting series of simultaneous equations are solved in this model by the iterative technique, line successive over relaxation.

For simulating vertical recharge to the aquifer, the complex physical processes have been modelled realistically using the flow mechanisms shown in Fig. 7.14.

The increases in storage in the three major vertical zones are obtained as follows:

surface storage = rainfall + overland flow + irrigation − runoff
− evapotranspiration − infiltration

with components dependent on land classification

aquitard storage = infiltration − evapotranspiration − vertical leakage
aquifer storage = vertical recharge − river leakage − lateral outflow − well abstraction.

The 1575 km^2 area is represented by 368 nodes and the grid sizes range from 5 km × 5 km to 1.25 km × 1.25 km. The orientation of the grid is based on the prevailing groundwater flow pattern and the boundaries defined by physical features. No flow is assumed across the northern and western boundaries since there is no aquifer sand outside the model limits. The eastern boundary following the course of the Nan River has a fixed lateral outflow in the aquifer continuing underneath the river and the southeastern boundary has been modelled as a fixed head boundary since knowledge of the very deep water levels is limited.

Within the study area, the base of the aquifer has been taken to be the lowest sand layer above the basement rocks or 75 m below sea level where the basement is unknown. The top of the aquifer is the base of the surface clay layer. The groundwater is confined at this level but, during pumping, a phreatic surface can result when the aquifer becomes unconfined.

Model calibration

There were difficulties in calibrating the model owing to the shortage of long-term data before pumping commenced in 1981. Thus the operation had to be carried out in stages. By using the best estimate of a pre-pumping piezometric surface and values of permeability and transmissivity, estimates of lateral inflows and outflows allowed only along the southeastern and eastern boundaries were obtained. Then the

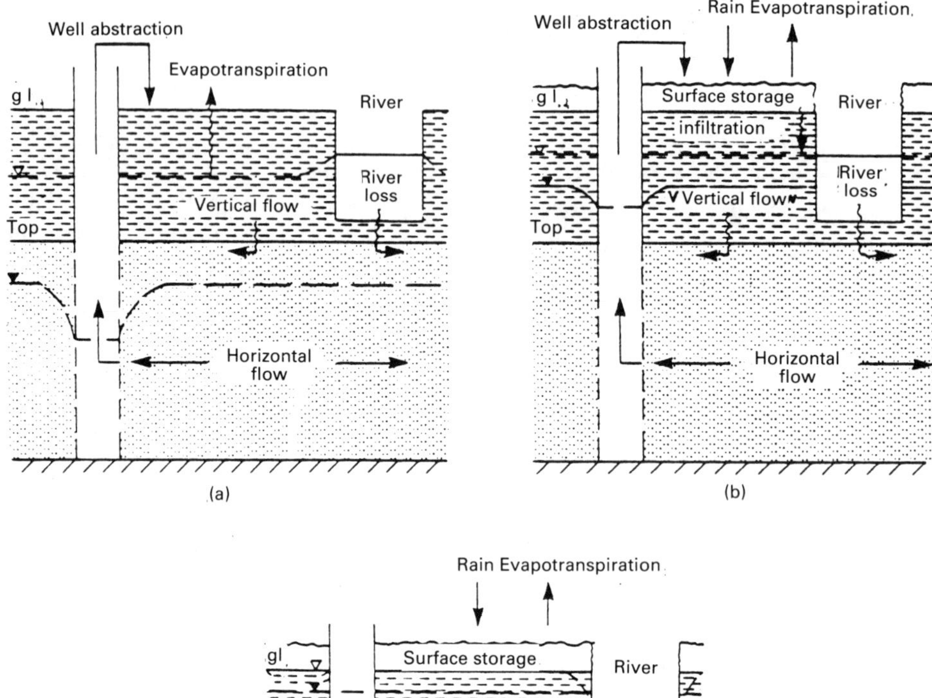

Fig. 7.14 — Examples of modelled flow mechanisms, (a) in the dry season, (b) at the start of the wet season and (c) in the saturated retaining layer; [symbol] a quitard layer; [symbol], aquifer; GL, grand lever; Top, top of aquifer, [symbol], base of aquifer; [symbol], prezometric surface in aquifer; [symbol], phractic surface in aquitard.

sensitivity of the aquifer system was tested according to changes in hydrological variables using rainfall data from 1980 to 1984 and observed well level fluctuations. Then long-term hydrometeorological data 1968-1984 were used to check the parameter values previously obtained. Numerous model runs were made to determine the sensitivity of the model to the various hydrological and hydrogeological parameters and the final calibrated piezometric surface used for all the pumping simulations was based on the following.

(1) Hydrological parameters.
  (a) Rainfall: Si Samrong weekly rainfall, 1968–1984.
  (b) Potential evapotranspiration calculated using modified Penman method for weekly data from Si Samrong, 1968–1984.
  (c) Maximum depth for evapotranspiration, 2.5 m below ground level.
  (d) Infiltration rate (soil), 0–2 mm day^{-1}.
  (e) Overland flow, 40, 50 and 150 mm depending on land use.
(2) Hydrogeological parameters.
  (a) Horizontal permeability (aquifer), 10–40 m day^{-1}; vertical permeability (aquitard), 0.001–0.0003 m day^{-1}.
  (b) Storage coefficient (aquifer) 0.001–0.005; specific yield (aquifer), 10%; storativity (aquitard) 3%.
  (c) Aquifer thickness, sand thickness, 0–90 m; aquitard thickness, surface clay, 5–50 m.

The main model program is served by five subsidiary programs defining the model structure, parameters and abstraction requirements and three output programs plotting hydrograph simulations at selected nodes, isopleths of required hydrogeological data and historic well data. In addition, 12 data files provide the input requirements.

Model application

Simulated abstractions for two configurations of wellfields in the two major development zones were made commencing with the input piezometric surface of December 1984 and applying 10 years of hydrometric data (1968–1977). Thus, in simulating operations for the following period 1985–1994, interannual variations in rainfall and crop requirements were incorporated. The annual combined total abstractions for two simulated abstraction rates are shown in Table 7.9.

The results of the pumping simulations were presented by piezometric and drawdown maps and by hydrographs at selected nodes. A summary of the findings at the end of the test period in 1994 is given in Table 7.10. The operation of the simulation model also provided information on the general availability of resources particularly with regard to the different methods of aquifer recharge. The small amount of lateral inflow into the area across the Nan River boundary amounting to 0.04 Mm^3 year^{-1} was considered negligible. From river bed infiltration, the model indicated that the Nan did not on average contribute water to the aquifer but a small amount, 0.2 Mm^3 year^{-1}, of groundwater leaked upwards to the river.

The downward leakage through the surface clay aquitard was induced by increasing abstraction rate from 16.33 Mm^3 year^{-1} with no abstractions up to 92.32 Mm^3 year^{-1} with the average abstraction rate of 90.5 Mm^3 year^{-1}. Infiltration from the surface into the aquitard was not significantly affected by pumpage.

Thus, over the 10 year period of simulation with an average annual rainfall of 1231 mm, there was sufficient recharge through the aquitard to sustain the higher average abstraction rate without affecting stored groundwater reserves. In the

Table 7.9 — Total simulated average abstraction rate

Year	Rain year	Abstraction rate I (Mm^3 $year^{-1}$)	Abstraction rate II (Mm^3 $year^{-1}$)
1985	1968	99.73	74.73
1986	1969	89.38	68.03
1987	1970	80.03	61.98
1988	1971	91.83	69.63
1989	1972	96.03	72.33
1990	1973	91.53	69.43
1991	1974	89.03	67.83
1992	1975	86.83	66.38
1993	1976	88.33	67.03
1994	1977	92.23	69.83
Average		90.5	68.7

Table 7.10 — Pumping simulation drawdowns, 1994

	Wellfield A, maximum drawdown (m) for the following average annual abstraction rates		Wellfield B, maximum drawdown (m) for the following average annual abstraction rates	
	90.5 Mm^3 $year^{-1}$	68.7 Mm^3 $year^{-1}$	90.5 Mm^3 $year^{-1}$	68.7 Mm^3 $year^{-1}$
Zone 1	32	18	27	15
Zone 2	20	17	20	17

future, more data will enable a more precise calibration of the aquifer system to be made.

8

Reservoir yield

INTRODUCTION

This chapter contains studies of water resource projects in which impounding reservoirs constitute the major storages in a scheme.

In the last century when a growing town or city wished to increase its water resources, a consultant water engineer was usually called in to advise. Then a convenient upland catchment area with an acceptable average annual rainfall was found, and depending on the supplies required, a site for a dam was selected to provide the necessary reservoir capacity. After the requisite site investigations, the design and construction of the dam proceeded, dependent only on the passage of a special Act of Parliament. The resultant reservoir served a limited community and water supplies were usually piped directly to the demand centre. This may or may not have necessitated some pumping but the water was usually of good quality and needed little treatment to bring it up to the specified standard for domestic supplies. The rising cost of such installations and the contemporary enthusiasm for preserving the natural environment has led to fewer but larger schemes funded from several sources. Improvement in engineering techniques has also meant a change in the function of impounded reservoirs. They are tending to become multipurpose, as flood controllers and as river-regulating storages, with water releases being made to the river to supplement low flows as well as for user abstraction at greater distances downstream nearer the demand centres. The question of water quality then becomes a major issue.

This progression of events relates primarily to development in the water supply industry in the UK but may be considered representative of the changing situations in many developed countries. In the developing world, the earlier stages in the progression may still be normal practice as will be seen in some of the overseas examples.

In the UK, regional responsibility for water resources and increased computer facilities, allowing more comprehensive initial hydrological analyses, have led to the broader evaluation of resources on a regional scale. With the benefit of long records of both precipitation and river flows, hydrologists have been able to make thorough assessments of the occurrence probabilities of dry periods, with their attendant threat of water shortages, and of the reliable yields of existing reservoirs. The calculated reliability of regional runoff parameters and reservoir yields enables authorities to plan further water storages and improved management of distribution to supply centes in the event of increased demands.

Similar computations for overseas schemes are generally hindered by lack of long records and recourse is usually made to simulating data using stochastic models. Catchment models are used to produce runoff and river discharge data when the historic data are limited to rainfall records. However, once good runs of monthly inflow and loss data have been established, the classical probabilistic methods of evaluating reservoir yields can be applied (McMahon & Mein, 1978).

The case studies in this chapter concentrate on the water resources of an impounding reservoir and the reader is reminded of the other important hydrological consideration in reservoir design, the evaluation of the design discharge for the dam spillway, dealt with in Chapter 6.

8.1 COW GREEN RESERVOIR

LOCATION Cow Green Reservoir is in Upper Teesdale, on the border of counties Durham and Cumbria, UK.

SOURCES Sandeman, Kennard & Partners (1965), Report on Further Augmentation of Water Supplies to the Tees Valley & Cleveland Water Board.

Kennard, M. F., & Reader, R. A. (1975) Cow Green Dam and Reservoir. *Proc. Inst. Civ. Eng.*, **P58** 147–175.

PROBLEM In the early 1960s, the Tees Valley and Cleveland Water Board were concerned about the rapidly increasing demand for water by the industries concentrated around the Tees mouth. Including the newly completed Balderhead Reservoir (Fig. 8.1), it was estimated that the net yield of all the sources in a critical dry year amounted to only 295.5 Ml day^{-1}. The existing demand (1966) was already of that order and the estimate for 1990 was given as 454.6 Ml day^{-1} of which 245.5 Ml day^{-1} would be required for potable supplies and 209.1 Ml day^{-1} supplied as raw water. The main abstraction points are at Lartington and Broken Scar where there are treatment plants and at Croft where raw water only would be available. It was therefore imperative that further storage was found to ensure adequate supplies in future years.

Solution

Prior to the construction of the Balderhead Dam, numerous sites had been investigated in the valley of the Upper Tees and its tributaries. Further considerations were given to four sites in the higher reaches of the river where there were favourable geological conditions and where sufficient storage could be obtained economically.

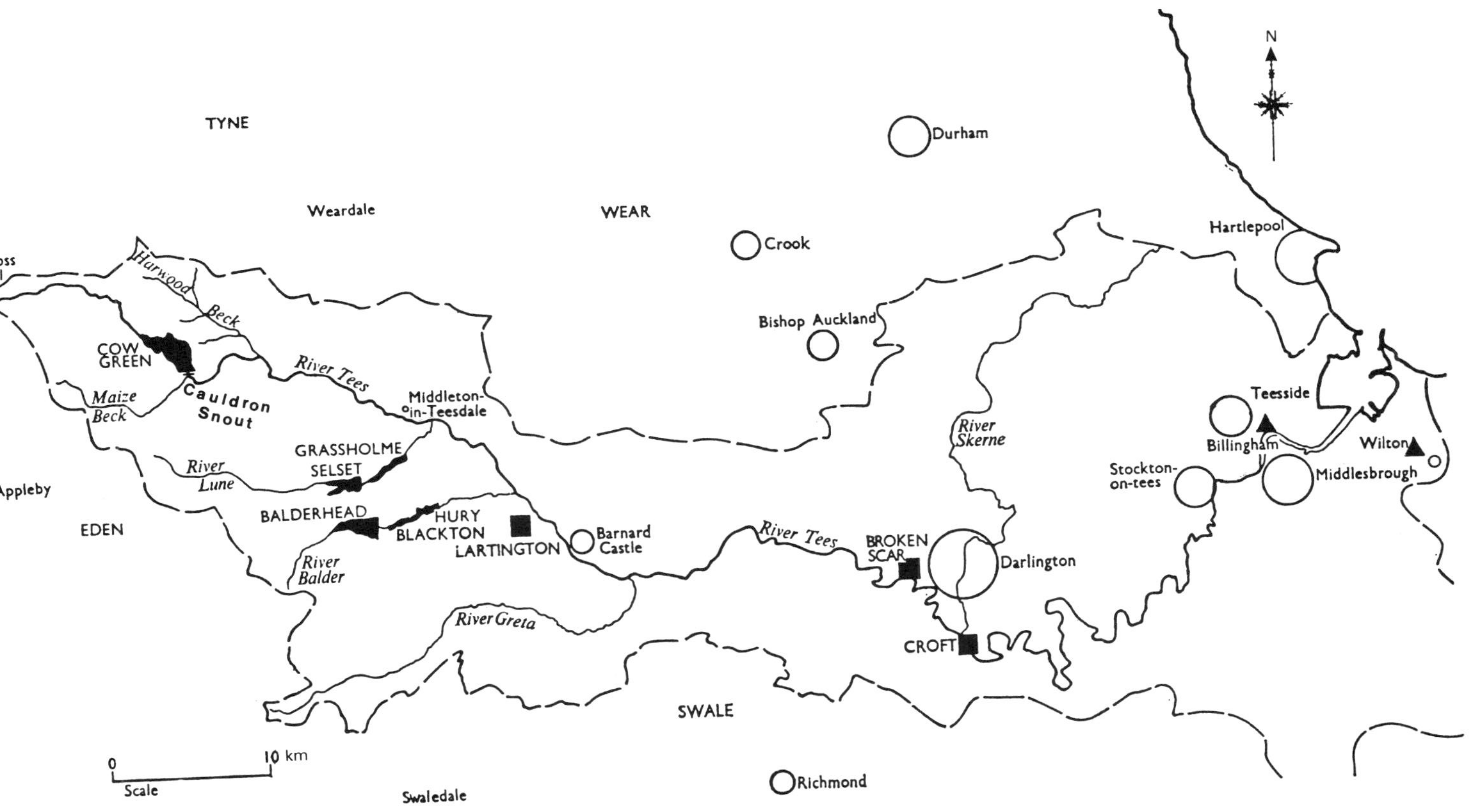

Fig. 8.1 — Tees catchment, reservoirs and abstractions: — —, boundary of Tees catchment; ○, population centres; ■, abstraction points; ▲, major industrial users.

The extensive area of uninhabited moorland upstream of Cauldron Snout waterfalls on the main river was the ideal site for a new reservoir. A catchment area of about 58 km^2 with an average annual rainfall of 1768 mm and evaporation losses of 320 mm would provide an average of 84 Mm^3 $year^{-1}$. In a dry year, the amount would be 50 Mm^3 assuming only two-thirds of the average rainfall. From topographical measurements assuming the dam site to be about 40 m upstream of a small bridge over the river above the waterfalls, the reservoir capacities shown in Table 8.1 were

Table 8.1 — Balderhead Reservoir capacities

Water level (m AOD)	Capacity (Mm^3)	Reservoir area (ha)
475	7.96	174
482	20.46	247
488	38.19	304
489	40.91	312
491	47.73	332

evaluated. In order to decide on the optimum reservoir capacity, an assessment of the water balance, yields versus demands, of the whole Tees Basin was necessary. The calculations were based on the minimum daily flows for the 6 month drought period of 1949 measured at the Broken Scar river gauging station where there was a prescribed flow of 0.789 m^3 s^{-1} (68 Ml day^{-1}) to be maintained. It was assumed that any inflow to the reservoir was added to any released water. From considerations of observed flows in the river and neighbouring tributaries, the quantity of 38.6 Ml day^{-1} for compensation water seemed reasonable.

The calculations proceeded as follows:

Net amount required to make up 1949 flows to the demand 409 Ml day^{-1} above Broken Scar	41.432 Mm^3
Add 22% to above for losses	9.115 Mm^3
Abstraction at Lartington 183 days at 145.5 Ml day^{-1}	26.627 Mm^3
Compensation water + inflow 183 days at 43.6 Ml day^{-1}	7.979 Mm^3
Gross amount to be released	85.153 Mm^3
Deduct runoff from subcatchments	6.524 Mm^3
Total water required from storage	78.629 Mm^3
Total net storage existing	42.732 Mm^3

Additional storage required	35.987 Mm^3
Allow for bottom water in reservoir	3.159 Mm^3
Allow for provision of freshets	1.818 Mm^3
Total new storage required	40.874 Mn^3

It was therefore recommended that a 40.91 Mm^3 capacity reservoir be constructed at Cow Green. This would necessitate a dam 25 m high with a crest length of 506 m and having a top water level of 489 m AOD. The reservoir area would be 3.12 km^2 representing about 5% of the catchment area.

Conclusion

Two major difficulties arose in the development of the scheme and in the detailed designing of the dam. Following the promotion of a parliamentary bill to obtain powers for the construction of the reservoir, there were serious objections by the environmental lobby against the endangering of rare plant species existing in a habitat unique in Great Britain. After detailed technical hearings by parliamentary committees and debates in both Houses of Parliament, the scheme was allowed to proceed.

The second problem arose from the complex geology of the reservoir site. The existence of pervious limestone strata and old mine workings necessitated careful investigations to make sure that the reservoir would be watertight. Detailed preparatory surveys of the dam site ensured that the designed dam, part concrete and part earthfill, was successfully constructed.

8.2 THIRLMERE: DESIGN DROUGHT SYNTHESIS

LOCATION Thirlmere is in the English Lake District.

SOURCES North West Water (1977) *Minimum runoff analysis and design drought synthesis*, Internal Report. Directorate of Resource Planning.

North West Water (1981) *RP Suite, computer programs for water resource studies*, Internal Report. Directorate of Planning.

PROBLEM In 1974, the North West Water Authority initiated probability analyses of long runoff records pertaining to the major sources of water. This study was necessary to form a consistent basis for the scientific planning and management of the water resources of the Authority's area (14 500 km^2). Nine records sampling catchments throughout the length of the area (Fig. 8.2) were analysed together with two comparable records from the Severn Trent Water Authority area.

The study of the Thirlmere record provides an example of the statistical methods used.

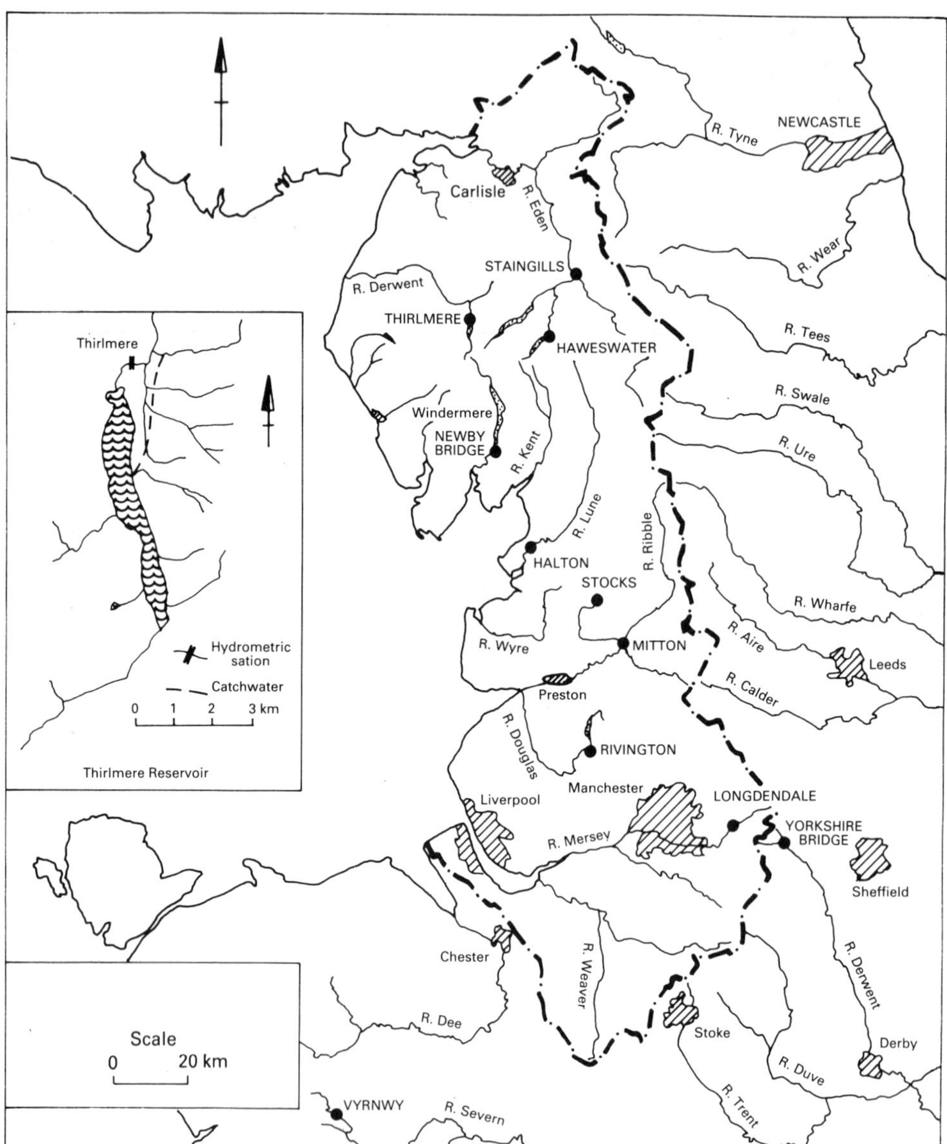

Fig. 8.2 — North West Water drought studies: ●, minimum runoff stations; —·—, North West Water boundary.

Thirlmere Reservoir

The Thirlmere Dam raising the capacity of the natural lake, was completed in 1894; later a catchwater was constructed from Mill Gill enlarging supplies and making a total catchment area of 40.9 km^2. Although supplying water since its completion, no flow records have been traced for the period before 1920 and only weekly inflow values calculated from abstractions, compensation water and changes in reservoir content are available from then. Complications to the records were added with the inclusion of Mill Gill in 1926. However, a daily data series including the Mill Gill diverted flow is available from 1939. The homogeneity of the series before 1954 was queried and it was estimated from a study of monthly rainfall-runoff relationships that discharges released through a scour valve were 20% too high in the earlier period. The corrected daily flow sequence from 1939 to 1974 in the form of monthly discharges (Ml) gave a satisfactory 35 years of data for analysis.(Recent evidence has led to a new review of the data.) Features of the record are given in Table 8.2.

Table 8.2 — Thirlmere Reservoir monthly inflows, 1939–1974

Month	Mean (Me)	Maximum (Me)	Year	Minimum (Me)	Year
January	10085	23403	1974	791	1963
February	7228	15125	1939	841	1963
March	6370	17893	1963	1682	1933
April	5491	13411	1947	936	1974
May	4470	12038	1967	1082	1946
June	3935	10165	1948	873	1957
July	4765	9888	1939	877	1949
August	6714	15584	1962	914	1947
September	8499	16179	1954	468	1959
October	8701	25799	1967	1268	1946
November	10591	21430	1954	2082	1945
December	10772	21107	1959	3460	1947
Yearly total	87622	127292	1954	54553	1973

Probability analysis

The first task was to investigate the occurrence of dry-month sequences and the rigorous manipulation of so much data is now readily carried out by computer. The Authority's special suite of computer programs was used for the analysis to identify critical periods up to 36 months duration.

The independent dry-month sequences are abstracted first. The record is scanned to find the lowest monthly discharge irrespective of year or season and this is repeated for successive remaining lowest values until individual monthly events have been identified for the number of years of record.Then the process is repeated taking the lowest 2 months together and then again for increasing durations up to 12 months. As the duration increases and according to the number of years record, the number of independent events is reduced. For each of the duration series, the events are ranked in ascending magnitude and given a probability of occurrence according to the formula

$$P(x) = \frac{m}{N+1}$$

where m is the rank number and N the number of years record. For durations between 12 and 36 months there are naturally overlapping seasonal events and thus the analysis takes one event finishing in each year. The events for each duration are similarly ranked and assigned a probability. For comparison with results from other records, the discharge totals were expressed as percentages of the average annual flow (AAF) over the 35 years of record. The series of events of durations less than a year constitute 'partial' series and may conform to one of many probability distributions. For simplification following trials, the independent events were plotted on log–log paper and the overlapping events on arithmetic probability paper and straight-line fits by eye were found to be satisfactory. Adjustments were made accounting for outliers, etc., to produce a family of lines on each plot.

The total discharges expected to occur 1 in 50 years on average were read off for each duration and plotted to form a minimum runoff curve and the historic events were added for comparison (Fig. 8.3).

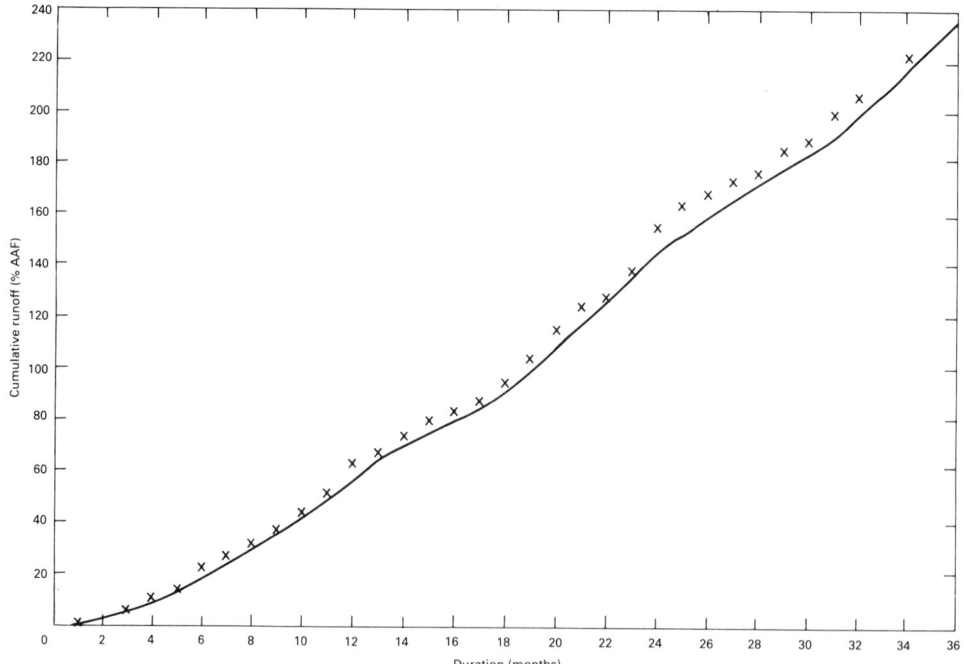

Fig. 8.3 — Minimum runoff diagram for Thirlmere Reservoir, 1939–1974 (AAF 1939–1974, 87 620 Ml; catchment area, 40.9 km^2): ——, 2% analysis; ×, minimum historic events.

Drought synthesis

In order to use the results of the probability analysis for assessing water availability or deriving control rules, a realistic daily calendar sequence of flows is required. The 1 in 50 years (2%) flows for 1 up to 36 months' duration were used following the method outlined by Twort *et al.* (1974). The procedure for Thirlmere is demonstrated in Table 8.3. According to the duration (column (1)), the cumulative 2% flow

Table 8.3 — Drought synthesis for Thirlmere (AAF, 87 620 Ml)

Cumulative sequence			Calendar sequence					
Duration (months)	2% flow (% AAF)	2% flow Increment	Month	2% flow (% AAF)	2% flow (Ml)	Historic Month	Historic Flow (Ml)	Design flow (Ml)
(1)	(2)	(3)	(4)	(5)	(6)	(7)	(8)	(9)
1	0.47	0.47	February	10.0	8762	February 1958	9351	8760
2	1.5	1.1	March	5.0	4381	March 1970	4400	4382
3	5.2	3.7	April	5.1	4469	April 1944	4473	4469
4	8.8	3.6	May	4.1	3592	May 1957	3554	3591
5	12.9	4.1	June	3.5	3067	June 1967	3023	3109
6	18.0	5.1	July	3.7	3242	July 1951	3214	3241
7	23.0	5.0	August	1.1	942	August 1947	913	936
8	29.0	6.0	September	0.47	416	September 1959	468	418
9	35.3	6.3	October	6.0	5257	October 1966	5464	5255
10	42.0	6.7	November	6.3	5520	November 1957	5555	5519
11	49.0	7.0	December	6.7	5851	December 1943	6291	5873
12	56.0	7.0	January	7.0	6133	January 1953	6205	6133
13	65.0	7.0	February	7.0	6133	February 1972	6487	6133
14	70.0	9.0	March	9.0	7886	March 1966	7860	7887
15	75.0	5.0	April	5.0	4381	April 1944	4473	4382
16	80.0	5.0	May	5.0	4381	May 1965	4400	4382
17	84.0	4.0	June	5.0	4381	June 1963	4423	4382
18	91.0	7.0	July	4.0	3505	July 1971	3482	3505
19	99.0	8.0	August	7.0	6133	August 1973	5518	6137
20	108.5	9.5	September	8.0	7010	September 1969	6991	7010
21	117.0	8.5	October	9.5	8324	October 1941	8132	8324
22	126.0	9.0	November	8.5	7448	November 1941	7455	7446
23	135.0	9.0	December	9.0	7886	December 1945	7932	7887
24	144.5	9.5	January	9.0	7886	January 1954	7850	7887
25	152.0	7.5	February	9.5	8324	February 1949	8423	8324
26	158.0	6.0	March	7.5	6572	March 1946	6587	6569
27	165.0	7.0	April	6.0	5257	April 1943	5305	5255
28	172.0	7.0	May	7.0	6133	May 1942	6273	6133
29	178.0	6.0	June	7.0	6133	June 1944	6205	6133
30	183.0	5.0	July	6.0	5257	July 1974	5327	5269
31	190.0	7.0	August	5.0	4381	August 1972	4250	4382
32	199.0	9.0	September	7.0	6133	September 1956	6009	6133
33	208.0	9.0	October	9.0	7886	October 1964	7946	7887
34	218.0	10.0	November	9.0	7886	November 1950	7546	7887
35	227.0	9.0	December	9.0	7886	December 1962	7355	7887
36	236.0	9.0	January	9.0	7886	January 1952	8055	7887

as a percentage of the AAF is taken from the probability plots (Column (2)). Column (3) gives the 2% flow increments between durations. From inspection of the monthly flow records, it was appreciated that September was the driest month and February flows were nearest to the average during a year. Therefore a monthly calendar sequence was started in February (column (4)) and in column (5) a 2% drought pattern was set up using the 2% increments in column (3) so that sequential combinations equalled the appropriate cumulative percentages in column (2). This is known as a 'stacked' drought. The corresponding monthly flow volumes from the AAF of 87 620 Ml are given in column (6). To obtain daily flows, the historic record was examined and the daily flows of the corresponding months with the nearest total to the 2% flows were abstracted (months and totals, columns (7) and (8)). These were examined and adjusted at the beginning and end of months to ensure an acceptable continuous flow pattern and, if necessary, flood peaks were modified to give closer total volume fit. The derived daily flow hydrographs for each month were also checked for a proper balance of high and low flows to ensure a realistic pattern of flows for the time of year. The monthly totals of the 3 year design daily flow hydrograph are given in column (9).

The cumulative volumes of runoff derived from column (2) were compared with the cumulative volumes from column (9) and the greatest departure of the final design drought from the required 2% drought was +3% for the 2 month duration event. For 26 out of the 36 durations there was no identifiable difference.

Conclusion

The assemblage of records for the other 10 locations was sometimes more difficult since statutory releases and pumped quantities of water had to be considered. However, some of the older reservoirs had much longer records on which to base the probability analysis, thereby sampling a greater number of drought conditions. The whole study provided a valuable tool in the planning and management of the water resources.

8.3 YIELD OF VYRNWY RESERVOIR

LOCATION Vyrnwy Reservoir is near the head of the Afon Vyrnwy, a major Welsh tributary of the River Severn.

SOURCES Jack, W. L. (1977) *Annual catchment losses — a practical method. J. Inst. Water Eng. Sci.* **31** No. 4, 280–284.

North West Water (1977) *Minimum runoff analysis and design drought synthesis*, Internal Report. Directorate of Resource Planning.

North West Water (1981) *Survey of Existing Surface Water Resources*, Internal Report. Directorate of Resource Planning.

PROBLEM During the development of public water supplies, reservoirs were built separately by local authorities and the individual yields were determined at different times by a variety of methods. The North West Water Authority, now responsible for the water resources and supplies of its large area, has adopted a consistent statistical method of analysis and made a review of all its sources. Both impounding reservoirs and sources with offstream storage have been studied but Vyrnwy Reservoir, an impounding reservoir has been taken as an example of the yield assessment (Fig. 8.4).

Vyrnwy Reservoir

Completed in 1891 by Liverpool Corporation for a reliable water supply to the city, Vyrnwy Reservoir has a catchment area of 94.3 km^2. The 454 ha reservoir is contained by a masonry gravity dam 44 m high and has a usable capacity of 55 150 Ml. While its major function remains as a water supply for Merseyside in the North West Water area, its ownership by the Severn–Trent Water Authority and its location in the Severn Basin has led to redeployment of its compensation water to assist in the regulation of the River Severn.

Catchment runoff

The first stage in the assessment of yield when measured flow records are not available is the determination of the long-term average annual rainfall from the catchment. The more recent standard 30 year period 1941–1970 adopted by the Meteorological Office has been used for the long-term averages.

The area contributing to the reservoir inflow needs careful definition. The watershed boundary is drawn on large-scale maps and the enclosing area planimetered. For the catchwater areas, the aqueduct capacity has to be considered and a scaled reduction in area may be made to allow for losses in peak storm flows. Thus the total effective catchment area is defined.

The average annual rainfall for the 1941–1970 period evaluated for most of the rain gauges in the region and the 1941–1970 isohyetal map published by the Meteorological Office form the bases for the determination of the AAAR over the catchment. For the Vyrnwy catchment, the value is 1897 mm.

To obtain residual rainfall, equivalent to the runoff, estimates of losses from evaporation and transpiration are required. Two methods have been used: the first from survey maps compiled by the former river authorities and the second applying the method used by the Welsh Water Authority (see first source cited above). The latter takes into account the elevation of the catchment, its exposure and its degree of afforestation and resulted in an average annual loss of 460 mm for the Vyrnwy catchment. For this catchment, comparing the results from the two methods and relating rainfall to the measured runoff, a value of 450 mm loss was assumed.

When there are no long-term flow records the average annual runoff is calculated from the rainfall assessment and estimated loss. The AAF (1941–1970) of 137 000 Ml derived from the Vyrnwy records gave an average annual runoff of 1447 mm over the 94.3 km^2.

Reservoir storage

The capacity of a reservoir is always assessed from height–area diagrams at the time of construction. The gross capacity of Vyrnwy was given as 59 650 Ml in the 1890s but

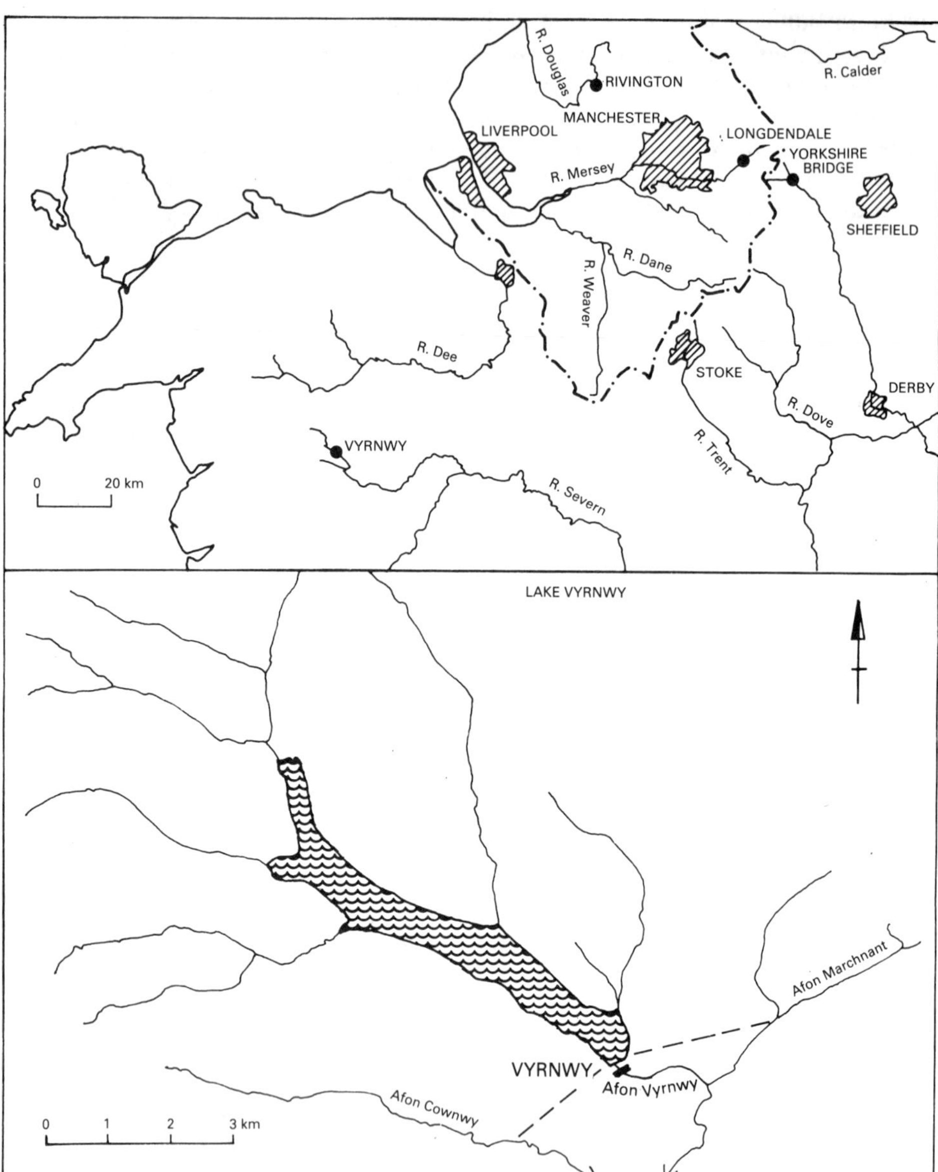

Fig. 8.4 — Lake Vyrnwy location and site: ≠, hydrometric station; — —, catchwater.

since then some siltation will have occurred reducing the volume. The reservoir was resurveyed for the Severn–Trent Water Authority in 1975. Allowances for dead water unavailable to the draw-off valves have also to be made and thus, deducting 10%, a conservative estimate of the available storage in Vyrnwy is now 55 150 Ml.

Yield assessment

A regional approach has been adopted by the North West Water Authority for the determination of a severe low-flow sequence, the 2% monthly drought sequence, in order to utilize the long-period records including the serious 1933–1934 drought. A composite cumulative 2% drought curve was constructed for the reservoirs of the Pennine region but for Vyrnwy its own 2% drought curve derived from the measured flow record 1879–1974 was used (Fig. 8.5).

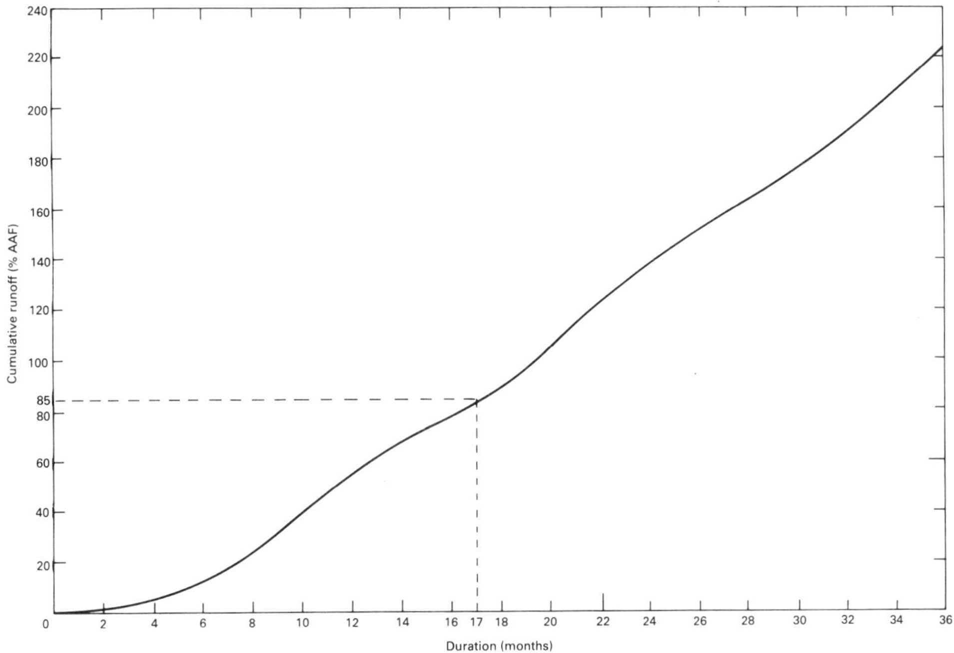

Fig. 8.5 — Minimum runoff diagram for Vyrnwy Reservoir 1879–1974 (AAF 1879–1974, 129 220 Ml; catchment area, 94.3 km^2): ——— 2% analysis.

The yield is obtained by dividing the sum of the inflow over a period and the available storage by the number of days in the period. This calculation is made for each of the durations 1–36 months using the 2% drought curve and the minimum yield is taken as the reliable gross yield. The duration at which this is found is then the critical period. To take into account droughts not conforming to calendar months, daily analyses led to a correction factor being applied to the monthly values. Incorporating this correction, a computer program was written to perform the yield analyses.

For a given reservoir size, the yields for the durations 1–36 months using the regional 2% drought curve were calculated and the lowest yield and critical period identified. This was repeated for a range of reservoir sizes and the results presented in non-dimensional form; storage as percentage of AAF and gross yield as percentage of ADF (Fig. 8.6). The increase in critical period with available storage is also shown.

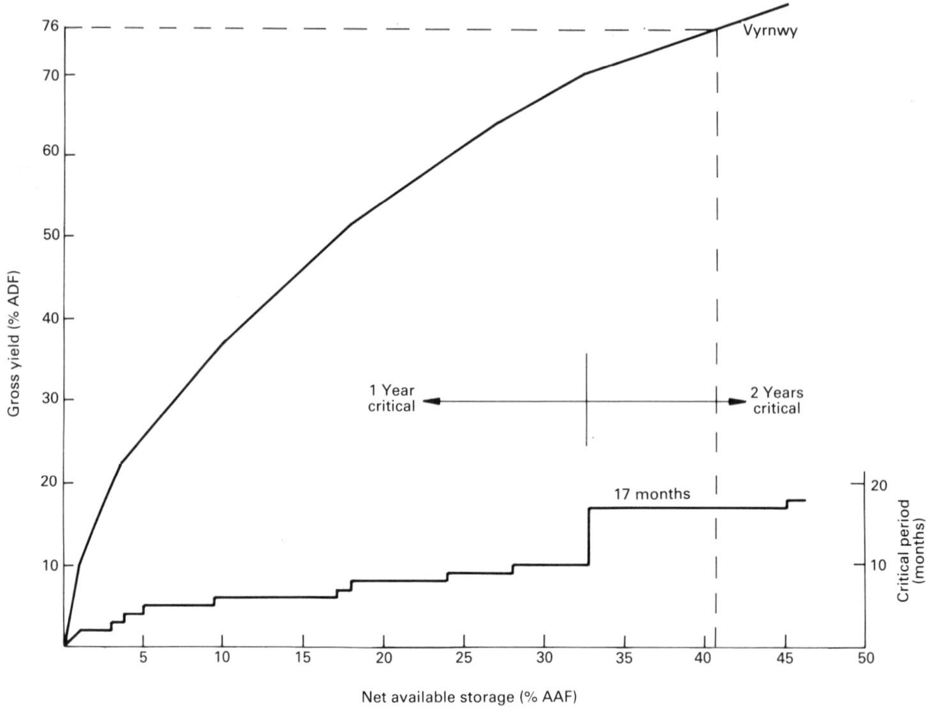

Fig. 8.6 — Yield–storage curve based on 2% regional drought.

Applying the generalized results (1941–1970) to Vyrnwy gives

$$\text{available storage as percentage of AAF} = \frac{55\,150}{137\,000} \times 100 = 40.3\%$$

$$\text{ADF} = \frac{137\,000}{365} = 375 \text{ Ml}$$

$$\text{estimated gross yield} = 76\% \text{ of ADF} = 285 \text{ Ml day}^{-1}$$

From Vyrnwy's own 2% drought sequence (1879–1974), the 17 month critical period has a cumulative runoff volume of 85% of AAF = 85 × 129 220/100 = 109 837 Ml. Thus

$$\text{gross yield} = \frac{109\,837 + 55\,150}{17 \times 30.4} = 314 \text{ Ml day}^{-1}.$$

The volume of water available for abstraction from an impounding reservoir, the net yield, is obtained by subtracting from the gross yield the quantity of compensation to be released downstream which is fixed by statute when permission for the dam construction is granted. Changes in release rates and the location of the rate measurements may cause difficulties in assessing daily values to be set against the gross yield. For Vyrnwy, 65 Ml day^{-1} is the equivalent daily compensation giving a net yield of 249 Ml day^{-1} based on the Vyrnwy drought.

Conclusion

The survey of water resources was carried out for over 60 sources. The regional approach produced a uniformly derived set of yields and the non-dimensional results, presented also in tabular form, give ready access to the gross yield expected from any of the very varied sources. Where an individual reservoir has a good long reliable record of inflows, as for Vyrnwy, then its own 2% drought sequence of flows was used in the analysis. The study forms the basis for updating yield evaluations to incorporate changes in reservoir systems, compensation requirements or management policy.

8.4 A RESERVOIR FOR CHIPATA, ZAMBIA.

LOCATION The town of Chipata is in southeast Zambia near the Malawi border.

SOURCES Gauff Ingenieure (1981) Consulting Engineers.
Hyde, D. I. (1985) *Reservoir capacity and spillway design; a case study in Zambia*, M.Sc. Dissertation. Imperial College, London.
Johnston, P. M. (1986) private communication.

PROBLEM To establish a secure water supply for Chipata, it was proposed to construct a new reservoir in the nearby catchment of the River Lutembwe (270 km^2 to the new dam site) to fulfil an estimated demand of 12 000 m^3 day^{-1} (Fig. 8.7). A good long record of rainfall was available at Chipata on the northern edge of the catchment but river flow records were fragmented and no correlation between rainfall and runoff was apparent. The patchy nature of hydrometric records over the catchment led the consulting engineers to request an independent assessment of the reservoir capacity required.

Investigations

The River Lutembwe has two major tributaries, the River Luntwere and the Nyamsech Stream. The main catchment is defined by hills rising to 1640 m which form the international boundary with Malawi and the undulating plain averages 1100 m in height. The savannah vegetation of tall grasses, bush and trees is interspersed

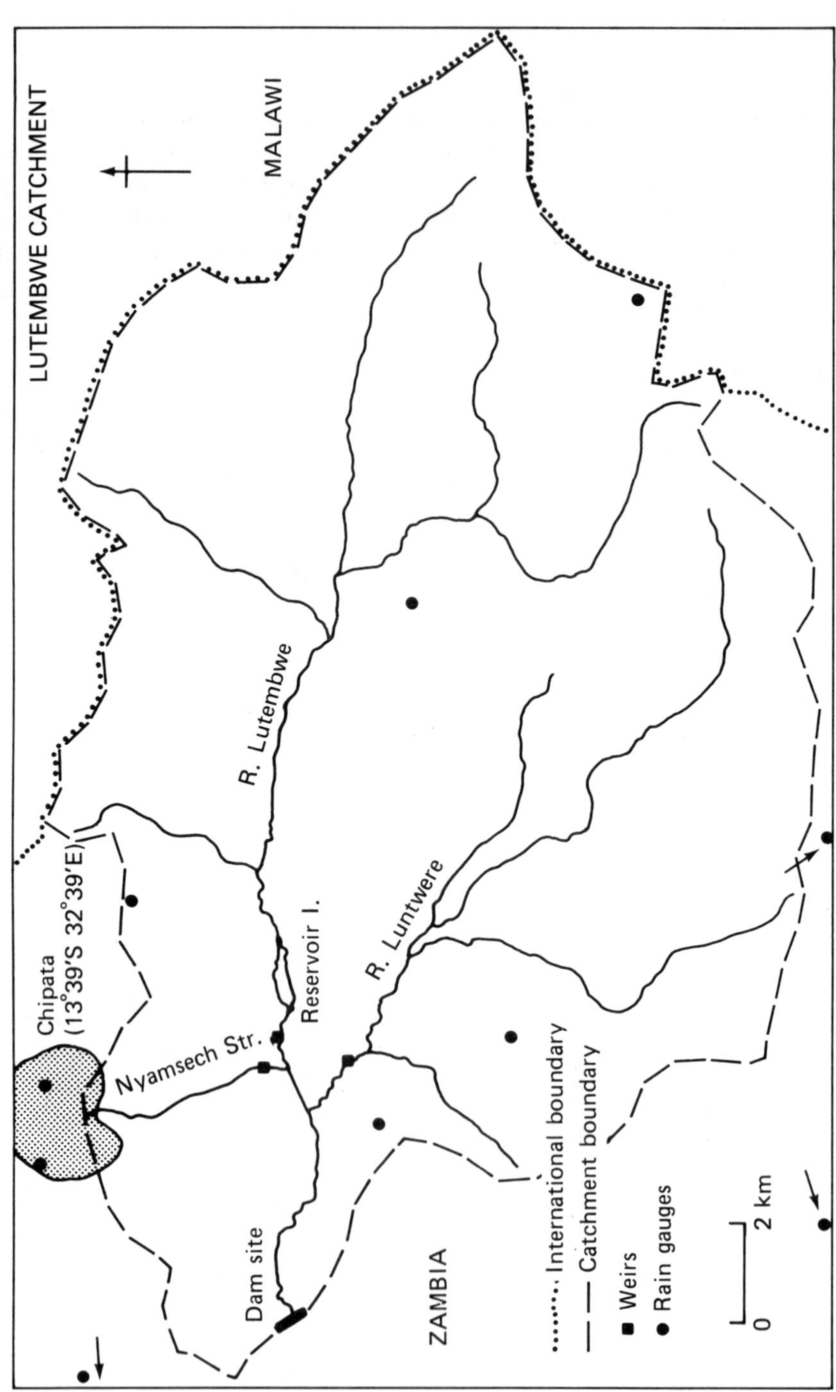

Fig. 8.7 — Lutembwe catchment, Zambia.

with some cultivated land. There are three well-defined seasons: the summer rains November–April, the cool dry months May–August and a hot humid spell in September– October. The catchment annual average rainfall is about 1025 mm and the Penman estimate of open-water evaporation is 1907 mm $year^{-1}$. For hydrometric data, there are 10 daily rain gauges in and around the catchment, three with over 10 years of records, and three weirs (on the main streams) where, from 1980, water level readings were made 3 or 4 times a day. At the Lutembwe River weir occasional daily readings were taken prior to 1980 but the whole record is irregular and further discontinuities were caused by the installation of replacement stage gauges at different levels.

Reservoir capacity design

Owing to the unsatisfactory nature of the catchment data, recourse was made to a detailed regional study made by the Institute of Hydrology (Drayton *et al.*, 1980) of floods and low flows in neighbouring Malawi. No fundamental regional differences were identified and therefore it was reasonable to extend the results across to the Lutembwe catchment. The standardized storage–yield curves for Malawi (Fig. 8.8) derived from the low-flow studies were used to construct a storage–yield curve for the Lutembwe.

The method of analysis proceeded as follows

(1) The ADF at the proposed dam site was evaluated from an amalgamation of the short stretches of water level measurements at Lutembwe Weir suitably converted to discharges and increased by an areal factor of 1.84 so as to represent stream flow from the whole catchment area. From 8 years of record the ADF was found to be 0.73 $m^3 s^{-1}$ from a range of daily flows from 0 to 19 $m^3 s^{-1}$ with a standard deviation of 1.52 $m^3 s^{-1}$.

(2) The flow duration curve (FDC) (showing the relationship between any given discharge and the percentage of time when this discharge is exceeded) was then constructed using the daily flows subdivided into 19 class intervals and by plotting the discharges at each lower class limit against the percentage of days exceeded. The plot on log-normal probability paper approximated a straight line which facilitated extrapolation beyond the range of the data.

(3) The gross yields, the flows exceeded for selected percentages of time, were taken from the fitted FDC so that

Q_{40} = 0.37 $m^3 s^{-1}$ is the flow exceeded 40% of the time
Q_{50} = 0.25 $m^3 s^{-1}$ is the flow exceeded 50% of the time
Q_{60} = 0.18 $m^3 s^{-1}$ is the flow exceeded 60% of the time
Q_{70} = 0.12 $m^3 s^{-1}$ is the flow exceeded 70% of the time
Q_{80} = 0.08 $m^3 s^{-1}$ is the flow exceeded 80% of the time
Q_{90} = 0.04 $m^3 s^{-1}$ is the flow exceeded 90% of the time

(4) The return period of failure of the reservoir was set at 10 years and thus a storage–yield relationship was determined with the chance of the reservoir running dry once every 10 years on average.

(5) Storages for T = 10 were obtained from the standardized Malawi curves (Fig.

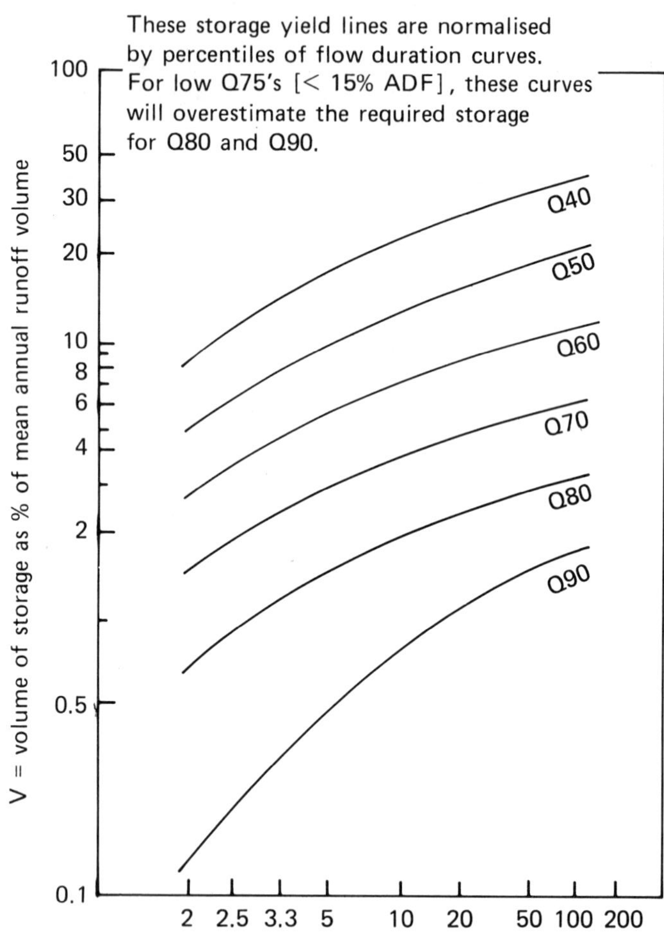

Fig. 8.8 — Standard storage–yield curves for Malawi.

8.8). For each of the gross yields, a percentage of the mean annual runoff volume (ADF $m^3 s^{-1}$ for a year) gave an equivalent storage (Mm^3) (Table 8.4).

Table 8.4 — Reservoir storages

Exceedence time	Equivalent storages (Mm^3)
Q_{40}	5.27
Q_{50}	2.86
Q_{60}	1.67
Q_{70}	0.87
Q_{80}	0.46
Q_{90}	0.18

(6) The derivation of net yields began with the estimation of evaporation losses. The surface area A (km^2) was determined for each equivalent storage for the gross yields from elevation volume–area curves at the proposed dam site. Then, first assuming the reservoir full, the evaporation losses were calculated from

$$\text{loss} = \frac{AE}{31\,500}\ m^3 s^{-1}$$

where E is 1907 mm. To correct any error from the assumption, the loss for each percentile state was multiplied by $\frac{2}{3}$, thus allowing for the average status of the reservoir being less than full. The gross yields minus losses gave net yields. (Table 8.5).

Table 8.5 — Derivation of net yields

Gross yield ($m^3 s^{-1}$)	Loss ($m^3 s^{-1}$)	Net yield ($m^3 s^{-1}$)	Net yield ($m^3 day^{-1}$)
0.37	0.10	0.27	23 330
0.26	0.06	0.20	17 280
0.18	0.03	0.15	12 960
0.12	0.02	0.10	8 640
0.08	0.01	0.07	6 050
0.04	0.01	0.03	2 590

(7) Storage–yield curve. The series of storages for the exceedence times were plotted against the net yields (Fig. 8.9). To ensure a steady supply of 12 000

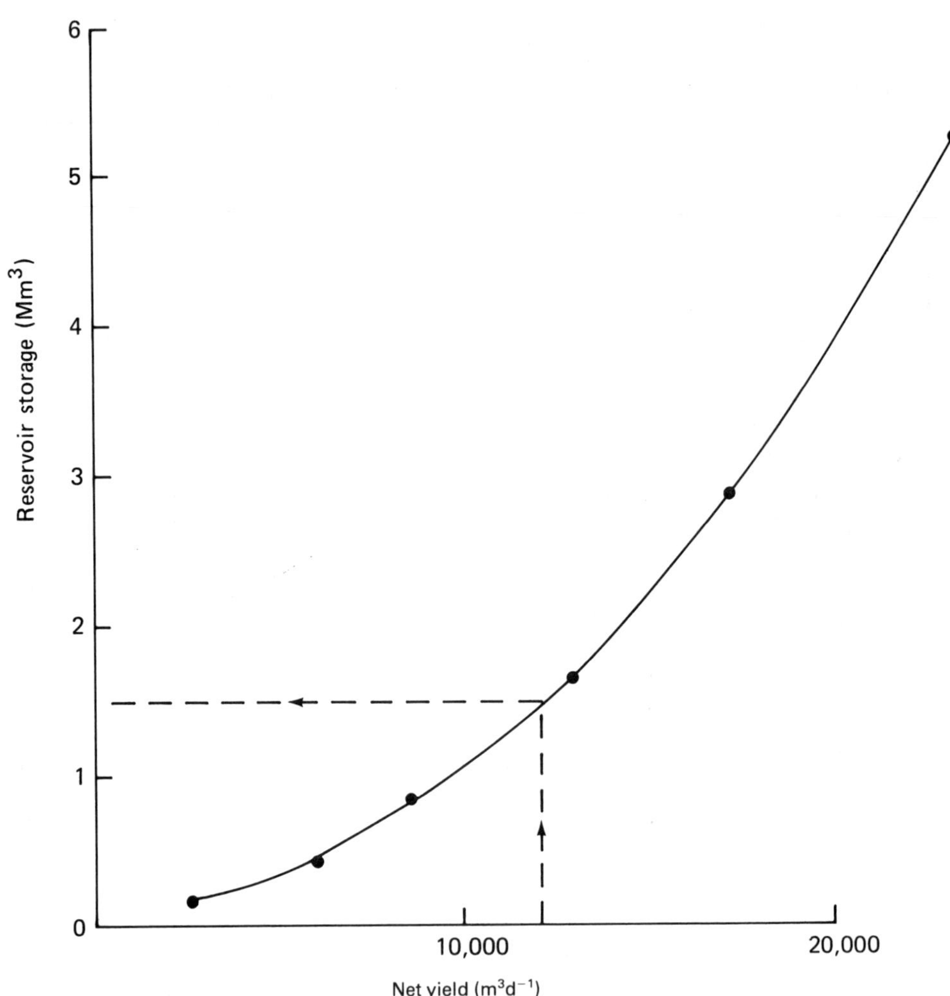

Fig. 8.9 — Storage–yield curve for the Chipata Reservoir (T = 10 years).

$m^3 \, day^{-1}$ for 9 years out of 10 on average, a storage of 1.5 Mm^3 would be required. In view of the uncertainty attached to the quality of the hydrometric data, a check on the result was made by adopting empirical methods to estimate the ADF and FDC before proceeding to derive the storage–yield curve. A smaller storage value of 0.8 Mm^3 was obtained.

Conclusion

The evaluation of the capacity of a reservoir to ensure a reliable supply of a specified quantity of water is a difficult problem even in regions with rainfall all the year round. In tropical climates with seasonal rainfall and greater losses, the difficulties are greater and the lack of reliable stream flow records makes the hydrologist's task complex. The present study resulted in a reservoir capacity of 1.5 Mm^3 for the estimated demand. The design engineer is thus well advised to have estimates derived by several analytical methods before using his engineering judgement in selecting a reservoir capacity.

8.5 OANOB RIVER DAM

LOCATION Oanob River Dam is near Rehoboth on the Oanob River, a headwater of the Orange River, SWA Namibia.

SOURCES Department of Water Affairs, SWA Namibia (1985) Internal Reports. Hydrology and Planning Divisions.

McMahon, T. A. (1981) *Reservoir capacity design and yield*, Short course. University of Pretoria.

Webster, P. (1986) Report compiled with permission of the Department of Water Affairs.

PROBLEM The development of the town of Rehoboth was based upon a reliable supply of groundwater from the alluvium of the nearby Oanob River (Fig. 8.10). However, rapid urbanization has resulted in a dramatic increase in water demand, which could only be met during droughts by the imposition of water restrictions. Detailed investigations by the Department have indicated that extensions to the existing groundwater scheme would barely guarantee supplies for a further 10 years and that the only feasible long-term solution would be the construction of a dam on the Oanob River. The Hydrology Division was therefore requested to investigate the available yield. Since the silt content of the flow is approximately 1.5% by volume, the Division was asked to investigate schemes with and without a silt-trap dam.

Investigations

The Oanob River has been gauged at Rehoboth (catchment area, 2730 km^2) since 1970. The 12 seasons of data collected at the time of the investigations were clearly inadequate for yield analysis. However, the data sequence was sufficient for the calibration of a monthly rainfall–runoff model. This permitted the synthesis of a runoff sequence of 52 years which was deemed adequate for yield analysis. The

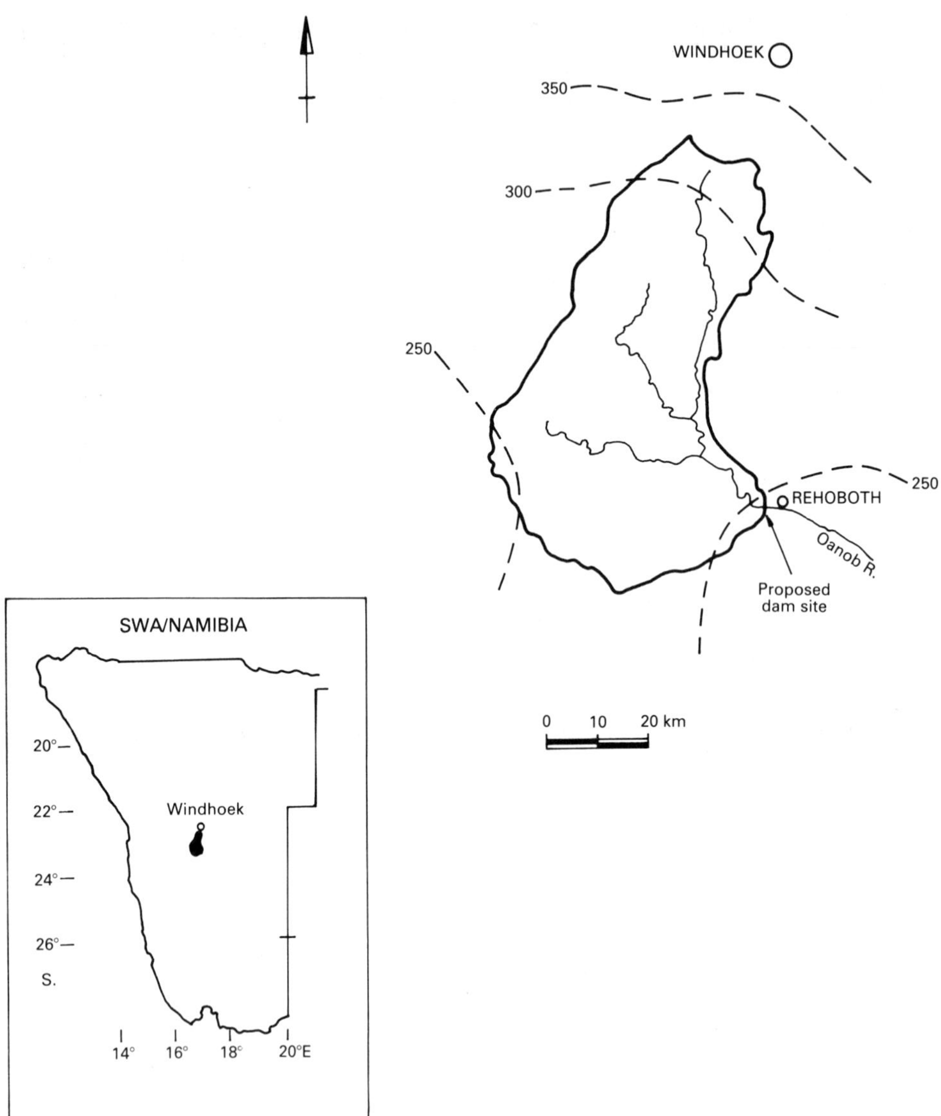

Fig. 8.10 — Oanob River catchment: ———, isohyets, the numbers give the rainfall in milimetres.

monthly and annual parameters derived from the runoff record are given in Table 8.6. It can be seen that the data have characteristics typical of ephemeral rivers in semiarid regions; it is highly seasonal and displays annual variability. Also noteworthy is the negative value of the annual lag-1 serial correlation which exerts a profound influence over certain aspects of yield analysis.

Table 8.6 — Monthly and annual flow parameters

Month	Mean (Mm^3)	Standard deviation (Mm^3)	Coefficient of variation	Skewness	Serial correlation
October	0	0	0	0	0
November	0.49	2.47	5.04	6.80	−0.01
December	0.62	1.73	2.81	3.35	0.13
January	3.43	7.13	2.08	3.36	0.36
February	4.15	9.02	2.17	3.26	0.48
March	4.67	7.27	1.56	1.59	0.68
April	0.86	2.04	2.37	2.72	0
May	0	0	0	0	0
June	0	0	0	0	0
July	0	0	0	0	0
August	0	0	0	0	0
September	0	0	0	0	0
Annual	14.21	20.97	1.48	2.62	−0.23
Standard error	2.91	2.06	0.33	0.33	0.14

Methods of yield analysis

Two methods of yield analysis have been favoured in recent investigations in Namibia and, together, they now form the basis of a standardized departmental approach.

(1) *Concatenated behavioural (CB) method* This is based on a standard behavioural analysis which has been modified such that the data sequence is concatenated upon itself and the results so derived are virtually independent of the initial storage (i.e. steady state).

(2) *Transition probability (TP) matrix method* Repeated application of a monthly water balance equation for various start-of-year dam storages leads to the development of a 1 year TP matrix. Repeated squaring of the matrix then yields steady-state storage probabilities which, when combined with a failure vector, give the probability of failure. The computer program for the calculations leans heavily on the work of McMahon & Mein 1978 and McMahon (see second source cited above). It has been modified to account for evaporation, dead storage, siltation and variable-draft operation.

The application of the two methods in tandem provides a powerful measure of the correlation structure of the data sequence. Whereas the TP method samples each year of data without reference to its sequencing, the CB method is sensitive to the data sequence. It could be argued that use of the standard equation for calculating the serial correlation coefficient could be used to arrive at the same conclusion. However, the fact that there is a high standard error associated with the use of this equation coupled with the fact that yield is sensitive to lags of greater than 1 year mitigate against its use. Many comparative yield analyses in Namibia confirm the presence of slightly negative annual serial correlation resulting in the higher yields obtained using the CB method. By giving preference to the results of the TP method, a measure of conservatism can be introduced.

Siltation

The effects of siltation can be handled by a steady-state analysis of a virgin dam basin (i.e. pre-siltation) and repeating the analysis on the dam at the end of its design life with the accumulation of silt within the basin (i.e. post-siltation). Prior estimates are required of the volume of silt that would accumulate during the design life and also of its distribution within the basin. For the Oanob Dam, it was estimated that 10 Mm^3 of silt would be deposited in the dam basin during its design life of 45 years. The silt distribution was assumed to follow a simple model calibrated with data from successive basin surveys of existing dams in Namibia.

Silt-trap dam

A suitable site for a silt-trap dam exists 11 km upstream of the main dam site. Should a silt-trap dam be required, its capacity is established by the volume (10 Mm^3) of silt that would be deposited during the design life. The incorporation of a silt-trap dam into the CB analysis poses few additional programmimg problems. For the TP analysis, the silt-trap dam is not viewed as an extra dam, but for computational convenience it is viewed as an extension of the main dam with due allowance for its different basin characteristics. An operating procedure was assumed for interdam transfers which maintained storage in the main dam as high as possible in order to minimize evaporation losses. For convenience, trap efficiencies of 100% were assumed and transmission losses between the silt-trap dam and the main dam were assumed to be neglible.

Results

Results are presented for pre- and post-siltation conditions and for schemes with and without a silt-trap dam. A yield–reliability table has been compiled from a number of yield–reliability diagrams (Table 8.7). The yield–reliability diagrams have been compiled from a number of computer analyses for which reliability has been calculated for specified pairs of draft–capacity.

The following points may be noted from Table 8.7.

(1) Yields from the CB method are consistently higher than those from the TP method. This was expected from the highly negative annual serial correlation.
(2) Silt-trap dam schemes produce lower pre-siltation yields than single dam schemes with the same gross storage since, to minimize evaporation losses, it is more efficient to store water in a single dam basin.
(3) Schemes which incorporate a silt-trap dam are only capable of producing comparable post-siltation yields to single dam schemes with the same gross storage.

(4) The effect of the silt-trap dam on yields declines as gross storage capacity increases. The silt-trap dam with its fixed storage of 10 Mm^3 makes up a decreasing percentage of gross capacity as capacity increases.

Recommendations

The following recommendations were made to the Planning Division of the Department of Water Affairs.

(1) The uncertainty over the authenticity of the largely synthetic data sequence leads one to opt for the results of the more conservative TP method in preference to those from the CB method.
(2) Schemes involving a silt-trap dam cannot be justified on the grounds of yield since they have a lower initial and only comparable post-siltation yield than schemes involving a single main dam with the same gross storage.

8.6 NAIROBI WATER SUPPLY

LOCATION Nairobi is the capital of Kenya, east Africa.

SOURCE Howard Humphreys (Kenya) Ltd (1986) *Third Nairobi Water Supply Project, short term plan to* 1995, Report to the Nairobi City Commission.

PROBLEM The Nairobi population of 0.859 million at the census of 1979 is expected to have increased to 1.2 million by 1985 based on the highest forecast rate and for the project period 1985–1995 to increase further by 6% per annum. While the proportion of the population supplied by the mains was a constant 89% over the years 1981–1984, the overall consumption per person per day has declined probably owing to the restrictions of the distribution system. In assessing future demands, it has been assumed that the 89% will rise to 95% of the total population served and that water unaccounted for will be reduced from the current estimate of 40% of production to 20% by 1995. With the highest forecast rate of population growth, the water demand is projected to grow from the 1986 figure of 203 to 366 (1000 m^3 day^{-1}).

The existing supplies come from four main sources in order of development: Kikuyu Springs, Ruiru Reservoir, Sasumua Reservoir and the latest Chania 2 scheme whereby releases from the Sasumua Reservoir regulate the River Chania for abstractions at Mwagu Weir (Fig. 8.11). However, to meet the 1995 demands in Nairobi, the needs of the other towns and irrigation demands in the region, the previous regional water studies have estimated a total demand of 10.95 $m^3 s^{-1}$ (946000 $m^3 day^{-1}$). The increase in local demands within the study area could be met by run-of-river flows but, for the increase projected for Nairobi, a further major storage is required.

Table 8.7 — Oanob Dam: 95% reliability yields

Capacity (Mm^3)	25	30	35	40	25	30	35	40
	Pre-siltation				Post-siltation			
Single dam yield								
CB (Mm^3 $year^{-1}$)	5.7	6.6	7.2	8.0	3.5	4.6	5.4	6.4
TP (Mm^3 $year^{-1}$)	4.3	4.8	5.2	5.5	3.0	3.6	4.2	4.6
Main + silt-trap dam yield								
CB (Mm^3 $year^{-1}$)	5.5	6.3	7.0	7.7	3.8	4.7	5.7	6.6
TP (Mm^3 $year^{-1}$)	4.1	4.6	5.0	5.4	3.1	3.7	4.1	4.6

Table 8.8 — Rainfall at Gethumbwini and evapotranspiration at Thika

Month	January	February	March	April	May	June	July	August	September	October	November	December	Total
Rainfall (mm)	40	45	109	260	176	39	25	19	32	90	169	86	1090
Evapotranspiration (mm)	151	159	152	131	110	95	82	94	124	136	129	140	1503

Source catchments
Within the Nairobi region extending from 60 km south of the city to 80 km to the north, the most reliable water resources come from the numerous rivers flowing in a general southeasterly direction from the Aberdare Mountains (Fig. 8.11). From extensive studies reported between 1980 and 1984 and consideration of comparative yields and costs of dams at several sites, the optimum development for this stage was found to be a dam on the River Thika (catchment area, 71 km^2; capacity, 70 Mm^3).

The complexities of the Thika catchment and its previously developed neighbouring catchment of the Chania are shown in Fig. 8.12. The distribution of average annual values for rainfall and potential evaporation is shown by the respective isopleths and the seasonal variations of rainfall and evapotranspiration are indicated in Table 8.8. The extent of stream flow information is indicated by the river gauging sites. Since the initial studies, further record years have become available including a drought period in 1884, to make the series up to 39 years, 1946–1984. Where there were missing data, infilling was carried out by correlation with adjacent stations using the monthly stream flow simulation model HEC 4 (US Army Corps of Engineers, 1971) The records were finally formed into 10 day series by comparison with recorded data in preparation for the system analysis.

Ancillary data
Before any analyses of the proposed Thika operation could be undertaken a considerable body of further data was required. The Sasumua Reservoir on one of the Chania headwaters already supplied Nairobi directly via the Chania 2 scheme and, to obtain naturalized flow data, evaporation losses from the reservoir, abstractions and downstream releases had to be taken into account. Irrigation abstractions in both catchments were assessed over the database period (1946–1984) in order to adjust flow data for the stations at the confluence of the two rivers. Annual mean 10 day daily abstraction rates were 1.51 $m^3 s^{-1}$ on the Chania and 0.80 $m^3 s^{-1}$ on the Thika for 1984. The derivation of flows in the Lower Thika River downstream of the Thika–Chania confluence was also necessary in order to ensure the irrigation demands of horticultural crops, coffee and pineapples in this productive area. Examples of highest annual demands are 1372 mm for horticultural crops, 664 mm for coffee and 262 mm for pineapples. In addition, processing of coffee and pineapples on the estates requires further supplies of water.

System simulation
The water resources and demands of the Chania–Thika catchment system were simulated by the reservoir operation by simulation study (ROSS) model (see Gambia study in section 9.3). The updated ROSS model for the Chania–Thika catchment has 76 reaches, 54 nodes and 29 offtakes. It was operated in two different modes: to assess the resources and to simulate the effects of making various additional demands on the system. There were three principal RUN forms of ROSS.

RUN1 simulated the Chania 2 scheme whereby water from the Sasumua Reservoir was diverted to the Chania River for abstraction at Mwagu for supply to Nairobi. The storage volume and diversions from the Thika and other headwaters were set to zero.

RUN2 simulated the operation of the Thika Reservoir without diversions to or

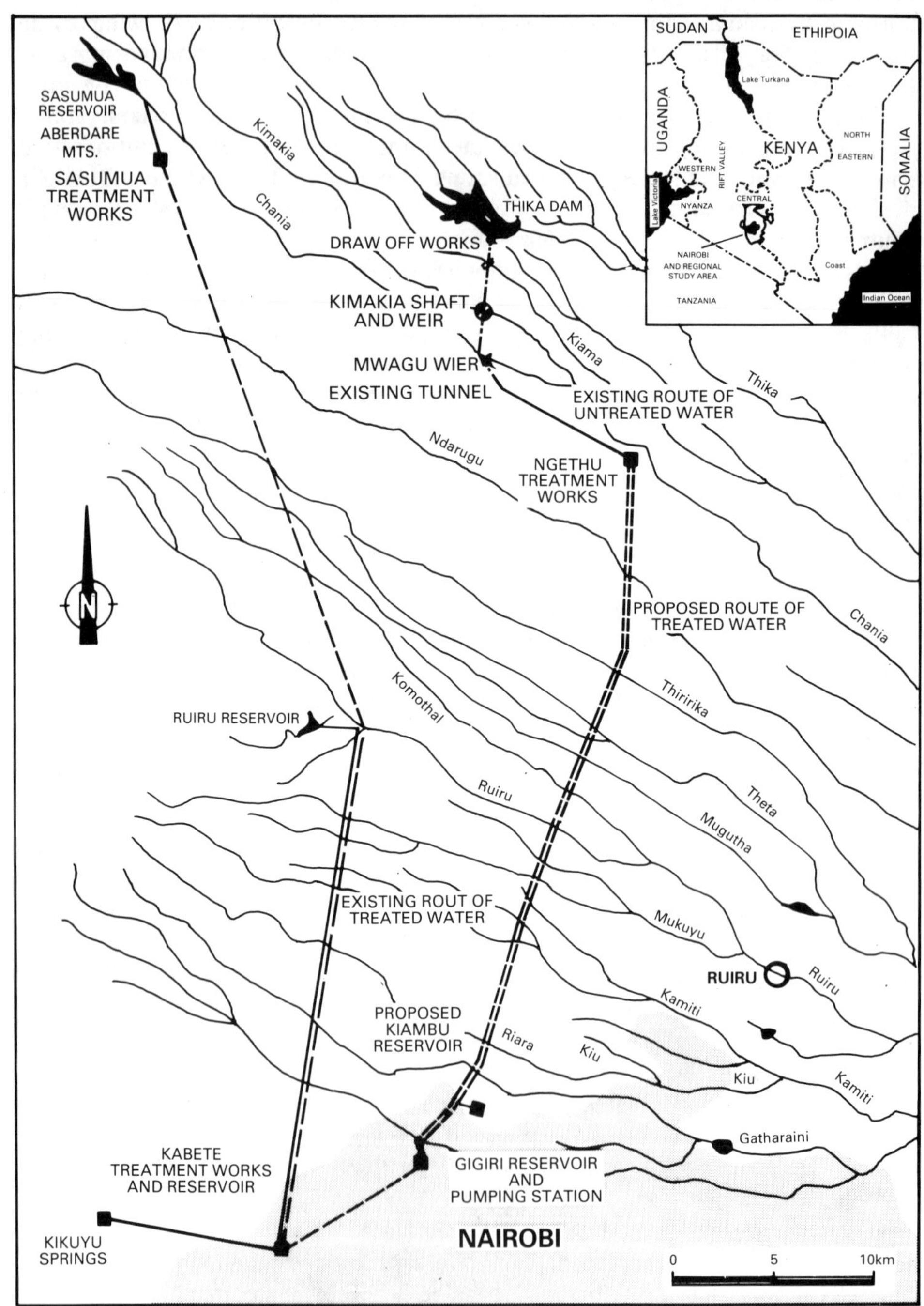

Fig. 8.11 — Nairobi water supply project.

repeat run put the net reservoir inflow into the system below the reservoir and all downstream demands not met by the natural flows were satisfied by drawing on the reservoir storage. Thus the reservoir releases represented the additional demand on the reservoir storage. The difference in the reservoir demands between the two runs represented the downstream demand met by the inflow into the Thika Reservoir.

RUN3 category of operations used the full ROSS model representation. Two major runs simulated the system with abstraction at Mwagu for Nairobi of 4.0 and 3.8 $m^3 s^{-1}$ corresponding to system reliabilities of 80 and 90% respectively. Reservoir volumes and spills, shortfalls and significant reach flows over each system time step of 10 days during the 39 years of records were printed out. Incidence of Thika Reservoir spills and Mwagu abstraction shortfalls taken from the detailed tables ares shown in Table 8.9 (p. 241). From subsidiary runs it was concluded that most of the agricultural demand is satisfied by natural inflows downstream of the Mwagu intake and Thika Reservoir site and it is only in drier than average years that upstream resources of the Chania and Thika Rivers are required and only for limited durations.

Reservoir yield

A statistical analysis of the Thika Reservoir yield was made by the computer program RESIZE part of the ROSS package. For this application, the procedure required the inflow data, details of the demand pattern and the surface area–evaporation relationship for the system. The program calculates the storage required to meet the demands at different risk levels by probability analysis of annual storage deficits. The storage requirement is calculated for each time interval and the maximum value of cumulative storage deficit is determined for each year. A Pearson type III probability distribution is applied to the annual series to determine the storage necessary to provide the required reliability. A sample of the output from the program is given in Table 8.10. With a minimum compensation flow of 0.3 $m^3 s^{-1}$ required below Mwagu, then the demand of 4.1 minus 0.3 gives a yield of 3.8 $m^3 s^{-1}$ with a reliability of 90% for a live storage of 69.8 Mm^3.

Conclusions

From the several simulation studies, for planning purposes it was concluded that the design yield at Mwagu intake should be 4 $m^3 s^{-1}$ following the construction of the Thika Reservoir of 70 Mm^3 capacity. This together with existing sources would produce 420 000 $m^3 day^{-1}$ total treated water, enough to satisfy projected needs for Nairobi until 2000. The main features of the recommended project are shown in Fig. 8.11. The Thika Dam is to be a conventional earthfill embankment maximum height of 63 m and the reservoir would have a surface area of 280 ha at full storage level. New tunnels to transmit the raw water from the reservoir to the Chania River upstream of the existing Mwagu intake and extensions to the Ngethu treatment works would be required. Transmission pipelines for the treated water, a new service reservoir at Kianbu and expansion of the distribution system would be included in this stage of development.

8.7 RESERVOIR YIELD IN NORTHEAST BOTSWANA

LOCATION Northeast Botswana is bounded by Zimbabwe to the north and the Republic of South Africa to the south.

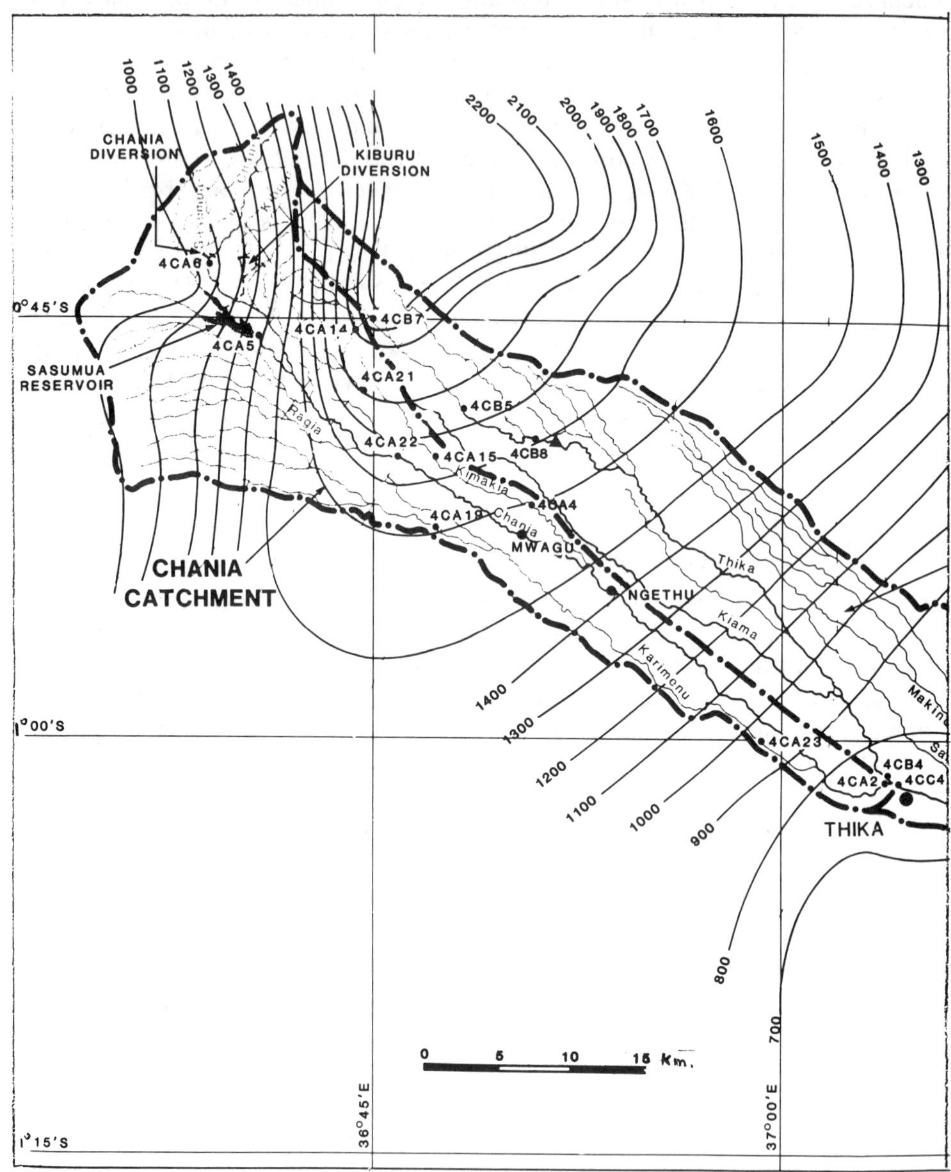

Fig. 8.12 — Chania and Thika catchments: —·—, catchments boundaries; ○, gauging stations with reference number; ▲, new gauging weir; ———, median annual isohyets, the numbers gave the rainfall in millimetres; ———, mean potential evaporation lines, the numbers give the evaporation in millimetres.

from the Thika. A first trial assumed an inexhaustible reservoir and all downstream demands were met either by the natural flows or by drawing on storage. With no spillage, the total releases were those that satisfied 100% of downstream demands. A

Table 8.10 — Thika Reservoir yield analysis (total volume, 70.0 Mm^3; maximum live volume, 68.5 Mm^3)

Downstream demand ($m^3 s^{-1}$)	Storage statistics			Live storage for reliability			
	Mean	Standard deviations	Skew	Median	80%	90%	95%
3.9	29.1	22.6	0.78	26.2	46.7	59.2	70.5
4.0	32.1	24.6	0.77	28.9	51.2	64.8	77.1
4.1	34.8	26.3	0.78	31.4	55.2	69.8	82.9
4.2	37.4	27.6	0.78	33.8	58.9	74.1	87.9

SOURCE Parks, Y. P., & Sutcliffe, J. V. (1987) The development of hydrological yield assessment in N.E. Botswana. *National Hydrology Symposium, Hull*. British Hydrological Society, 12.

PROBLEM The economic development of emergent nations in arid and semiarid regions of the world is dependent on the efficient use of scarce water resources. In this region of Botswana, several investigations of resources have been carried out over the past 20 years. In the beginning, assessments of yield were necessarily simple owing to lack of hydrological records but, with increasing lengths of records and the establishment of gauging stations at potential development sites, more realistic methods of analysis have become practicable.

Northeast Botswana

The study area drained by two major tributaries of the River Limpopo is shown in Fig. 8.13. Rainfall increases locally from south to north and occurs in the hot summer season when evaporation is at a maximum. Evaporation exceeds average rainfall in all months. The typical seasonal distribution is demonstrated in Table 8.11 for Francistown which has an annual average rainfall of 463 mm and a Penman estimated annual open-water evaporation of 1886 mm.

In such a climate, the estimated average annual runoff is between 10 and 50 mm and, since the convective storms result in only a few spates during the rainfall season, the sandy river beds are dry most of the time. Satisfactory river gauging sites are difficult to find but three weirs were built in 1962 on rock outcrops near Francistown and data from the reservoir site at Shashe were also available. Initial studies of rainfall-runoff relationships demonstrated the vital importance of river flow records, and gauging stations more generally based on rated sections were established at potential reservoir sites.

Mean runoff estimation

For the most recent studies commissioned in 1986, there were river flow records for periods of 15–23 years as well as the inflow and outflow records of the Shashe Dam.

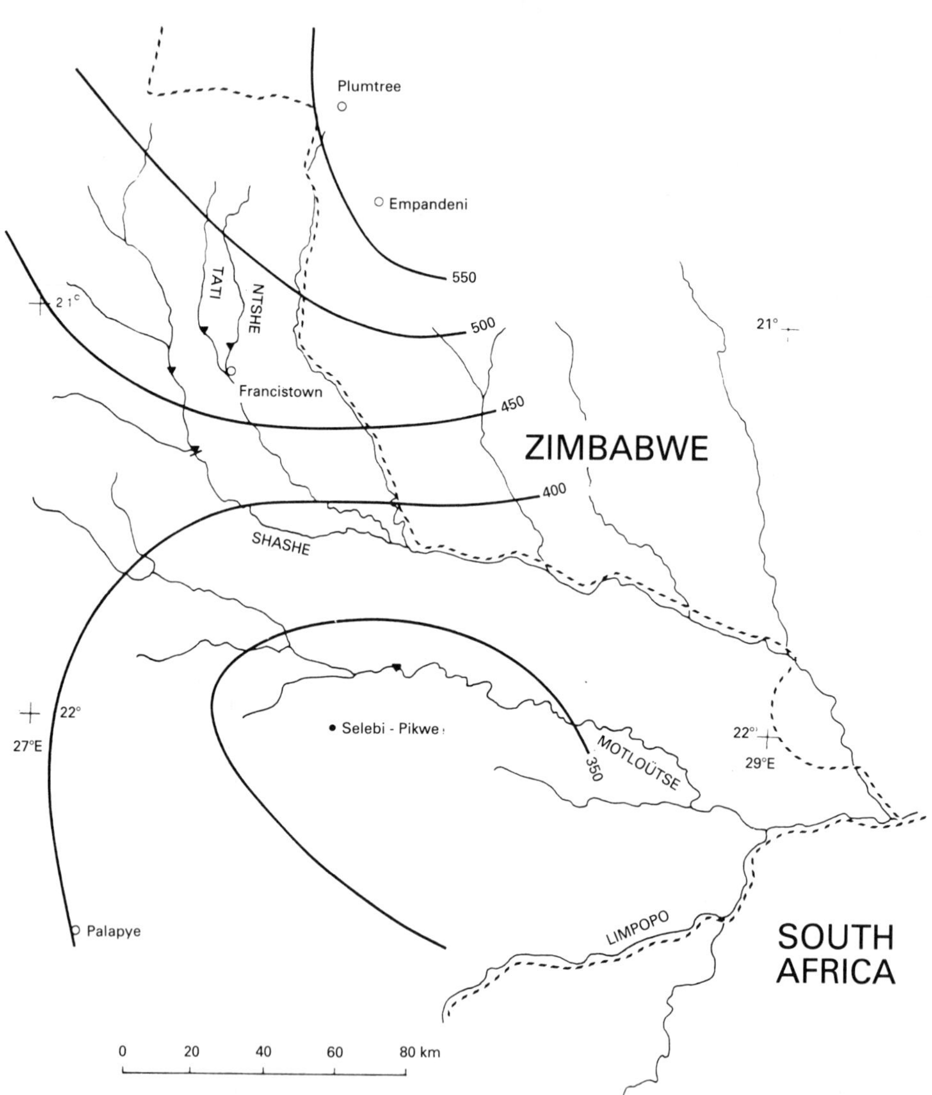

Fig. 8.13 — Northeast Botswana: ---, international boundaries; ▼, river gauging stations; ———, isohyets, the numbers give the rainfall in millimetres; ○, long-term rainfall stations; +, Shashe Dam.

From long-term rainfall records at Francistown, Serowe and Palapye and some nearby stations in Zimbabwe, long-term monthly rainfall series were derived for each gauged basin. Using the Pitman (1973) model to generate monthly river flows with the parameters chosen to give satisfactory fits for the observation periods of rainfall and runoff, the river flows were extended back to 1922 using the basin rainfall

Table 8.9 — Operation of Thika Reservoir, 1946–1984

	Number occurring in 10 day period											
	January	February	March	April	May	June	July	August	September	October	November	December
Demand, 4 $m^3 s^{-1}$												
Spills	30	17	10	20	38	47	52	28	8	2	12	27
Shortfalls	6	9	13	5	2	0	2	8	9	9	7	7
Demand, 3.8 $m^3 s^{-1}$												
Spills	39	21	14	20	40	53	60	35	11	3	16	34
Shortfalls	6	6	10	2	0	0	0	3	7	9	7	7

Table 8.11 — Average rainfall and open-water evaporation at Francistown

Month	January	February	March	April	May	June	July	August	September	October	November	December
Rainfall (mm)	104	82	61	24	7	3	1	1	6	27	59	88
Open water evaporation (mm)	208	178	180	136	105	78	85	124	171	216	204	201

series. Where necessary, these synthesized long records were related to potential dam sites by applying ratios based on differential area and mean runoff predicted from mean rainfall. These long-term runoff records were used to assess reservoir yields at the required locations.

Reservoir yield

The potential yield of an impounded reservoir in this arid climate is greatly affected by evaporation losses. A water balance analysis of 12 years of records of the Shashe Reservoir gave an annual evaporation estimate of 1820 mm which compared favourably with the Francistown Penman estimate of 1886 mm. This adverse factor results in the need for large storages to ensure required supplies.

The appraisal of critical historic droughts and drought estimates from regional analysis are two of the earlier methods used to assess reservoir yield. With the availability of a long sequence of runoff records, the storage yield assessment can be made by simulating the operation of the reservoir for different yields and a range of capacity storages. Several such 'behaviour' methods have been developed and three have been applied to the Shashe Reservoir records in order to optimize yields.

(1) *Simple failure counting method* The synthesized long runoff record as reservoir inflows I was run through a simple water balance model:

$$S(n+1) = S(n) + I + R - \mathrm{E}_0 - \mathrm{Y} - \mathrm{SPILL}$$

where S are storages according to month n, R, E_0 and Y are monthly rainfall, open-water evaporation and yield respectively. For a particular yield, the years (or months) in which a failure occurred were counted and expressed as a percentage of the total number of years (or months). Thus the return period of failure is estimated. However, several hundreds of years of records are needed to give precise results and the 59 year series from 1922 gives only approximate results for the long-term return period values required.

(2) *Deficient volumes method* This is also based on a water balance equation but the storages are replaced by deficits so that the failures occur when the deficit D increases during a drought until it reaches the maximum net reservoir storage. The water balance equation is given by

$$D(n+1) = D(n) - I - R + Y + \mathrm{E}_0 + \mathrm{SPILL}$$

with variables as previously defined. Following the application of the equation to the long record series, the monthly deficient volumes are ranked in decreasing order and are plotted using a log-normal plotting scheme (Parks & Gustard, 1982), total deficits or storage required versus reduced variate (return period). Once the failure occurs, the deficit is constant and the point of failure can be precisely defined. The procedure is repeated for different yields. Thus the method combines the simulation of the reservoir behaviour with the probability of failure. Yields with specific probabilities of failure for a given reservoir capacity are produced.

(3) *The Gould method* The matrix method developed by Gould is described in McMahon & Mein (1978). A monthly water balance is used to produce a transition matrix describing the likelihood of ending a year with a given storage, conditional on the storage at the start of a year. The reservoir is divided into N states of equal storage and each separate year of inflow is routed through the reservoir starting in each of the N states and recording the finishing state. When all the data years had been processed, the total number of times each end-of-year state was reached depending on the starting state were collated and these were transformed into a probability transition matrix $\boldsymbol{T}$. The number of times that the reservoir fails or spills according to starting state are stored in vectors **F** and **S** respectively. Then with a vector **P*** (derived from $\boldsymbol{T}$) of likely starting states the long-term likelihood of failure and spilling are given by

$$\sum_{i=1}^{N} \mathbf{P}_i^* \mathbf{F}_i \quad \text{and} \sum_{i=1}^{N} \mathbf{P}_i^* \mathbf{S}_i.$$

The steady-state probability vector of the reservoir state (storage contents) can be determined from the transition matrix $\boldsymbol{T}$ and the starting conditions of the reservoir. If $\mathbf{P}_1$ is the vector of the starting conditions then for the second year

$$\mathbf{P}_2 = \mathbf{T} \times \mathbf{P}_1.$$

With succeeding years

$$\mathbf{P}_{t+1} = \mathbf{T} \times \mathbf{P}_t$$

and gradually $\mathbf{P}_t$ reaches a steady state as the effect of the initial conditions $\mathbf{P}_1$ becomes negligible. Then $\mathbf{P}_t = \mathbf{P}^*$ which describes the likelihood of being in any of the N states. The probabilities of failure on an annual or monthly basis can be derived as in the previous method for the range of yields. One of the disadvantages of the Gould method is the individual treatment of each year's data without carryover effects but corrections to the method are described for series with high annual autocorrelation in McMahon & Mein (1978).

Results

The three reservoir yield methods were applied to the problem of the resources of the Shashe Dam. The present reservoir capacity of 84.5 Mm^3 was tested to derive the reliable yields for 20, 30 and 50 year return periods (Table 8.12). In presenting the results, an average yield from the three methods was considered suitable.

The capacity was increased to 122.0 Mm^3 by a proposed 2 m raising of the dam and to 176.0 Mm^3 by raising the dam 4 m and the analyses repeated. A final summary of yields for Shashe Reservoir according to capacity is given in Table 8.13. A smooth plot of these results (Fig. 8.14) provides the basis for an economic analysis to determine the optimum size of the reservoir.

Table 8.12 — Shashe Reservoir yield by three methods

Return period (years)	Yield (Mm^3 $month^{-1}$) by the following methods of analysis			
	Simple failure counting	Deficient volumes	Gould	Mean
20	1.95	2.04	1.60	1.86
30	1.75	1.85	1.44	1.68
50	1.59	1.67	1.27	1.51

Table 8.13 — Shashe Reservoir yields

Capacity (Mm^3)	Yield (Mm^3 $month^{-1}$) for the following return periods		
	20 years	30 years	50 years
84.5	1.86	1.68	1.51
122.0	2.13	1.93	1.77
176.0	2.37	21.5	1.93

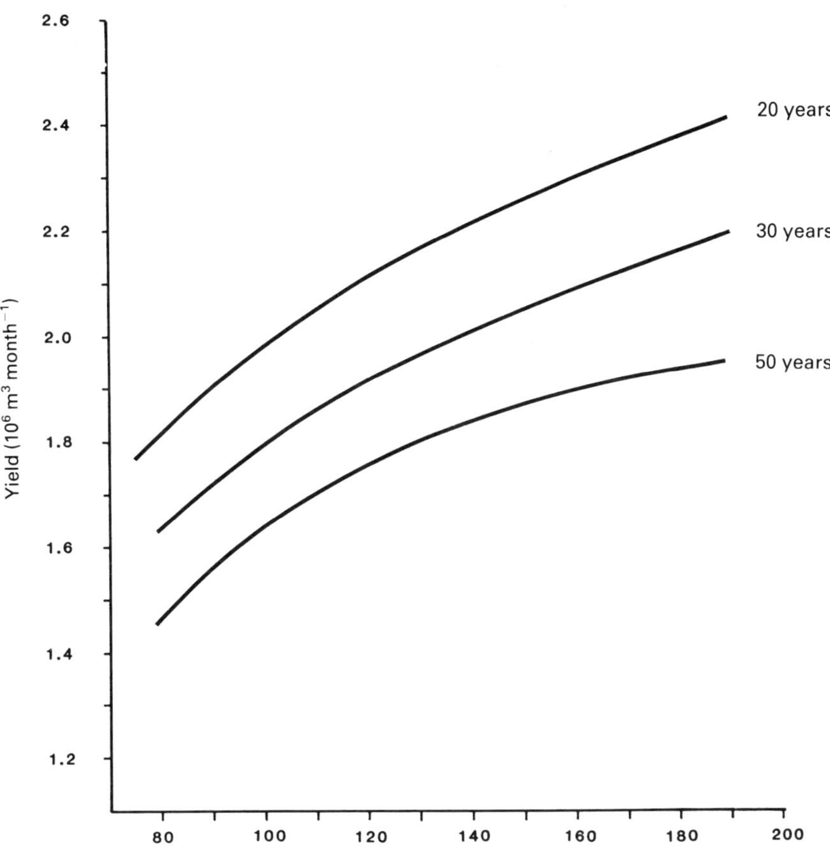

Fig. 8.14 — Shashe Dam yield–capacity curves.

9

River basin development

INTRODUCTION

The water resources case studies exemplified in the previous chapters have been concerned with supplies for domestic or industrial consumption but another important use of water is for the generation of electricity. Hydropower is the main source of electricity in mountainous areas such as in the European countries of Norway and Switzerland. In the establishment of the New World of the Americas and Australasia, the swift rivers from the high mountain ranges and the spectacular natural waterfalls were soon harnessed to supply power.

The development of hydrological knowledge in association with such schemes was clearly demonstrated by the work of Captain McClean in the Scottish Highlands. The North of Scotland Hydroelectric Board based many of its schemes on the pioneer river gaugings and rainfall measurements made by Captain McClean and his staff in the early decades of this century. Hydrological studies in all such undertakings are essential in the initial planning stages and in the subsequent management of the installations. These are admirably shown by the work of the Norwegians in their attention to the measurement of winter snowfall on the upland catchments and to the monitoring of the subsequent delay until the melting season before the swollen rivers affect reservoirs and power stations. Also in this context, the Swiss are noted for having the most comprehensive and efficient network of hydrometric stations in the world. For a country controlling the headwaters of many of the major rivers of Europe, the Swiss have their priorities right.

In the development of new lands or the modernization of originally primitive communities, it is now technically expedient and good practice to consider the potential water resources of a river basin as a whole. The needs of the people for good-quality piped domestic water supplies, for improved agricultural production assisted by irrigation, and for existing and future industrial usage are all aspects to be considered in planning the utilization of water resources. In countries without alternative sources of energy, the hydroelectric power scheme has top priority in the objectives to be realized in a river basin.

The planning of comprehensive water supplies for a river basin is, however, governed by the size of the area. All inclusive studies of large rivers passing through several countries, such as the Mekong in southeast Asia or the Danube in Europe, are quite impracticable and their catchments cannot be considered as a whole. It requires protracted international collaboration before agreement on plans can be reached over the development of catchment areas which extend over and beyond national boundaries. In this category, the Gambia River basin development extends over nearly 80 000 km^2 but clear objectives have been laid down by the three contributing countries. The other examples more concerned with hydropower development relate to much smaller catchments contained within a single national boundary.

The advent of high-speed large-capacity computers has allowed great expansion in the range of hydrological techniques available for analysing the water resources of a river basin. Taking into account all the possible movements of surface water in a catchment area, powerful computer models such as the ROSS model can simulate the continuous functioning of a river system (Simpson & Thorpe, 1982). Long records of river flows with a selected time interval are required and if necessary may be synthesized from rainfall records by means of rainfall-runoff models or the unit hydrograph method. The ROSS model operates on a system of nodes connected by reaches in which the nodes are computational points. These may be calculating releases from a reservoir given the demands to be met and the resources available or receiving the flow from a tributary and routing the joint discharges down the next reach. By feeding in the time series of river flows from the several contributing headwaters and entering the functioning constraints and demands, the resultant reliability of the resources during dry periods and the capability of the system to cope with floods during wet spells can be investigated. For the designing of the reservoir spillway, the methods outlined in Chapter 6 are usually employed with the derivation of probable maximum floods from calculations of probable maximum precipitation being applied in tropical extreme rainfall regions to reinforce results from statistical methods.

9.1 CLUTHA POWER DEVELOPMENT

LOCATION The Clutha River flows from the Southern Alps of Otago Province, New Zealand.

SOURCE Jowett, I. G. & Thompson, S. M. (1977) *Clutha power development, flows and design floods*, Report by the Power Division, Ministry of Works and Development, New Zealand.

PROBLEM In the development of the catchment of the Clutha River for hydroelectricity, from the numerous proposals advanced, a scheme involving the construction of five dams was approved by the Government. The sites of two earth dams, Upper Pisa and UC8, on the Upper Clutha between Cromwell and the outlets of Lakes Wanaka and Hawea, two concrete dams on the Kawarau River, K7/2 and K9, and the remaining concrete dam DG3 on the Clutha downstream of Cromwell are shown in Fig. 9.1. The hydrolo-

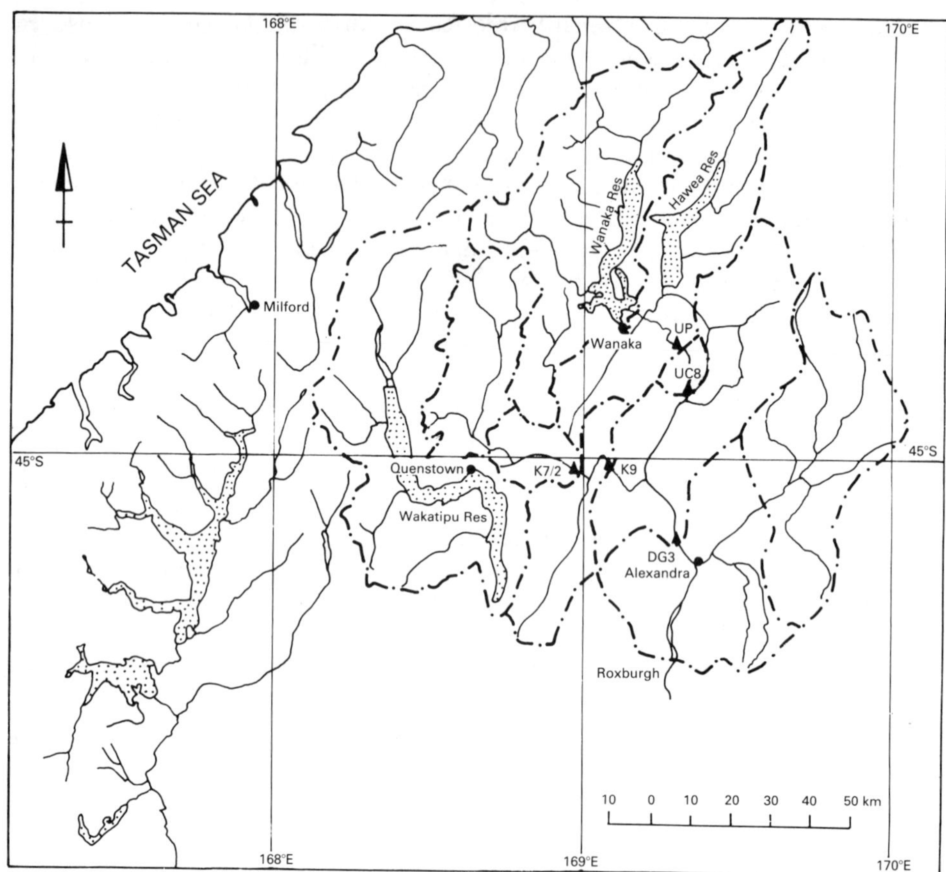

Fig. 9.1 — Clutha catchment: ●, towns; ▲, river gauging stations.

gical studies required of the Investigation Section of the Power Division, Ministry of Works and Development, fell clearly into two parts. The first was concerned with assessing the reliability of the river flows and analyses of annual and seasonal variations were necessary for planning the power generation. The other vital studies concerned with the dam designs were of the flood flows that would need to be passed by the dam spillways.

Investigations

The Clutha catchment area of 15857 km^2 is defined by the site of Roxburgh Power Station completed 1956. Fourty-four per cent of the area drains into three large lakes, Hawea, Wanaka and Wakatipu, which total 637 km^2 in water surface area. The altitude ranges from 2823 m above sea level, with the main divide averaging 2000 m, down to 126 m at Roxburgh but there are many subsidiary ranges rising to 2000 m throughout the catchment. The mean annual rainfall follows the topographical pattern and varies from about 3800 mm on the mountain range down to about

400 mm along a sheltered stretch of the valley. Most of the rain is brought by depressions and their associated fronts travelling in the westerly winds with the greatest falls occurring when slow-moving fronts are crossing the mountains.

Rainfall measurements are well provided over the catchment by about 50 rain gauge stations and for long-term analysis Queenstown has 94 years of record. River flow data are available for the outlets of the three lakes, six of the major tributaries and at Roxburgh. While five of the 10 records each provide well over 40 years of records, considerable work was needed in the initial stages of the studies to check on the homogeneity of the data, particularly with regard to datum levels and ratings and to infill gaps in the records.

River flow variations

As a basis for estimating flows from ungauged catchments, the runoff (flow per unit area) was calculated from the annual mean flow for each gauged catchment. Comparison of the gauged runoffs and the calculated areal precipitations gave estimates of the evaporation losses which could be related to the precipitation of the ungauged areas to give estimates of their runoffs and hence mean flows. Combination of the relevant gauged and ungauged portions of the areas draining to the proposed dam sites resulted in the required mean annual flows shown in Table 9.1.

Table 9.1 — Mean annual flow at the dam sites

	Area (km^2)	Mean flow ($m^3 s^{-1}$)
Kawarau River		
K7/2	4623	198.9
K9	5528	211.3
Clutha River		
Upper Pisa	4619	259.6
UC8	4736	260.5
DG3	12100	471.6
Roxburgh (gauged)	15857	490

The sum of the component area flows (500.3 $m^3 s^{-1}$) compared with the gauged total at Roxburgh gave a discrepancy of 2.1%, well within expected margins of error. Using the long reliable records of Wakatipu, Wanaka and Hawea (mean flows, 1560, 190 and 63 $m^3 s^{-1}$ respectively), maximum inflows, outflows and lake levels were analysed. The annual maximum inflows and outflows for 1–8 day durations were fitted by least squares to a Gumbel distribution and values for return periods of 2, 15 and 500 years were calculated. A sample of the results excluding Hawea where the outflow is controlled is given in Table 9.2. The comparable maximum flows at

Table 9.2 — Maximum lake inflows and outflows

Return period (years)	Wakatipu (1926–1975)				Wanaka (1930–1975)			
	Inflow ($m^3\,s^{-1}$)		Outflow ($m^3\,s^{-1}$)		Inflow ($m^3\,s^{-1}$)		Outflow ($m^3\,s^{-1}$)	
	1 day	4 days	1 day	4 days	1 day	4 days	1 day	4 days
2	1249	687	350	342	1643	909	473	460
15	2389	1228	570	554	3158	1516	749	725
500	4144	2062	909	882	5490	2450	1174	1132
Maximum recorded	2935	1625	694	675	4463	1786	769	752

Roxburgh were as follows for 1 and 4 day durations: 2 year return period, 1160 and 1083; 15 year return period, 1945 and 1794; 500 year return period 3152 and 2889; maximum recorded, 2415 and 2190.

Annual minimum monthly mean flows were fitted to normal distributions using simulated Hawea outflows and 2, 10 and 100 year return period values calculated (Table 9.3). Comparison between the records showed that low flows from the

Table 9.3 — Annual minimum monthly flows

Return period (years)	Annual minimum monthly flow ($m^3\,s^{-1}$)		
	Sum of three lakes		Roxburgh
	Inflow	Outflow	
2	148	195	266
10	115	153	204
100	89	118	150

downstream catchment areas tend to occur at different times from low flows from the lakes. Seasonal variations of the three lake inflows, taking into account the winter snow cover, and the variations over the whole catchment are shown in Table 9.4. Low is the Manuherikia River representing the relatively low altitude dry southeast areas of the catchment with maximum snow most in September and minimum flows in January and February due in part to irrigation demands. In the middle areas represented by the Shotover River, a secondary maximum occurs in March to May

Table 9.4 — Seasonal flow patterns (multiples of the mean)

	Mean flow ($m^3 s^{-1}$)	Variation in flow ($m^3 s^{-1}$)											
		Jan	Feb	Mar	Apl	May	Jun	Jly	Aug	Sep	Oct	Nov	Dec
Three lakes	409	1.15	1.60	1.01	1.01	0.92	0.66	0.61	0.66	0.94	1.26	1.39	1.27
Middle	38	0.92	0.66	0.90	0.82	0.87	0.68	0.55	0.79	1.42	1.53	1.71	1.53
Low	12.7	0.25	0.25	0.29	0.58	0.82	1.28	1.19	1.75	2.36	1.85	1.47	0.33

caused by higher rainfall and decreasing autumnal evaporation. River flow variations at the dam sites can be evaluated from the statistics provided for the gauged catchments.

Flood flows

The dam designers required the 1000 year flood flows for the earth dams on the Upper Clutha River and the 500 year events for the remaining concrete dams. Flood hydrographs, the behaviour of the flood flow with time, were needed in addition to the peak discharges.

From the isohyetal maps of 36 major storms, the rainfall distribution pattern showed little variation due to the predominantly orographic influences and the dominant northwest winds in such events. A Gumbel analysis of 1 and 3 day rainfall totals in records for 33 stations provided values for 15, 500 and 1000 year return periods and from the resulting isohyetal maps, catchment 3 day design rainfalls were obtained. (Table 9.5). Storm maximization using meteorological observations resulted in a PMP of 336 mm over the three lakes area.

Table 9.5 — 3-day design rainfalls

	Rainfall (mm) for the following return periods		
	15 years	500 years	1000 years
Three lakes, total	209	313	357
Shotover	166	230	279
Upper Pisa		176	
UC8		163	
K7/2		138	
K9		124	
DG3		146	
Roxburgh	62	94	107

From a study of the rainfall-runoff relationships over the years of the gauged catchment records, the direct (excess) runoffs were obtained by subtracting base-flows and thence 6 h unit hydrographs (25.4 mm) were calculated. Comparable unit hydrographs for the ungauged catchments were obtained from catchment characteristics using Snyder's method.

To determine the flood hydrographs at the required dam sites, a flood model was used (Fig. 9.2). Although flood flows from the lakes may contribute 50% of the peak flows at Roxburgh, it is the Shotover tributary that accounts for nearly all flood flow downstream of the lakes and there is a time lag of about 12 h between the Shotover and Roxburgh peaks. Thus the catchment contributions calculated from the design rainfalls and unit hydrographs had to be routed downstream to account for the lags.

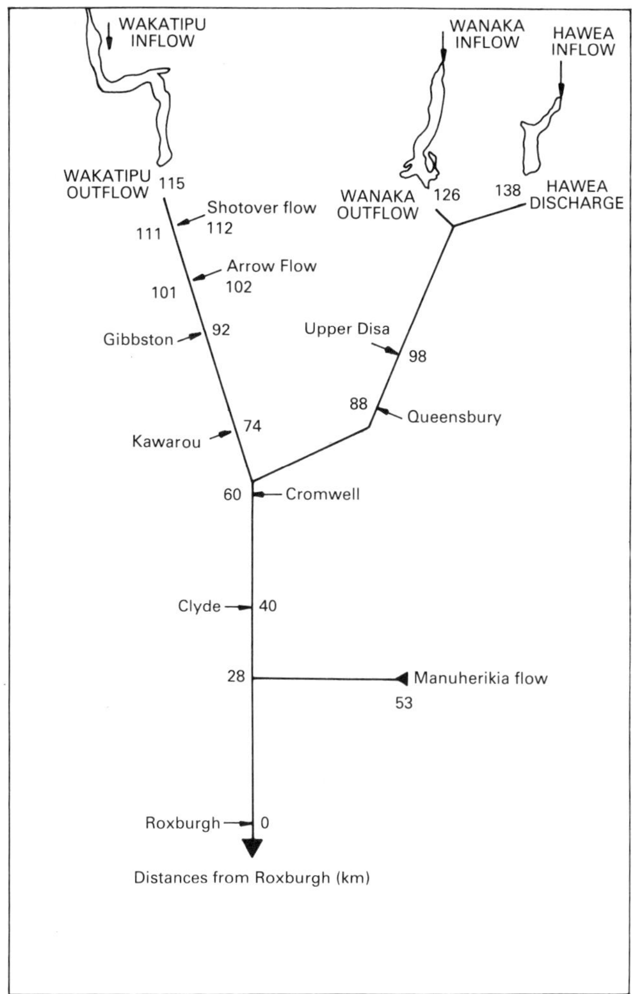

Fig. 9.2 — Flow prediction model.

The peaks were translated downstream without attenuation as kinematic waves with a speed of $2\,m\,s^{-1}$. The model was tested against the three largest floods and the three most recent floods with good agreement. From comparison of the results from the model with the peaks obtained from Gumbel analysis of the gauged catchment records and taking into account initial lake levels, state of the catchments together with optimum storm duration, a final table of recommended design floods was compiled (Table 9.6). The design hydrographs are shown in Fig.9.3.

Table 9.6 — Dam site design floods

	Mean flow ($m^3 s^{-1}$)	Peak discharge ($m^3 s^{-1}$) for the following return periods			PMF ($m^3 s^{-1}$)
		15 year	500 year	1000 year	
Upper Pisa	260	750		1610	2587
UC8	261	750		1610	2587
K7/2	200	950	1730		2400
K9	211	1020	1930		2760
DG3	472	1800	3200		6820
Roxburgh	490	2000	3600		7830

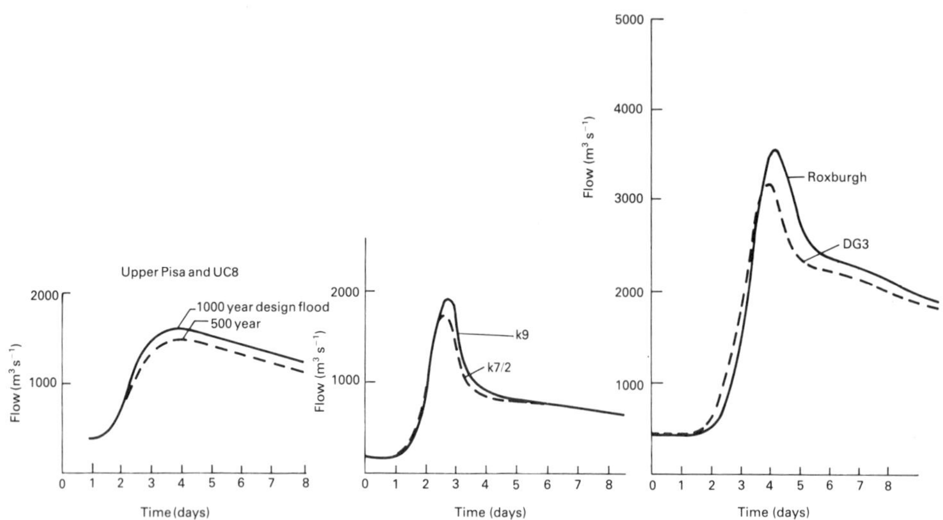

Fig. 9.3 — Design flood hydrographs.

9.2 MAMBILLA HYDROELECTRIC POWER: GEMBU SCHEME

LOCATION The Gembu Reservoir is on the River Donga in eastern Nigeria.

SOURCE Diyam Consultants in association with Binnie & Partners (1984) *Mambilla hydroelectric power development feasibility studies: Gembu scheme,* Report to the National Electric Power Authority, Lagos, Nigeria.

PROBLEM Progress in the modern economic development of Nigeria had entailed early investigations of the water resources of the Upper Benue River Basin for a multitude of purposes. In the late 1970s, the potential of the left-bank tributaries for hydroelectric power began to be realized and Diyam Consultants in association with Binnie & Partners was commissioned to make a feasibility study of the Donga Basin. The Gembu scheme was one of several suggested schemes for hydroelectric power development in the region. The consultants were able to build on previous hydrological studies to assess catchment yields, flood flows and possible operational practice for the several scheme configurations proposed.

Donga Basin

The River Donga is a left-bank tributary of the River Benue, the major tributary of the great Niger and it drains from the northern slopes of the Cameroun Mountains (Fig. 9.4). The Donga Basin down to the Benue has a catchment area of about 20 000

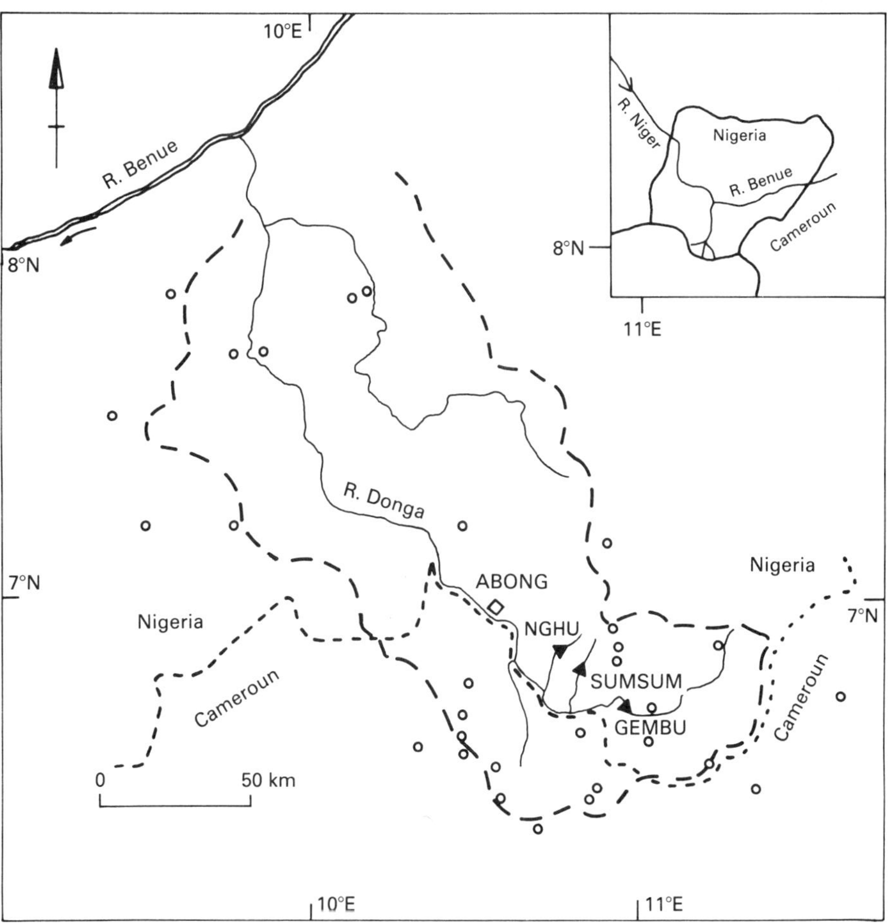

Fig. 9.4 — River Donga Basin: ○, rain gauges; ▲, dams; □, power station.

km^2 but the project is concerned with smaller areas in the higher reaches of the river. The Upper Donga and its tributaries occupy deeply incised valleys in the Mambilla plateau, a high level peneplain at 1300–1400 m. Below the proposed Gembu Dam at 1220 m, the river falls nearly 1000 m to the gently sloping plain comprising two-thirds of the basin. The igneous rocks of the plateau were once covered by high forest but much of this has been cut down to be replaced by montane grassland now heavily overgrazed. The climate of the plateau, 6–7°N of the Equator, has equable temperatures throughout the year but its marked seasonal characteristics result from the rainfall pattern. The dry northeast winds of November–March are replaced by the moist southwest monsoon winds with the northward migration of the intertropical convergence zone in the summer months. This frontal zone between the contrasting air masses produces heavy thunderstorm rainfalls as it passes over the Donga Basin in April–May and September–October. The salient climatic features are shown in the monthly mean temperatures and rainfalls for the climate station at Gembu where the average annual rainfall is 1938 mm (Table 9.7).

Hydrology

Analyses of rainfall and river flow data provided the initial information on potential catchment yields and on flood frequencies required to assess dam SDFs.

Rainfall

Although there were several rainfall stations in the study area, only Gembu had a long record. A double-mass analysis with long-term stations outside the catchment from 1950 indicated high values for Gembu in 1972 and 1973 but subsequent years appeared consistent with the earlier record. A standard period 1959–1980 was selected in order to derive catchment rainfalls for comparison with river flows. The 1959–1980 annual averages were obtained for six long stations and the shorter-period rainfalls of 1977–1980 (available for many of the plateau stations) were related to the standard averages. A mean percentage of standard of 95.6% was obtained from the long-term stations for the shorter period. Then the standard mean annual rainfalls were calculated for all the catchment gauges and a mean annual isohyetal map plotted. From the map, the AAAFs for the catchments to the proposed reservoir and intake sites were estimated. Sample values are given in Table 9.8. Annual rainfall frequency analysis indicated great variations for specific years over the Donga Basin and that the drought sequence of the Sahel (1970–1973) was not experienced in this region.

River flow

Six river gauging stations, five with autographic recorders, provided data from 1977 for the investigations. The most important record was from a point on the River Donga 300 m upstream of the proposed Gembu Dam site. Charts from the other stations were also analysed to provide the size and timing of floods from the subcatchments particularly with regard to intake sites. For the required inflows into the proposed reservoirs and the divertible flows from tributary intakes, long unbroken sequences of monthly flow values were needed. The 6 years of records at the plateau stations were related to a long record back to 1959 from a gauging station on the Lower Donga and extended to a standard period 1959–1981. Flows at

Table 9.7 — Climate features at Gembu (1463 m altitude)

Month	January	February	March	April	May	June	July	August	September	October	November	December
Temperature (°C)	19.6	20.6	21.1	20.6	20.0	18.8	18.0	18.1	18.4	18.7	18.9	18.8
Rain (mm)	6	25	117	182	208	231	254	233	313	278	81	10

Table 9.8 — 1969–1980 average annual catchment rainfalls

River	Site	Catchment area (km^2)	AAAR (mm)
Donga	Gembu Dam	1920	1625
Sumsum	Dam site	156	1905
Nghu	Dam site	42	2060

ungauged sites were extended by scaling the nearest extended flow record using catchment areas and AAAR. Considerable work was involved in this stage of the studies since numerous broken sequences in the records required infilling by cross-correlation between stations and by attention to significant storm events. The derived mean annual flows at the proposed dam sites in the basic Mambilla scheme are given in Table 9.9. An additional yield of 60.3 $m^3\,s^{-1}$ (1900 Mm^3) from 1631 km^2 of tributary catchments was estimated assuming maximum possible flow captured.

Table 9.9 — Mean annual yield from the direct catchments

Dam site	Catchment area (km^2)	AAAR (mm)	Mean annual runoff			Implied losses (mm)
			($m^3\,s^{-1}$)	(Mm^3)	(mm)	
Gembu	1920	1425	57.8	1823	949	676
Sumsum	156	1905	4.9	154	990	915
Nghu	42	2060	1.4	45	1070	990
Total	2118		64.1	2022		

Floods

The lack of long period records provided problems in the assessment of flood frequencies and previous regional flood studies were consulted. Using the peaks-over-threshold analysis for the 6 years of records from three of the gauging stations, peak discharges for a selection of return periods T were obtained. After comparisons with regional flood results, the values subsequently derived for the three dam sites (Table 9.10) were considered acceptable for preliminary designs.

However, the spillway designs required estimates of more extreme floods. A number of trials deriving the PMF via estimates of the PMP were made with pertinent data. 1 day PMP values were obtained from selected long-term rainfall stations and 510 mm at Gembu was considered representative for the catchment

Table 9.10 — Peak discharge estimates for different return periods

T (years)	River Donga, Gembu Dam		River Sumsum at dam		River Nghu at dam	
	Peak discharge ($m^3 s^{-1}$)	Peak discharge per unit area ($m^3 s^{-1} km^{-2}$)	Peak discharge ($m^3 s^{-1}$)	Peak discharge per unit area ($m^3 s^{-1} km^{-2}$)	Peak discharge ($m^3 s^{-1}$)	Peak discharge per unit area ($m^3 s^{-1} km^{-2}$)
$\bar{Q}$	410	0.214	113	0.72	49	1.17
2	425	0.211	117	0.75	51	1.21
5	530	0.276	146	0.93	63	1.51
10	610	0.318	168	1.07	73	1.74
25	720	0.375	198	1.27	86	2.05
50	800	0.417	220	1.41	96	2.28
100	880	0.458	242	1.55	105	2.50

extreme floods studies. For the response of the Gembu catchment, a 3 day PMP design storm of 750 mm was calculated which became 562 mm after applying an estimated 75% area reduction factor. Storm profiles were analysed at recording rain gauges and one with an extreme initial hourly fall was chosen. A theoretical unit hydrograph was derived for the complex Gembu catchment and after assuming small losses of 1.5 mm h^{-1} it was convoluted with the storm profile to give an estimated PMF of 34 800 $m^3 s^{-1}$. Simple triangular unit hydrographs were used for the other two catchments and the PMF estimates were 4760 $m^3 s^{-1}$ for Sumsum and 1900 $m^3 s^{-1}$ for Nghu. The flood estimates were presented with great reservations since no rainfall maximization studies were available and there were no records of major storms on which to base realistic areal reduction factors.

Reservoir operation

The plan of the Gembu scheme is shown diagrammatically in Fig. 9.5. The main storage is the Gembu Reservoir from which water is to be sent through unlined tunnels and the flow controlled by radial gates at the outlet into Sumsum Reservoir or by a turbine of about 15 MW capacity. From Sumsum to Nghu Reservoir the flow would be controlled by the hydraulic gradient along the connecting tunnel. The Abong Power Station is at 300 m and thus the generating head from Nghu Reservoir is about 950 m. In order to maintain the required daily flow to Sumsum during the driest recorded period, a minimum operating level in Gembu Reservoir can be calculated and the retention levels in Sumsum and Gembu can be optimized. From the optimization studies, the recommended levels are indicated on the diagram. The AAFs are also shown. The yields from the direct catchments are all absorbed in the three reservoirs and 99% of the AAFs in the five principal indirect catchments are diverted. The reservoir operation was simulated by computer using data from 1959 to 1981. The estimated monthly inflows from direct and indirect catchments were the main inputs. Reservoir levels in Sumsum and Nghu were held constant and the effects of rainfall and evaporation on each of the reservoir surfaces were accounted for in routing the flows. For given inputs and minimum operating level in Gembu Reservoir, the simulation model calculated the firm continuous flow for power, necessary reservoir storage and maximum flow in the Gembu to Sumsum tunnel.

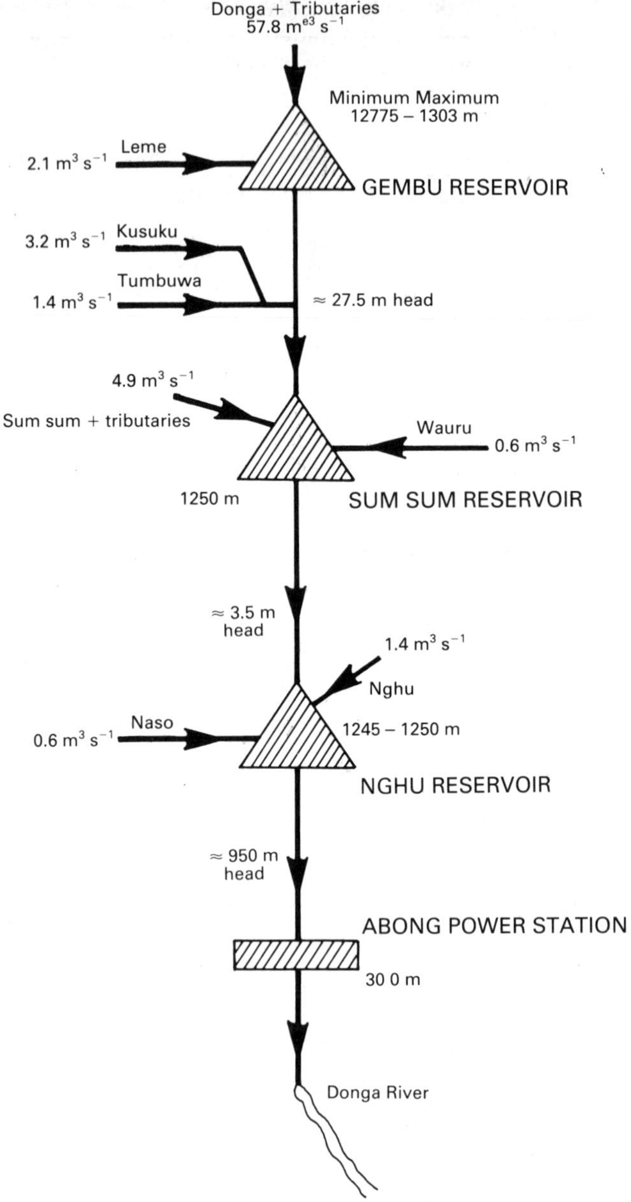

Fig. 9.5 — Gembu Basin scheme.

The reservoir operation study was carried out for six possible schemes but only the results for the first basic scheme are presented here as an example. With a minimum operating level of 1277.5 m in Gembu Reservoir, a maximum retention level of 1303 m ensured a reliable continuous flow to the turbines of 69 $m^3 s^{-1}$ with a maximum of 68 $m^3 s^{-1}$ from Gembu to Sumsum. To allow for sedimentation, a maximum level of 1305 m was recommended for Gembu.

The energy yield at Abong for the basic scheme was 4950 GW h $year^{-1}$ which together with a small yield of 60 GW h $year^{-1}$ at Sumsum gave a total yield of 5010 GW h $year^{-1}$.

Further studies tested the capabilities of the system under drought conditions and during periods of maintenance when the linking tunnels may be closed for inspection.

9.3 GAMBIA RIVER BASIN DEVELOPMENT

LOCATION Gambia River is in west Africa

SOURCES Agrar und Hydrotechnik Gmbh and Howard Humphreys & Partners (1986) *Kekreti Reservoir project, feasibility study,* Report for Gambia River Basin Development Organization and Federal Republic of Germany.

Howard Humphreys & Partners (1985) *Gambia River Basin study,* Fact Sheet.

Simpson, R. W., & Thorpe, G. R. (1982) Use of a generalised computer program for resource systems optimisation in developing countries. Optimum allocation of water resources, Exeter Symposium, IAHS Publication No. 135. International Association of Hydrological Science, pp. 285–298.

PROBLEM The development of the resources of the Gambia River Basin has been a long-term project commencing in the early 1970s. The main aims of the member countries of the Gambia River Basin Development Organization — The Gambia, Senegal and Guinea — are the stimulation of employment for the rural population and the promotion of agro-industrial development with the expectation of a more equitable distribution of revenues. From the numerous basic studies which included the hydrology of the basin and an estuary salinity barrage has evolved the major Kekreti Reservoir project whereby a multipurpose dam becomes the focus of an integrated development programme for the Gambia River Basin. After a basic economic evaluation of a number of options, the study concentrated on the following aims.

(1) To provide irrigation water for up to a maximum of 70 000 ha from a reservoir containing up to 3500 Mm^3.
(2) To produce sufficient hydroelectric power to meet the needs of the region.
(3) To control salt water intrusion in the lower tidal reaches of the Gambia River.

An account of the reservoir operation studies undertaken by the consultants describes how these aims may be met.

Hydrological data

The boundary of the project area of 78 000 km^2 is shown in Fig. 9.6. The river

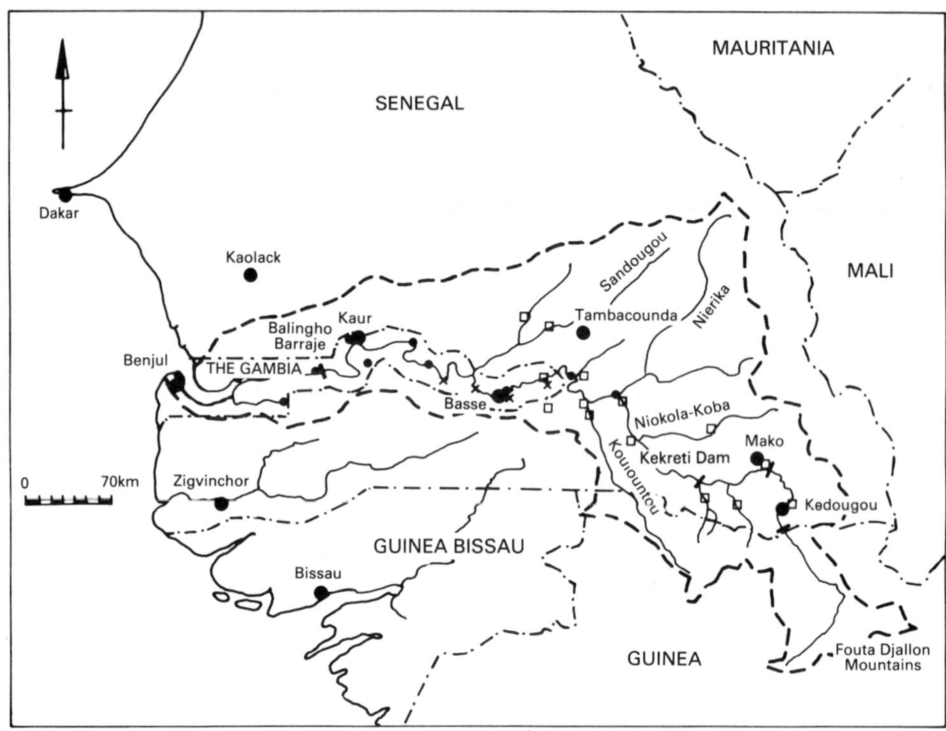

Fig. 9.6 — The Gambia River Basin: —·—, international boundaries; — — —, project boundary; +, probable dam sites; ●, water levels; ×, tidal discharges; □, river gauging stations.

gauging stations on the main river and on the major tributaries were established during the earlier studies. Their particulars are given in Table 9.11. Where there were only water level records, estimates of mean annual runoff were made by areal proportioning of neighbouring area unit runoffs. The records of a majority of the stations began in 1970 but there were several stations which began later or which had missing records. In order to assemble 10 years of continuous records at each station, missing data periods were infilled by correlation with adjacent stations using the monthly stream flow simulation model HEC-4 (US Army Corps of Engineers, 1971). The 10 day flow series required for the reservoir studies were generated where necessary using observed 10 day mean flows.

At the proposed Kekreti Dam site, there was no river gauging station and thus continuous monthly and 10 day flow series were synthesized using records upstream and an estimate of local inflows. The 10 day series of mean flows at Kekreti for the required period were generated from the addition of Mako, Thiokoye and Diarha

Table 9.11 — Particulars of selected gauging stations

Main river	Catchment area (km^2)	Catchment to confluence (km^2)	Mean annual flow ($m^3 s^{-1}$)	Mean annual unit runoff ($m^3 s^{-1}$ per 1000 km^2
Kedougou	7 550		70.1	10.5
Mako	10 450		93.7	9.0
Simenti	20 500		129.2	6.3
Tributaries				
Thiokoye	950	1 250	11.5	9.2
Diarha	760	860	8.1	9.4
Nikola-K	3 000	4 700	10.5	2.2
Nieriko	Level only	11 450	12.3 (estimate)	1.07
Niaouli	1 230	1 580	0.36	0.23
Sandougou	Level only	11 950	3.5 (estimate)	0.29

recorded flows and the estimated local inflows. The mean annual flow at Kekreti adopted for the study was 116.2 $m^3 s^{-1}$.

With regard to the sedimentation of the reservoir, sediment load measuremen*s were available for two stations on the Gambia River and by interpolation a relationship between total sediment load in 10 days at Kekreti and mean 10 day discharge was obtained. A mean annual total suspended sediment load of 240 000 tonnes was obtained. Adding 10% for bedload gave a total of 265 000 tonnes which was estimated to represent a total volume of sediment of 220 000 m^3 $year^{-1}$.

Evaporation losses from the reservoir also required consideration. The annual mean open-water evaporation Eo calculated by the Penman formula with data from Kedougou was 2698 mm and this was adopted as representative of conditions to be experienced over the Kekreti Reservoir. To evaluate the expected net monthly evaporation, E the runoff proportion of the rainfall (25%) flowing into the reservoir was deducted from the rainfall to be balanced against the calculated open-water evaporation. The computations are explicit in Table 9.12.

Modelling the Gambia River system

The complexities of water resources and the demands on a river system can now be modelled and operated on a digital computer. A range of system designs and operation decisions can be readily made with the aid of numerical simulation and optimization techniques.

The ROSS model conceived in 1975 has gradually developed into a general and powerful computer program for river basin and water resource simulation.

The system is represented by a series of nodes connected by reaches. The types of node are shown in Fig. 9.7. These are the computational points. The reaches, the

Table 9.12 — Net reservoir loss by evaporation

	Mean annual totals (mm)	Net reservoir loss by evaporation (mm)											
		Jan	Feb	Mar	Apl	May	Jun	Jly	Aug	Sep	Oct	Nov	Dec
Open-water evaporation Eo	2698	216	238	309	306	312	226	198	175	170	186	173	189
Rainfall	1236	0	0	0	0	47	178	274	284	292	141	13	1
75% of rainfall	927	0	0	0	4	35	133	205	213	219	106	10	0
Expected evaporation E	1867	216	238	309	302	277	93	0	0	0	80	163	189

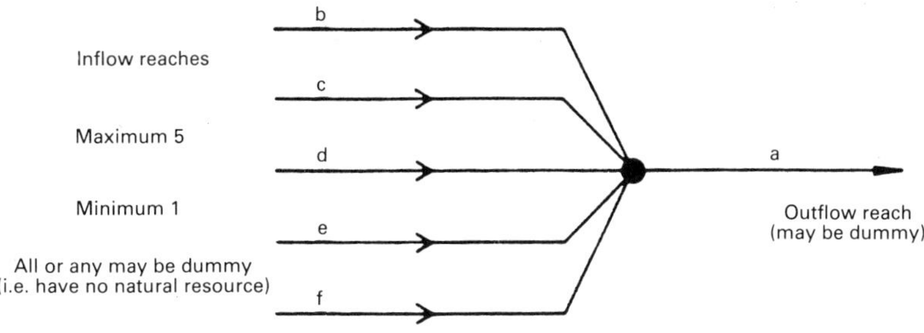

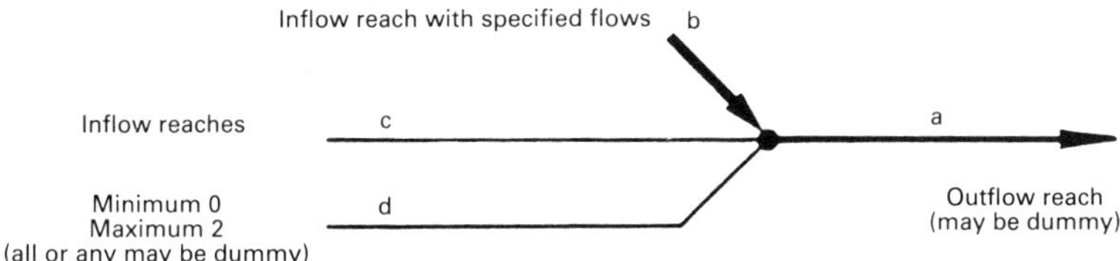

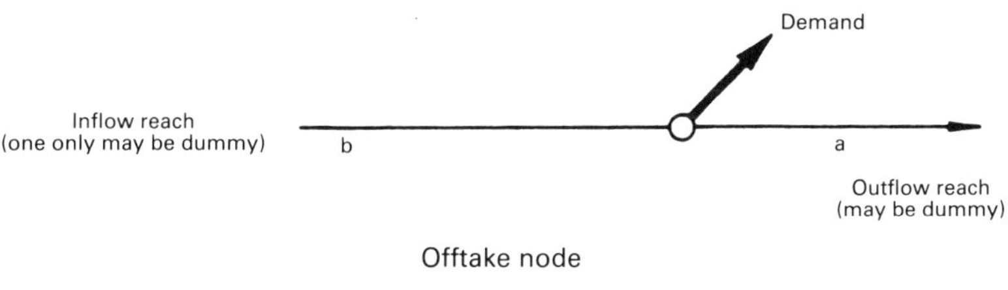

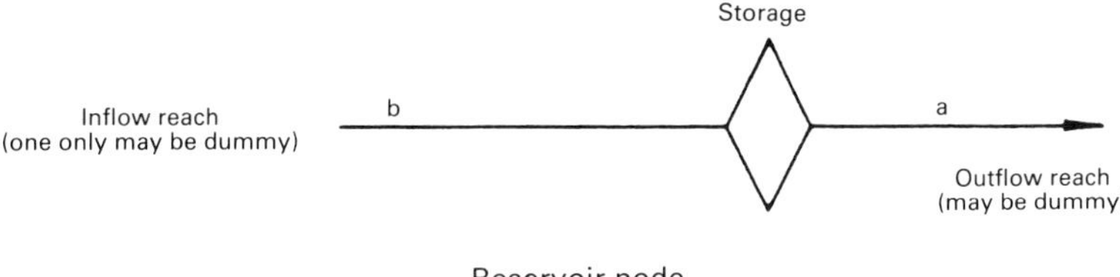

Fig. 9.7 — Node types for the ROSS simulation model.

links in the system, are usually the natural watercourses whose discharge including natural lateral inflows is computed at the immediate upstream node. The ROSS model network for the Gambia River system contained 56 reaches including the tributaries and 43 nodes on the main stream. Evaporation losses, irrigation offtakes and power generation releases were located at 28 nodes. At a final node, the last water requirement in the system was evaluated to maintain the level at the proposed Balingho Barrage, the defence against saline intrusion. An additional reservoir at Kouya upstream of Kekreti was incorporated into the model since the effects on Kekreti of the proposed Kouya Dam needed to be studied.

Model operation

The network description of nodes, offtakes and confluences with associated demands and control rules in downstream order is kept on file. The required details of the input data and output information are also filed.

Input data

Time series of the natural river flow resources for all the reaches have already been derived. Not all available water may be used at an offtake since flashy flood peaks can be lost downstream; so a proportioning factor is applied.

Each offtake node has its own demand pattern which may be constant with a mean annual demand divided into proportions for each period of the year or a series of the variable demands during the simulation period.

Compensation flow takes priority at reservoirs, diversions and offtakes and any spillage is taken into account in assessing compensation releases. For reservoir releases, the stored volume X may be related to the volume Y available for release by the following equations:

linear:	$Y = A+BX$
exponential:	$Y = A+\exp(X+B)$
logarithmic:	$Y = A+B\ln X$
power curve:	$Y = AX^B$.

For controlled releases, monthly values of A and B must be given. Evaporation losses from the reservoir are accounted for in each time step according to the previously calculated monthly net values.

In water use, upstream consumers have priority over downstream users and this forms the default option in the program. However, this can be altered by giving priority specifications under separate rules and constraints. Incoming drainage to the system from returning irrigation water, domestic effluent or surplus industrial uses can be accounted for by giving delay intervals at the offtakes according to the abstraction use.

Output data

There are three printing routines providing the following.

(1) Diagnostic printout detailing every computation.
(2) Details of flow throughout the network for each time period.

(3) Annual summary of flows, reservoir releases, offtake demands, supplies and shortfalls.

From initial optimization studies using an ancillary program RESIZE, it was concluded that a reservoir volume for Kekreti of 3500 Mm^3 would meet the irrigation demand for 70 000 ha and any greater reservoir would not be cost effective. In addition, a capacity of 3500 Mm^3 would be needed to ensure a maximum power yield with a necessary dead storage volume of 250 Mm^3.

Model results

The ROSS model simulated the operation of the Kekreti Reservoir and the downstream irrigation systems over the period 1971–1980 using the following data.

(1) The synthetic 10 day mean flows in the Gambia River at Kekreti.
(2) Recorded flows in all tributaries downstream of Kekreti adjusted to the river confluences.
(3) Mean 10 day water demands for each of 26 irrigation areas in Senegal and The Gambia (previously computed).
(4) Surface area–water level–reservoir volume relationships in Kekreti Reservoir for evaporation calculations during the simulation period.
(5) Water required to maintain the Balingho Barrage water level using calculation of evaporation losses.
(6) Assumption that drainage return flows were 20% of the total abstracted volume for the irrigation areas and time lagged by one 10 day period.

The results of the model run with the reservoir producing hydropower only and with releases to irrigate the maximum 70 000 ha is shown in Fig. 9.8. It shows that the irrigation development of 70 000 ha can be supported with a reliability of ever 80% since there were only two indications of reservoir failure in the 10 years of simulation. The detailed results showed that in meeting these irrigation demands the minimum reservoir volume in a 10 day period was 575 Mm^3, well above the prescribed dead storage of 250 Mm^3.

Reservoir operation

To investigate the operational developments of the Kekreti Reservoir, several model runs were made first with power only and then gradually increasing irrigation demands for 30 000, 50 000 and 70 000 ha. For the power only, the releases are at a constant rate throughout the year of 102 $m^3 s^{-1}$. This is the most efficient mode of operation for the turbines and the irrigation demand is expected to exceed this when about 10 000 ha have been developed. However, with increased irrigation demands, the releases through the power station must increase to meet the demands. For all the irrigation areas simulated, the annual yield of the reservoir exceeds the irrigation demand and therefore extra releases for power may be made during the wet season.

In each of the four operation runs of the model, the reservoir failed twice in the 10 year period and spilled three times (Fig. 9.8). However, as the water demands become more seasonal with increasing irrigation area, the efficiency of the reservoir operation will decline. This would be caused by increases in spillage and evaporation

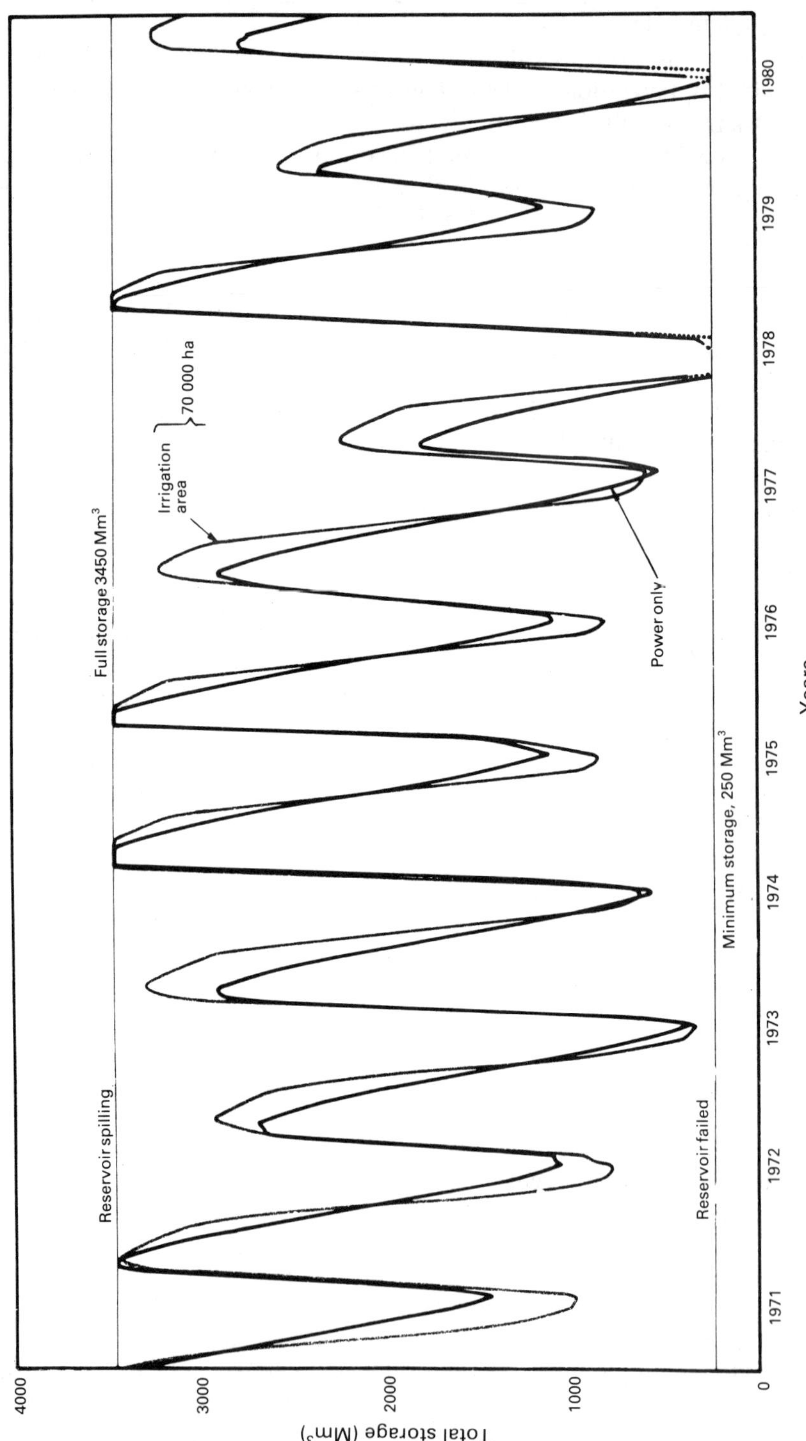

Fig. 9.8 — Results of reservoir operation study.

although some of the extra spillage could be harnessed for power during periods of low irrigation demand.

After including the proposed upstream Kouya Reservoir in the Kekreti operation investigations, the construction of the Kouya Dam could only be justified on an increased energy generation of 4%.

10

Irrigation

INTRODUCTION

Agricultural production in semiarid and arid regions of the world has long benefited from irrigation, the application of water to the fields from a large perennial river or a storage reservoir. The exploitation of the seasonal floods of the Nile over many centuries from the times of the ancient Egyptian civilization is well known but the early reservoirs at old dam sites in Sri Lanka have been appreciated comparatively recently.

The pattern of irrigation depends on the climate, the amounts of rainfall expected and whether or not it arrives in well-defined rainy seasons. In some countries, crops are grown entirely with irrigated water but in some areas, even when rainfall occurs all the year round, irrigation may be practised to enhance crop yields in the summer season. Thus there is a very wide range of irrigated crops from those grown in local subsistence farming to those produced for commercial gain from large plantations. They can benefit from irrigated water in regions extending from the tropics to lands with a temperate climate.

In the planning and design of modern irrigation schemes, the hydrologist is engaged for two main functions, the investigation of water resources and the evaluation of the crop water requirements. The studies may begin with assessing the resources and then deciding what crops can be grown or, alternatively, the decision on the required crops may come first and the necessary resources have to be found. Many other factors come into the decision-making process of which soil quality is the most important but adequate competent labour and the economics of the scheme must also be considered. A vital factor was overlooked in some of the earlier designed schemes which only came to light after many years of operation. That was the inadequate drainage of the land being irrigated. Saturated soils became impacted and accumulations of harmful salts made the land infertile. This lesson has been well learnt by successive planners and attention is now focussed on the adequate drainage of agricultural land to be irrigated (see Chapter 2).

The following case studies come from tropical Africa and the Mediterranean island of Cyprus. The development of new irrigated areas , the extension of existing schemes and the improvement in crop yields by irrigation are demonstrated in the examples. The evaluation of resources is the major hydrological topic and where reservoirs are concerned the secondary functions of hydroelectric power development and domestic water supplies are considered. River flow records are the necessary basic data for assessing the adequacy of run-of-river flows or perhaps the need for storage to augment low flows. The techniques of catchment modelling to derive runoff from rainfall or where there are sufficient historic records the methods of statistical analysis may be employed. For estimating crop water requirements, the cropping pattern is needed and then rainfall and evaporation values on a monthly or 10 day basis must be established. Long-term records are analysed to estimate frequencies of occurrence of required amounts to ensure optimum crop yields. Probability calculations are also necessary for reservoir yield and spillway design as described in previous chapters. Hydrological studies concerned with irrigation have been practised and the results applied for many years in the USA in the drier states of the south and west. The methods of the US Soil Conservation Service (USDA, 1967) are applied worldwide and the Blaney–Criddle formula (FAO, 1977) is also used in many areas. Hydrologists emanating from Europe tend to use the evaporation formulae of Penman or Turc (Shaw, 1988) in conjunction with crop coefficients for assessing crop water requirements.

10.1 RUBINGAZI IRRIGATION FEASIBILITY STUDY

LOCATION The Rubingazi River, in central Kenya, is a headwater of the Tana River.

SOURCE Booker Agriculture International Ltd and Binnie & Partners (1980) *Rubingazi Irrigation Project: feasibility study*, Report to the Tana River Development Authority, Government of Kenya.

PROBLEM As part of a wider study of land use development, the consultants were requested by the Tana River Development Authority with support from the UK Ministry of Overseas Development, to investigate the feasibility of irrigating approximately 2000 ha extending on both banks of the Rubingazi River, south of Embu (Fig. 10.1). The feasibility study was to take into account all aspects of land development from soil surveys, appropriate agricultural crops and their marketing and the engineering of the irrigation works to the cost–benefit analysis and the human impact on the population. In this account, the contribution of the hydrologists is focussed on the problem of water resources for the project.

Rubingazi catchment

The Rubingazi River forms part of the deeply incised radial drainage pattern from Mount Kenya, an extinct volcano. The long narrow catchment of about 200 km^2 extends from an elevation of 4500 m on the mountain to about 1050 m at the confluence of the Rubingazi with the Nyamindi River. The project area is just

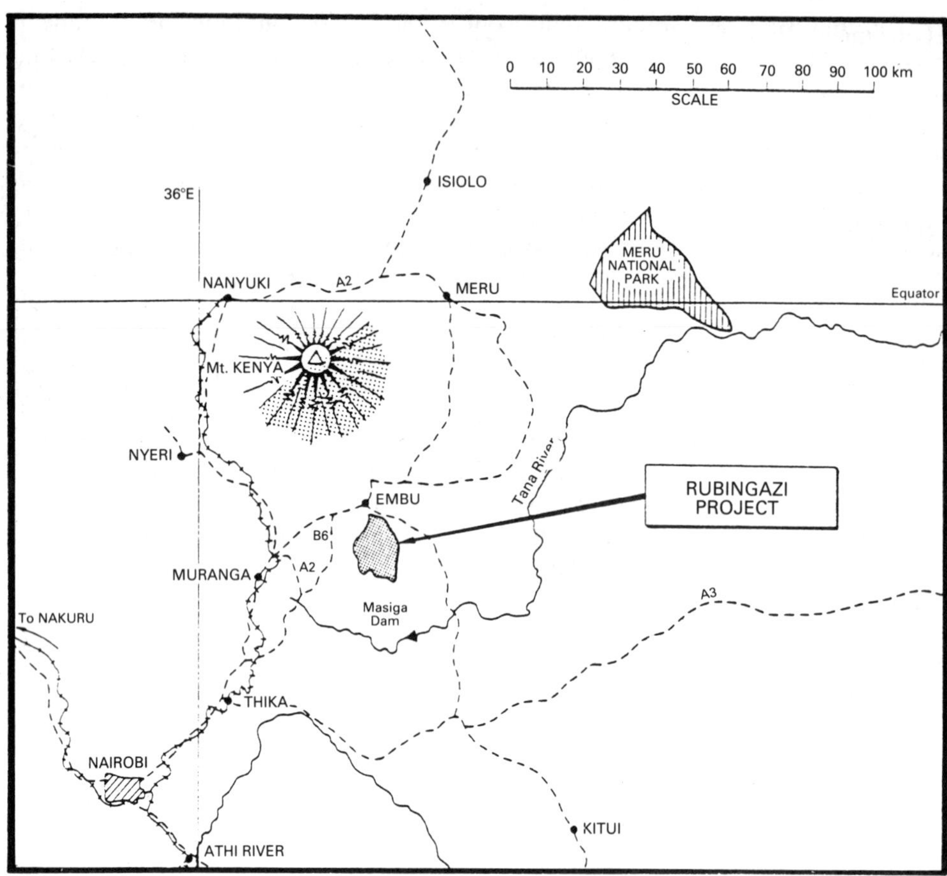

Fig. 10.1 — Rubingazi irrigation project: – – –, trunk roads, ―+―+―, railways; •, major towns.

upstream of this confluence at 1050–1300 m. The wide variety of volcanic rocks support three vegetation zones, the high alpine moorland belt down to about 3200 m, the continuous forest and bamboo belt from 3200 to about 1800 m and below this the savanna and cultivated areas. The expected equatorial-type climate is modified by the high mountains and the monsoonal systems of the Indian Ocean. In the transitional periods between the southeast monsoon of June–September and the northeast monsoon, December–March are the long heavy rains of March–June and the short rains October–November. Table 10.1 of monthly mean rainfalls demonstrates some seasonal differences between the upper catchment and the project area. At the river gauging station 6 km upstream of the project area there is an estimated mean annual runoff of 830 mm with maximum mean flow of 9.62 $m^3 s^{-1}$ in May after the long rains and a minimum of 2.33 $m^3 s^{-1}$ in March at the end of the dry season. There is a tendency for flows to be sustained by groundwater inflows in the recession periods particularly in August and September.

Hydrometric data

The evaluation of the detailed temporal distribution of the water resources for irrigation required the analysis of hydrometric data for rainfall, evaporation and river flow. Considerable effort was afforded to the assemblage and appraisal of existing information.

Rainfall

The coverage of the rain gauge network (Fig. 10.2) was good except in the upper catchment. The longest record at Embu from 1908 to 1972 had few interruptions and the record for Mwea experimental farm, a climate station just southwest of the catchment, was nearly complete from 1965. Of the 17 stations, nine were visited by the project hydrologist and, after quality control and some infilling of gaps, eight of the records were considered acceptable. The mean of the monthly rainfalls at Embu and Mwea were taken to represent the seasonal pattern over the project area. For other rainfall values, a factor of 1.033 was needed to adjust Mwea data to the project area and 0.844 to transpose rainfall from Embu. For example, the long Embu record was used to assess the 50% and 80% probabilities of exceedence of the annual rainfalls and these transposed values provided the plotted isohyets (Fig. 10.2).

Evaporation

Although knowledge of the frequency of high-temperature damage to maize seedlings is required from climatological data, the main value of the climate stations is to supply the hydrologist with information on evaporation. Two stations Tebere (height, 1159 m) and Mwea (height, 1150 m) near the project area and Kinderuma 40 km away provided published data. It was found that Penman estimates of open water evaporation E_0 and potential evapotranspiration ET, using albedos of 0.05 and 0.25 for short grass respectively, from Mwea data could be transposed satisfactorily to the project area.

River flow

There were a number of permanent river gauging stations in the neighbourhood of the project area. The most valuable stations for the studies was DC3 on the Rubingazi (catchment area, 207 km^2) and DC6 on the Kapingazi (30.6 km^2) which together measured most of the flow available to the project (Fig. 10.2). Of the 16 stations in the region only one downstream on the main Thiba River had a water level recorder; the remainder had staff gauges read once or twice on working days with rating by current meter measurements on up to 14 occasions per year at important stations. The stations likely to be useful in the studies were visited and the quality of their records assessed. DC3 had a stable control but DC6, although a poor site and bypassed by ungauged flows, was more satisfactory than its upstream neighbour and therefore used in analysis. These two records were fully processed: preliminary ratings were used to give estimates of daily mean flows; then the hydrographs were checked, erroneous data identified and corrected and gaps infilled. A sample of monthly evaporation and river flows is given in Table 10.2. In addition, river water samples were taken from the Rubingazi for water quality analysis and, while the existing supplies were chemically suitable for irrigation, poor bacteriological quality

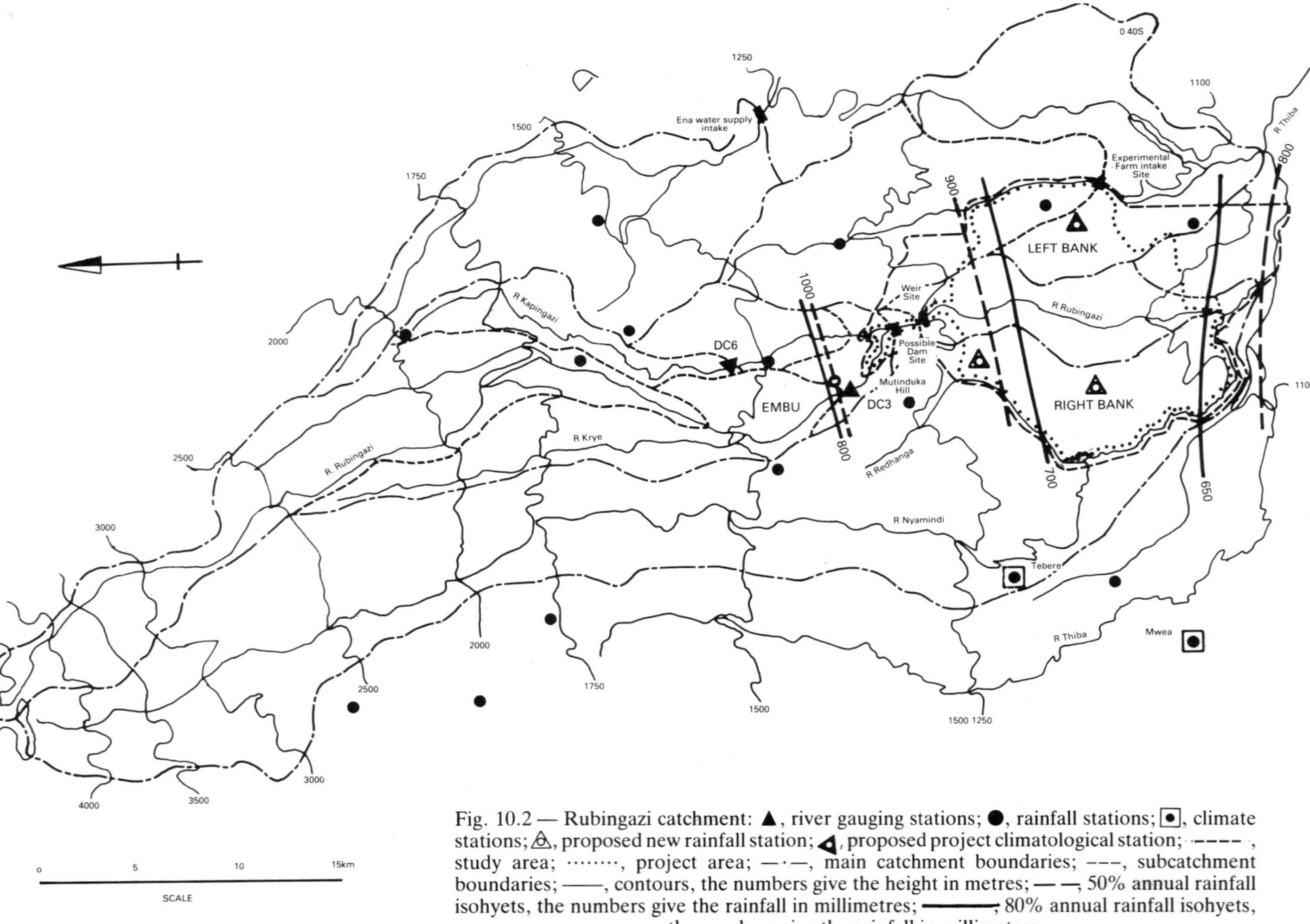

Fig. 10.2 — Rubingazi catchment: ▲, river gauging stations; ●, rainfall stations; ⊡, climate stations; △, proposed new rainfall station; ◀, proposed project climatological station; ----, study area; ·······, project area; —·—, main catchment boundaries; ---, subcatchment boundaries; ——, contours, the numbers give the height in metres; — —, 50% annual rainfall isohyets, the numbers give the rainfall in millimetres; ━━━, 80% annual rainfall isohyets, the numbers give the rainfall in millimetres.

indicated that they were unsuitable for drinking without treatment. From sediment ratings for the Thiba River, an estimate of the annual sediment load in the Rubingazi was calculated.

Hydrological studies

Phase I

The hydrological studies were divided into two parts according to differing plans for the project development. In phase I, it was envisaged that irrigation water as run-of-river flows should be available all the year and led by lined canals from a weir offtake to feed suitable blocks of land totalling 1800 ha. The irrigation method recommended was to have sprinkler application on a 24 h basis to ensure maximum use of the water resources, night storage being difficult and expensive to provide.

(1) In assessing crop water requirements, two cropping patterns were investigated (Table 10.3). Monthly and 10 day values of rainfall and evapotranspiration for a frequency of 1 in 5 years (80% probability) were used in the calculation of crop water use. The effective rainfall was computed by the US Soil Conservation Service method (USDA, 1967) and actual crop evapotranspiration ET_c calculated for each crop in each cropping pattern by multiplying evapotranspiration values by the appropriate crop factors. Irrigation requirements were then obtained by subtracting effective rainfall from the ET_c. The project area water requirements estimated for the two cropping patterns, allowing for an irrigation efficiency of 70% and a conveyance efficiency of 85%, are compared with the water available at the proposed intake site given by the flows at DC3 and DC6 scaled up for increased catchment area (Table 10.4). On a monthly basis, only March appeared a critical period but assessing needs on a 10 day basis revealed higher rates required in October. These shorter-period calculations provided the design discharges of 1.48 and 2.69 $m^3 s^{-1}$ for the main canals in the two cropping pattern options.

(2) Flood estimates for several return periods were needed for the design of the intake weir, canal crossings and other bridges. No reliable flood flows or river levels were available for the Rubingazi catchment since twice daily staff gauge readings rarely coincided with the peaks. An analysis of peak flows at 12 river gauging stations in the Tana River Basin produced the relationship of MAF ($m^3 s^{-1}$) to area A (km^2):

$$\text{MAF} = 1.11A^{0.72}.$$

The ratio of the 100 year flood to the MAF was reasonably independent of catchment area with an average of 3.34. Thus, for the station DC3, the MAF would bc 52 $m^3 s^{-1}$ and the 100 year event 174 $m^3 s^{-1}$. From a more widely derived East African flood model, the 100 year flood for DC3 should be in the range 300–370 $m^3 s^{-1}$. A conservative estimate of 300 $m^3 s^{-1}$ was adopted and scaled to the proposed intake gave 353 $m^3 s^{-1}$. Peak flows for return periods of 5 and 25 years for catchments draining to culvert and bridge sites were calculated from $Q_5 = 2.56A^{0.88}$ = and $Q_{25} = 3.53A^{0.88}$ with A in square kolimetres.

(3) The third study in phase I was a simple model to simulate the operation of the

Table 10.1 — Monthly mean rainfall

	Mean rainfall (mm)											
	January	February	March	April	May	June	July	August	September	October	November	December
Upper catchment	105	96	122	332	488	167	205	183	132	260	200	127
Project area	26	27	79	216	145	25	18	20	21	104	174	48

Table 10.2 — Monthly mean evaporation and river flows

Month	January	February	March	April	May	June	July	August	September	October	November	December
Open-water evaporation[a] E_0 (mm day^{-1})	6.2	6.5	6.6	6.0	5.5	4.7	4.2	4.7	6.2	6.6	5.6	5.7
Evapotranspiration[a] ET (mm day^{-1})	4.7	5.1	5.1	4.7	4.2	3.7	3.3	3.7	4.9	5.2	4.3	4.3
River flow												
DC3 (m^{3}s^{-1})	3.28	2.52	2.33	4.93	9.62	5.74	4.71	4.57	4.17	4.20	7.39	4.17
DC6 (m^{3}s^{-1})	0.29	0.18	0.15	0.88	1.37	0.74	0.57	0.40	0.28	0.35	0.83	0.53

[a]Penman estimates

Table 10.4 — Water requirements for 1800 ha at 80% probability and water availability

	Water (m^{3}s^{-1})											
	January	February	March	April	May	June	July	August	September	October	November	December
Water required, cropping pattern 1	0.58	0.77	0.97	0.67	0.68	0.59	0.46	0.23	0.35	0.71	0.51	0.67
Water required, cropping pattern 2	0.73	0.86	0.99	0.35	0.80	0.93	0.58	0.57	1.10	1.15	0.50	0.67
Available water	1.54	1.30	1.13	1.64	4.76	3.60	3.35	3.10	2.94	2.83	3.59	2.46

Table 10.3 — Project cropping pattern areas

Crop	Area of cropping pattern 1 (ha)	Area of cropping pattern 2 (ha)
Food crops (maize and pulses)	283	212
Cotton	850	425
Tobacco	283	—
Vegetables (including green beans)	284	638
Sweet corn	—	425
Fruit trees (Avacado)	100	100
Total	1800	1800

scheme and its affect on river flows. A monthly time basis was used although this could mask some of the river flow variability. Eleven years of historic data 1969–1979 with some estimated infillings in the record gaps were related to the project area and proposed weir intake. Allowances were made for upstream demands of 0.41 $m^3 s^{-1}$ and a minimum flow of 0.12 $m^3 s^{-1}$ to pass the intake. Both cropping patterns were tested and 1976 conditions proved to be critical. Shortfalls of 0.12 $m^3 s^{-1}$ occurred for both cropping patterns in the March when the river flow was estimated to be of 20 year return period.

Phase II

In order to develop more of the potentially irrigable land on both sides of the river, a regulating reservoir would be required to overcome the variability of the river flows. The net storage volume required for four possible combinations of cropping patterns over 1800 ha on the left bank, with 1490 and 300 ha on the right bank were determined by simulation studies using the 11 years of data and with the previous upstream flow and minimum release assumptions. The results are summarized in Table 10.5. The proposed dam site is shown in Fig. 10.2. It was considered that the reservoir should pass safely a 10 000 year flood and this was estimated to be about 600 $m^3 s^{-1}$. Trial reservoir flood routings were carried out to test spillway provisions and to assess extra flood storage required. A height of 38 m was determined for the dam at the proposed site to ensure the necessary storage and allowances for sedimentation and flood freeboard.

Conclusion

Only a general outline of the hydrological studies has been given and as so often occurs in practice they have been based on inadequate data. Therefore additional hydrometric stations were recommended. These studies formed an integral part of the feasibility of the Rubingazi irrigation project relying upon the findings of the soil scientists and agricultural economists and providing a basis for the detailed engineering design phase of the scheme.

Table 10.5 — Rubingazi phase II storage

Cropping pattern Area (ha)	Crop option	Net storage (Mm^3)	Dead storage (Mm^3)	Total (Mm^3)	Depth (m)	Surface area (ha)
1800	1					
1490	1	5.8	2	7.8	31.9	59
300	Fodder					
1800	2					
1490	2	7.0	2	9.0	33.6	66
300	Fodder					
1800	2					
1490	1	6.5	2	8.5	32.9	66
300	Fodder					
1800	1					
1490	2	6.4	2	8.4	32.8	63
300	Fodder					

10.2 KHRYSOKHOU HYDROAGRICULTURAL DEVELOPMENT

LOCATION The Khrysokou River is in western Cyprus.

SOURCE Ministry of Agriculture and Natural Resources, Cyprus (1981) *Hydro-agricultural development*, *Khrysokhou area*, Feasibility Report, Annexe 3, Water Resources Development. UNDP, Food and Agricultural Organization.

PROBLEM The Khrysokhou watershed irrigation project began in 1979 with the principal aim to increase agricultural production and employment opportunities by converting low-yielding subsidized agriculture to modern irrigated farming practices. The project area is in western Cyprus covering 930 km^3 in the Polis region (Fig. 10.3). The essential problem for the hydrologist is the assessment of the water resources and their reliability for the irrigation of the lowlands (31 km^2) along Khrysokhou Bay and the uplands (12 km^2) between 200 and 400 m elevation up the Khrysokhou River valley. This account is confining attention to the provision of irrigation water for the lowlands. The study was carried out by the project's Water Resources Team led by Dr I. M. Goodwill and the project managed jointly by Mr J. H. Visser, Food and Agricultural Organization and Dr C. A. Christodoulou, Republic of Cyprus, whose organizations kindly permitted this publication.

Investigations

The average annual precipitation over the area is about 600 mm of which an estimated 70% evaporates leaving 180 mm, approximately 167 Mm^3 of water per

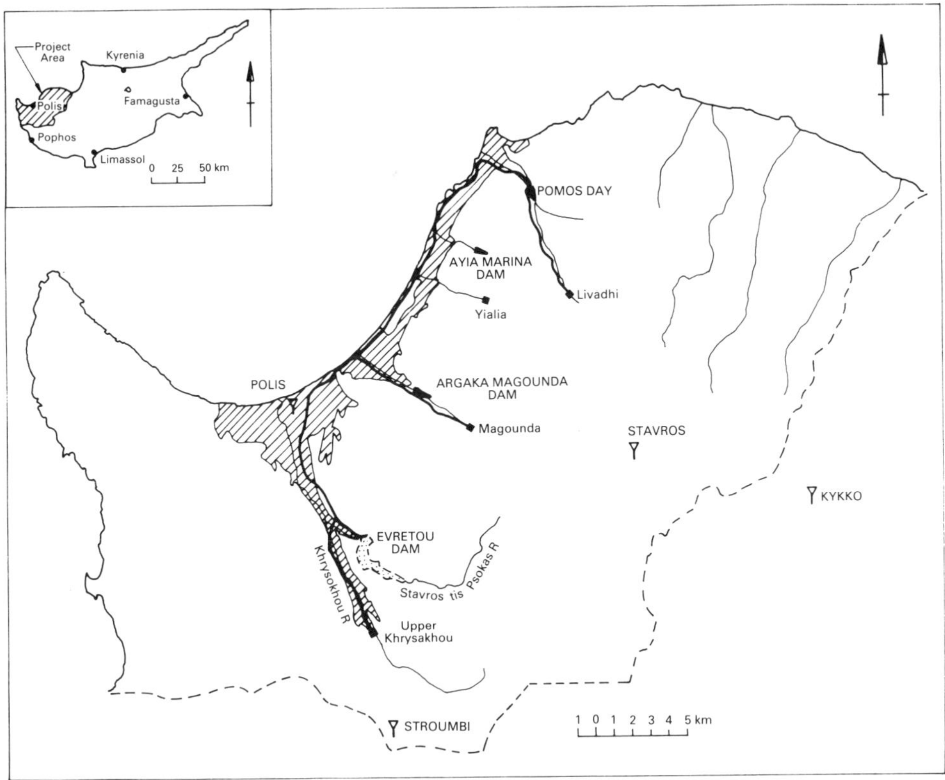

Fig. 10.3 — Project location map: ▶, existing reservoirs; [symbol], proposed reservoir; ■, proposed intakes; ——, conveyor; [hatched], irrigation area; Y, rainfall stations.

annum available for development. Of this average total, 58% is surface runoff, the remainder going to groundwater recharge. Rainfall records from 1916 and river discharges from 1965 were available for analysis. The rainfall occurs in the winter months and is highly variable on a monthly and annual basis so that storage of surface runoff must be provided to serve dry years as well as the regular summer seasonal demands. Temperatures allow cropping throughout the year. Existing storages total about 2 Mm^3 in the three small reservoirs Pomos, Ayia Marina and Argaka Magounda built in the 1960s.

Development demands

To realise the full potential of the soils suitable for irrigation, estimates of the crop demands were made using the Blaney–Criddle method. Cropping patterns for perennial crops such as citrus and deciduous fruits and nuts and annual vegetables were determined by the project agronomist and economist. Then irrigation water demands were evaluated on a monthly basis. To assess the annual variability of the irrigation requirements, stochastically generated series of hydrometeorological data of 49 years length based on historic records were used as input for the Blaney–Criddle formula to obtain monthly water demand values by weighting each crop demand to give an average over the irrigated area. The statistics for one series run of the formula for the lowland irrigation area are given in Table 10.6. During the months of

Table 10.6 — Statistics of water demand from 49 years' simulated data

Month	Mean (mm)	Standard deviation (mm)	Coefficient of variation	Maximum (mm)	Minimum (mm)
November	17.0	11.1	0.7	35.8	0.2
December	4.3	6.6	1.5	23.9	0.0
January	2.1	4.2	2.0	20.2	0.0
February	4.6	7.5	1.6	36.8	0.0
March	16.5	10.9	0.7	47.5	3.1
April	34.1	16.2	0.5	79.0	15.5
May	87.7	20.5	0.2	131.4	50.0
June	144.4	4.7	0.0	154.0	129.9
July	149.0	6.4	0.0	160.7	132.6
August	126.2	5.8	0.0	136.8	108.2
September	85.4	3.8	0.0	92.9	72.1
October	41.6	15.1	0.4	64.6	8.5
Year	713.0	56.7	0.1	883.0	615.7

the dry season, May–September, which account for 95% of the irrigation demand, the variation in demand is very low. The maximum annual demand of 883 mm was the highest from 20 series of 49 years' generated data, thus representing an infrequently occurring very dry year which had only 40% (186 mm) of the average annual rainfall. The non-agricultural demands for domestic water supply, tourist development and possible industrial use, totalling annually 1.8 Mm^3 are expected to be met from groundwater resources so that it is the total of 22 Mm^3 $year^{-1}$ required on average for irrigating the 31 km^2 of the lowlands area that is to be provided by the project.

The supply scheme

A new large reservoir (capacity, 25 Mm^3) contained by the Evretou Dam 71 m high on the Stavrostis Psokas River forms the major storage for the scheme. The three small reservoirs are incorporated into the supply system and four river intakes Livadhi, Yiallia, Magounda and Upper Khrysokhou augment the natural inflow to the Evretou Reservoir. The intakes are sited at a higher level than the reservoir and hence all water is conveyed by gravity to the irrigation areas and to the main storage via two-way flow pipes ranging from 300 to 900 mm in diameter. The total length of the main conveyor pipeline from the Pomos Dam to Evretou Reservoir is 29.8 km. The conveyance system is composed of pressure pipes (i.e. an enclosed system) so that water can be automatically controlled and sent to wherever it is needed. The irrigation headworks consist of nine storage ponds automatically fed from the main conveyor and acting as constant-head supplies to the irrigation system. Groundwater sources, most important in dry years, are connected into the system at the level of the irrigation headworks.

The system has been planned to make its operation on a daily basis as automatic as possible. Diversion to Evretou from the small stream intakes, mainly in the winter months, takes place whenever there is stream flow; part of the flow *en route* feeds the local irrigation storage ponds. The Evretou Reservoir can supply the irrigation requirements of the lowlands area for a whole year but the conveyor pipeline sizes have been designed according to peak irrigation demands when supply can be afforded by the smaller reservoirs, thus reducing required flow from Evretou. The operation of the system in an average year is demonstrated in Table 10.7.

Table 10.7 — Average irrigation water supply to the lowlands

Month	Total demand	Evretou		Small reservoir		Groundwater	
	(Mm^3)	Supplied	(%)	Supplied	(%)	Supplied	(%)
January	0.0	—	—	—	—	—	—
February	0.1	0.1	100	—	—	—	—
March	0.5	0.5	100	—	—	—	—
April	1.1	1.1	100	—	—	—	—
May	2.7	2.4	89	0.1	4	0.2	7
June	4.4	3.6	82	0.4	9	0.4	9
July	4.6	3.4	74	0.6	13	0.6	13
August	3.9	2.9	74	0.5	13	0.5	13
September	2.6	2.1	81	0.2	8	0.3	12
October	1.4	1.2	86	0.1	7	0.1	7
November	0.5	0.5	100	—	—	—	—
December	0.1	0.1	100	—	—	—	—
Year	21.9	17.9		1.9		2.1	

System simulation

The long-term operation of the system was tested with a simulation model (Fig. 10.4). The model simulates the behaviour of the reservoirs connected together by pipes and is designed for a time step of 1 day or 1 month according to the stream flows. When one or more of the stream flows are high, the time step is 1 day to cope with the diversion water but in dry spells a 1 month time step saves computer time with little loss of accuracy. In conjunction with the stochastically generated series of irrigation demand data, similar 49 year series of rainfall and stream flow data were generated. Of 20 generated series, based on historic rainfall and stream flow records, six were selected by inspection for input into the simulation model to produce

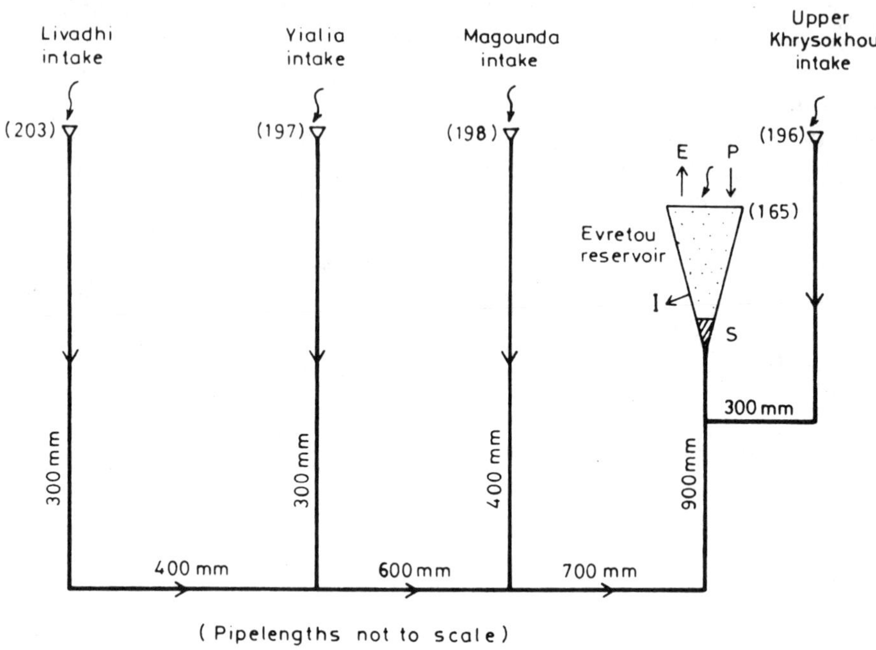

Fig. 10.4 — Layout of lowlands main conveyor simulation model: P, precipitation; E, evaporation; I, irrigation supply; S, sediments; ∇, diversion structure, the numbers give the elevation in m (asl); ←, river runoff.

detailed analyses of annual sequences likely to produce shortages of irrigation water. The results from one series is given in Fig. 10.5. The irrigation demand takes 10 years to build up from commencement of the scheme and thereafter its fulfilment fluctuates according to the resources. Over these 49 years, the average annual data for Evretou Reservoir assuming irrigation demand for 2500 ha was as in Table 10.8. From the model runs made, it was shown that the reservoir will spill, on average, in 1 in 3–4 years. In most years, the demand is fully met; in the dry years the deficit is quite small and only in a few extreme years the deficit approaches 50%.

From a subsequent analysis of the 20 series, the information in Table 10.9 for an average series was taken from cumulative frequency curves of the number of shortage years to be expected in a 49 year period.

Conclusion

Several possible development schemes were considered, many to enhance existing local systems of irrigation. However, the regional development of the lowlands and uplands would provide the greatest benefits for the capital invested with low running costs. The establishment of the lowlands scheme would ensure the greater reliability of meeting irrigation demands from the very variable water resources.

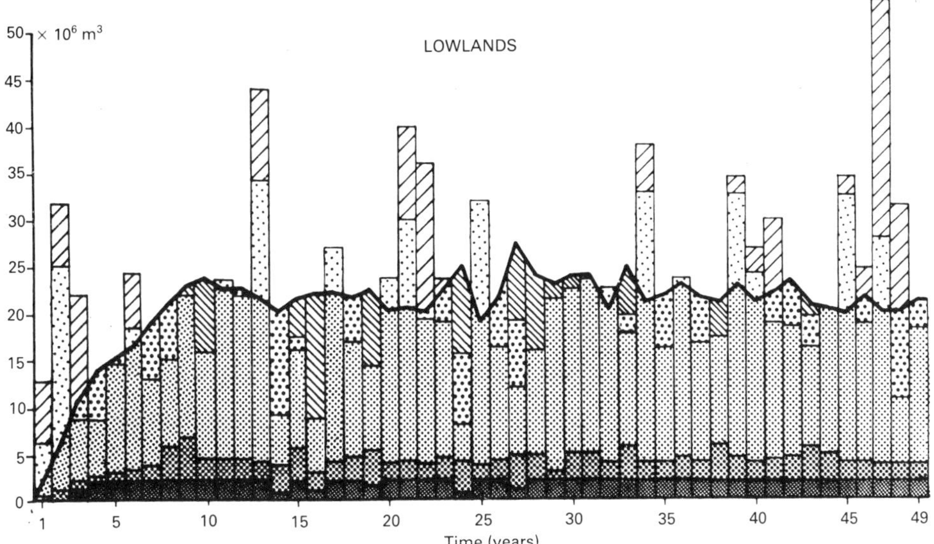

Fig. 10.5 — Graphical representation of system behaviour: ▨, spill; ⊡, international storage; —, irrigation demand; ▧, shortages; ▦, demand satisfied from interannual storage; ▦, demand satisfied from annual flow; ▦, demand satisfied from groundwater; ▦, demand satısfied from small reservoirs.

Table 10.8 — Average annul data

Inflow (Mm³)	12.9
Diverted inflow (Mm³)	5.5
Irrigation demands (Mm³)	17.8
Total spills (Mm³)	2.4
Inflow not spilled (%)	86
Diverted water retained (%)	88
Deficits (Mm³)	2.0

Table 10.9 — Number of shortage years

Shortages (%)	Number of years
⩾45	2
⩾30	4–5
⩾15	10

10.3 SENEGAL RIVER BASIN IRRIGATION

LOCATION The Senegal River in west Africa flows through Mali, Mauritania and Senegal.

SOURCE Hargreaves, G. L., Hargreaves, G. H., & Riley, J. P. (1985) Irrigation water requirements for Senegal River Basin. Am. Soc. Civ. Eng., *J. Irrig. Drain. Eng.* **111** No. 2, 113–124.

PROBLEM As a contribution towards attaining self-sufficiency in food supplies in the developing countries of Mali, Mauritania and Senegal, a scheme to irrigate 274 805 ha by 2030 AD in the Senegal River Basin has been proposed. Two dams on the river are intended to provide the water resources for irrigation, domestic and industrial supplies, hydropower and flood control. The present study assesses the crop water requirements for a variety of crops and comments on other factors influencing crop yields.

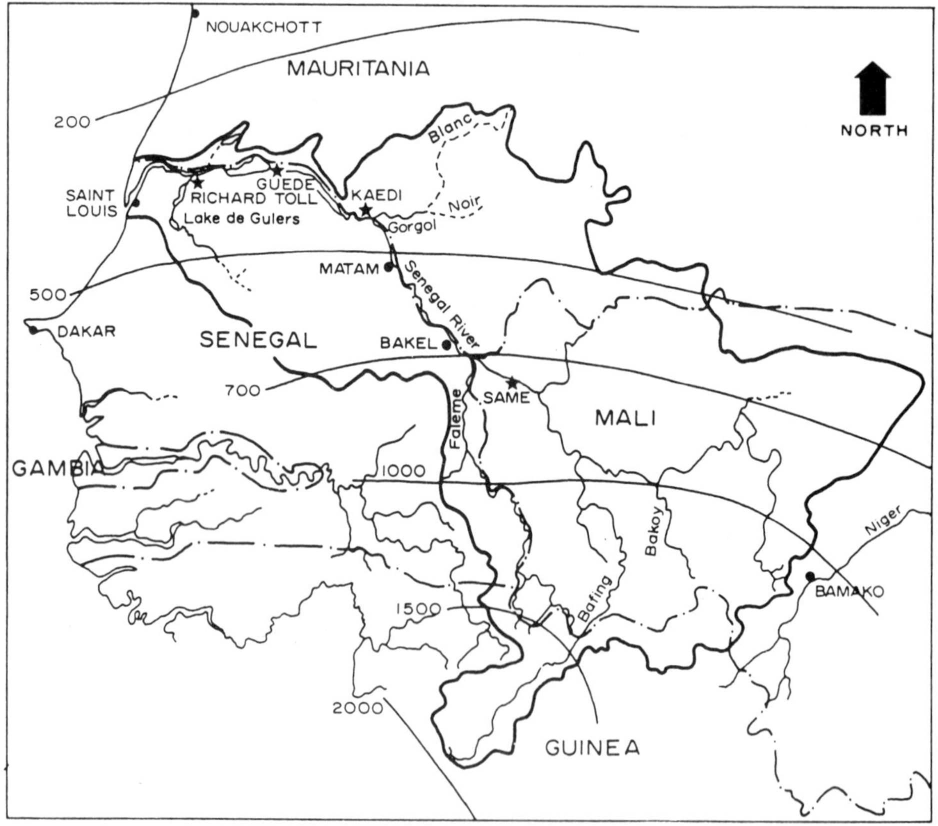

Fig. 10.6 — Senegal River Basin, map showing average annual precipitation ★, climatological stations used for this study; ——, isohyets, the numbers give the average annual precipitation in millimetres.

Senegal River Basin

The main tributary of the Senegal River, the Bafing River, rises in the mountains of Guinea but the greatest extent of the basin is contained in the territories of Mali, Mauritania and Senegal (Fig. 10.6). The area lies within the tropics with an average annual temperature of 29°C and a small monthly range from 25°C in January to 32°C in June. Rainfall decreases northwards from an annual 2000 mm near the headwaters to about 300 mm at the river mouth. There is a marked seasonal distribution; the rainfall of June–October constitutes the wet season and the rest of the months are dry. Four climatological stations Richard-Toll, Guede, Kaedi and Same provided data on temperature, radiation and rainfall for the proposed irrigation area.

Evapotranspiration

The first problem was the determination of the monthly potential evapotranspiration ET_0 from a reference crop. From previous studies,

$$ET_0 = 0.0135RS(T+17.8) \text{ mm} \tag{10.1}$$

where RS (mm of evaporation) is the solar radiation at the surface and T (°C) the mean monthly temperature. This formula produced good results with perennial ryegrass but tended to overestimate cool-season grass evapotranspiration for coastal locations and to underestimate for highly advective conditions. Using the measured RS and mean temperatures in equation (10.1), the monthly ET_0 were calculated (Table 10.10).

Table 10.10 — Estimated ET_0 for Senegal River Basin irrigation

Month	January	February	March	April	May	June
ET_0 (mm)	129	151	179	196	203	192
Month	July	August	September	October	November	December
ET_0 (mm)	186	171	173	167	132	129

In developing countries, identifying the ground cover of climatological sites is difficult and additionally values of RS are sometimes not available. However, the latter can be estimated from the extraterrestrial radiation RA and sunshine hours or the daily temperature range TD. With TD given by the mean maximum minus mean minimum temperatures in degrees Celsius averaged from the four climate stations,

$$RS = 0.16RA \times TD^{0.5} \tag{10.2}$$

Combining equations (10.1 and 10.2) results in the following:

$$ET_0 = 0.0022RA(T+17.8)TD^{0.5}. \tag{10.3}$$

The coefficient in equation (10.2) was tested on published data from worldwide locations and was found to increase in coastal areas and decrease at high elevations where there is an increase in TD. This tendency counterbalanced the bias in equation (10.1) and thus the coefficient 0.0022 in equation (10.3) was found to be remarkably uniform. Equation 10.3 was recommended for use in the scheduling of irrigation.

The potential crop evapotranspiration ET_c was calculated from a standard equation

$$ET_c = K_c ET_0 \tag{10.4}$$

where K_c is a crop coefficient which has been derived by experiment for a variety of irrigated crops (FAO, 1977). Crop coefficients were established on a monthly basis but in this application seasonal values were used.

Water requirements

The limited rainfall of the wet season provides some of the crop water needs and this must be taken into account when assessing irrigation quantities. Some of the rainfall is lost through immediate evaporation and, depending on the intensity, some forms direct runoff. Analysis of the rainfall records led to the choice of the 50% probability occurrence amounts for the monthly totals and these were reduced by a standard method to give the effective rainfall P. The evaluation of the crop water requirements $I = ET_c - P$ is demonstrated in Table 10.11. The length of season varies

Table 10.11 — Crop water requirements

Crop	Wet season					Dry season				
	ET_0	K_c	ET_c	P	I	ET_0	K_c	ET_c	P	I
Rice	829	1.20	995	207	1788	858	1.20	1030	0	2030
Tomato						655	0.90	590	0	590
Wheat						588	0.90	529	0	529
Cowpea						784	0.90	706	0	706
Sugarcane	889	1.05	933	215	718	1119	1.05	1175	0	1175
Sorghum	697	0.85	592	207	385	655	0.85	557	0	557
Maize	722	0.90	650	208	442	588	0.90	529	0	529
Cotton	889	0.90	800	215	585					

according to crop and forage crops have the same demand pattern as sugar cane. An extra 1000 mm for the rice crop is allowed for percolation and flooding the rice fields.

An irrigation efficiency of 60% was assumed and, for the total water requirements, the quantities represented by I were increased proportionally. For example, the sugarcane irrigation demand was assessed as 1197 mm and 1958 mm for the wet and dry seasons respectively.

Irrigation scheduling

There is a delicate balance between irrigation and crop yields. For some crops over watering can be as detrimental to yield as lack of water. Due attention should be paid to the water management to optimize crop production. The method and timing of application of the irrigating water influence the yield and quality of the crop. For relating application to crop demand, equations (10.3) and (10.4) may be used for irrigation scheduling in the developed areas using the data given in Table 10.12.

Table 10.12 — Senegal River Basin irrigation area data

Month	Mean T (°C)	RA (mm day^{-1})	TD (°C)	Actual P (mm)	Effective P (mm)
January	25	11.9	15	0	0
February	28	13.2	17	0	0
March	28	14.7	16	0	0
April	31	15.6	16	0	0
May	31	16.0	15	0	0
June	32	15.9	14	8	8
July	29	15.9	14	54	50
August	29	15.7	12	96	80
September	30	15.0	12	80	70
October	30	13.8	14	7	7
November	28	12.3	13	0	0
December	26	11.5	15	0	0

11

Water resource–river management

INTRODUCTION

The responsibilities of practising hydrologists have expanded greatly in recent years. No longer are they concerned solely with hydrometry and derivations of runoff, although these remain the foundations of the subject. Hydrologists are now called upon to study and advise on the management of water, a task which has much wider implications than those of the old water engineers in the organization of piped water supplies. The new connotation of water management also includes the attempt to control the resource when it is still in its natural state in a free-flowing river. In the furtherance of these expanded duties, a basis for study is the development of a river basin (Chapter 9). The UK government made this approach an operational reality in forming the regional water authorities from combinations of natural catchments. They were given statutory responsibility for the whole land phase of the hydrological cycle from the measurement of rainfall, evaporation and river discharges through all the uses and abuses of water down to the sea outfalls.

The technical discipline which makes possible attempts to manage the water resources of a river basin is that of systems analysis. The application of systems analysis implies that some decision-making process from optimization techniques is part of an overall modelling procedure. In developed countries where major water resource systems are already operational, large-scale systems analysis may not be appropriate (Rogers & Fiering, 1986). Modifications to an existing system required by changes in circumstances can be incorporated more simply. However, the sophisticated mathematical and dynamic techniques and other processes contained in systems analysis may be valuable tools in the evaluation of water resources related to requirements in developing countries. In such studies, while the analytical variables are mainly hydrological, man-made variables such as measures of costs and benefits can also be included in the models. The optimizing methods may be applied to the available water resources given a fixed storage system but they can also operate on the benefit-to-cost ratios in the selection of the 'best' scheme. In some of the most advanced systems analyses, consideration of intangible factors can be handled to give weight in the decision-making process.

The case studies assembled in this chapter are mainly concerned with the operation of existing systems. Data generated by several stochastic models were fed into a simulation model of Lake Nasser in order to devise a strategy for testing the operation of the lake under a wide range of flow conditions. The management of the River Torrens floodplain in Australia takes account of flood damage costs weighed against the costs of providing different standards of protection. The operation of the Kaudulla rice irrigation scheme is mainly concerned with ensuring that potential irrigation water is not wasted in the rainy seaon and that sufficient storage is provided to serve the fields in the dry season. The four examples from the UK cover a range of problems in the operation or management of water resource systems. Each has its own distinctive problem to solve and under differing circumstances a variety of techniques are employed. Hydroelectric power from Kielder is an additional asset from the original water resources scheme and only the Roadford Reservoir system is as yet incomplete. The operational model for this scheme has been constructed to assist in the management of the system to satisfy the requirements of a wide range of interested authorities.

In this dominating age of the computer, the hydrologist is able to apply ever more sophisticated techniques in helping to solve problems. Certainly, the design engineer and manager is now provided with sometimes an embarrassing number of options from which to choose the best economic or most expedient practical solution. However, good reliable hydrometric data remain the mainstay for providing the bases for sound engineering decisions. Hydrologists must be mindful of the physical limits of the natural world and be ever attentive to the dangers of accepting extrapolated results.

11.1 HYDROLOGICAL SIMULATION OF LAKE NASSER

LOCATION Lake Nasser is the reservoir contained by the Aswan High Dam on the River Nile in Egypt.

SOURCES Todini, E., & O'Connell, P. E.(eds.) (1979) *Hydrological Simulation of Lake Nasser.* IBM Italia Scientific Centers – Institute of Hydrology, UK.

Curry, K., & Bras, R. L. (1978) *Theory and applications of the multivariate Broken Line, disaggregation and monthly autoregressive streamflow generators to the Nile River*, Technology Adaptation Program. Massachusetts Institute of Technology, Cambridge, Massachusetts 416 pp.

PROBLEM The completion of the Aswan High Dam in 1964 represented man's greatest effort to control the water resources of the Nile which have been the basis of human development throughout history. To ensure the optimal operation of the dam to give the maximum benefits to the economy of Egypt, a Master Water Plan Project was set up in 1978 by the Ministry of Irrigation.

While the High Aswan Dam has created massive reservoir storage to protect Egypt from the worst ravages of floods and droughts, the reliability of Lake Nasser supply is still vulnerable to long-term fluctuations in Nile flows. A framework was required for assessing the impact of possible further fluctuations in Nile flows on alternative water resources exploitation plans and reservoir operating policies. Initial studies by the University of Cairo and Massachusetts Institute of Technology suggested a number of stochastic models suitable for generating synthetic inflows to Lake Nasser. Furher models were developed and implemented during this study, which has addressed itself primarily to evaluating the most suitable inflow generating model and to the development of a simulation model of the Lake Nasser system which could be used to explore the reliability of Lake Nasser under different demand scenarios and operating rules.

Lake Nasser

Lake Nasser is one of the largest man-made lakes in the world. It has a capacity of 162 000 Mm^3, of which about 130 000 Mm^3 is live storage and it extends upstream from the Aswan High Dam for 500 km with the upper 200 km in the Sudan. An agreement between Egypt and Sudan defines Egypt's share of Nile water as 55 500 Mm^3, $year^{-1}$. The operation of the reservoir has to provide annual flood control, meet irrigation requirements and generate hydropower. To prevent downstream degradation from large releases, the Toshka spillway has been constructed to divert excess flood water to a depression in the Western Desert. The major components of the River Nile system are shown in Fig.11.1.

Nile flows

The inflows to Lake Nasser are governed by the contributions from three main catchment areas.

(1) The equatorial lakes.
(2) The southern Sudan catchments.
(3) The Ethiopian plateau.

The flow is a minimum in May sustained primarily by areas (1) & (2) with only 20% on average coming from the Blue Nile which drains the Ethiopian plateau. The peak flow occurs in August–September and comes mainly from the Blue Nile and Atbara with only 10% on average coming from the White Nile. Of the several reservoirs constructed upstream, the Owen Falls Dam in 1948 is the largest but its benefit to Egypt will not be realized until the Jonglei Canal bypasses the swamps of the Sudd region.

Some research was carried out into the homogeneity of the flow record for the Nile at Aswan which showed an abrupt drop around 1902. A double-mass plot of the January and July gauge readings for Wadi Halfa and Aswan for the period 1890–1920 indicated that the pre-1902 published monthly Aswan discharges should be reduced by 8%. Further corrections to the published flows from 1938 onwards required an assessment of losses in Gebel Aulia Reservoir and the calculation of the regulation effects of this and the reservoirs on the Blue Nile and Atbara. The final corrected natural annual series for the period 1871-1975 at Aswan is shown in Fig.11.2. Some statistics of the natural flows are given in Table 11.1. Of special note is the range in the shape of the frequency distributions of the individual months with the wet months

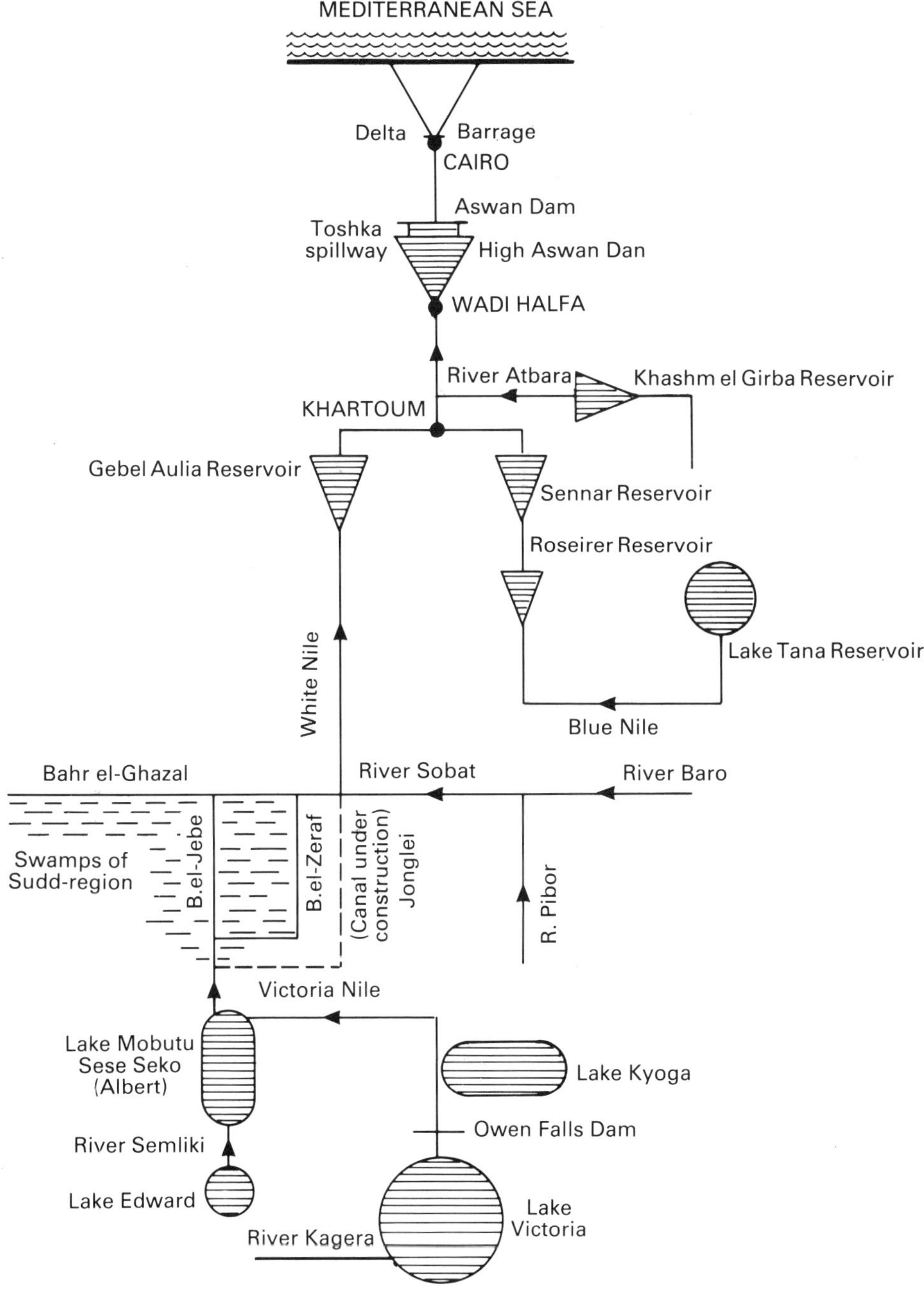

Fig. 11.1 — River Nile system.

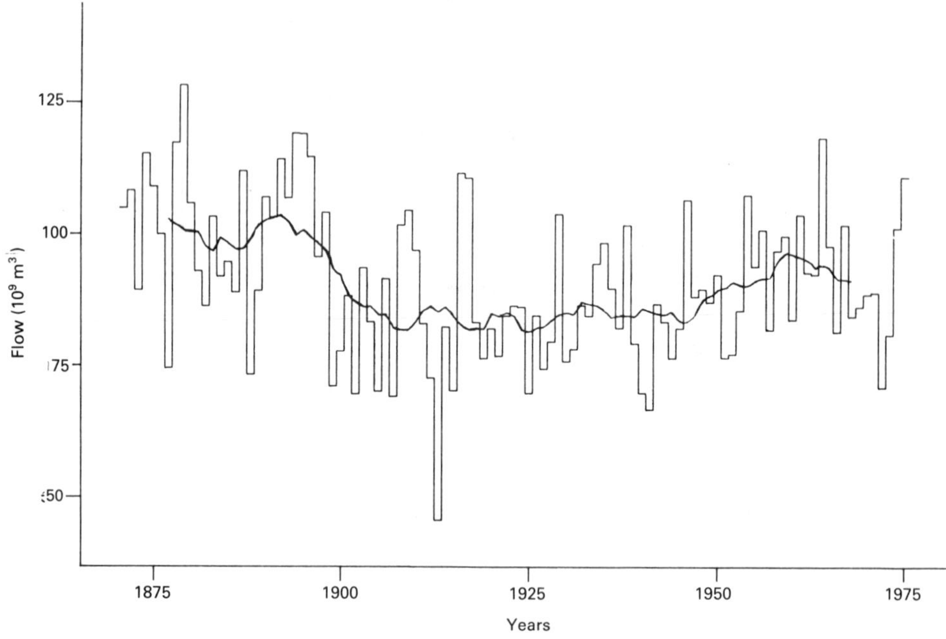

Fig. 11.2 — Annual natural flows at Aswan, 1871–1975, after applying an 8% correction to the pre-1902 flows, with a 13 year centred moving average.

being near normal and the more variable dry months of March to June being positively skewed. The lag-1 correlation coefficients between successive months showed that correlation is weakest between June and July and highest when the Nile flood is in recession. This has a bearing on model application and it was deemed appropriate to define the year from July round to June. For the River Nile, a further statistic of importance is the Hurst coefficient h which gives a measure of long-term persistence in the natural time series which needs to be represented in any stochastic model. After investigating a number of methods of estimation, a least-squares estimate of h=0.813 for the annual inflows was adopted.

Stochastic models

There were four basic models available and two variations, making a total of six models used in the evaluation study.

Model 1: Seasonal lag-1 autoregressive model (normal distribution)

This is a simple and widely known model for simulating synthetic stream flows (Thomas & Fiering, 1962). If $Y_{t,j}$ are the flows in year t and month j then the model is represented by

$$Y_{t,j} = \mu_j + b_j(Y_{t-1,\,j-1} - \mu_{j-1}) + \sigma_j\sqrt{(1\text{-}\rho_j^2)}\ \varepsilon_{t,j}$$

where μ_j and μ_{j-1} are the mean flows for months j and j-1, σ_j is the standard deviation of the flows for month j, ρ_j is the cross-correlation between flows in months j and j-1,

Table 11.1 — Statistics of the Aswan natural flows, 1871–1975

Month	Mean (Mm^3)	Standard deviation (Mm^3)	Coefficent of variation	Skewness	Minimum (Mm^3)	Maximum (Mm^3)
January	4 211	944	0.22	0.57	1 720	7 075
February	2 975	926	0.31	0.63	1 150	5 969
March	2 377	922	0.39	1.57	1 021	6 336
April	1 788	755	0.42	2.11	788	5 281
May	1 634	582	0.36	2.23	801	4 177
June	1 843	605	0.33	1.74	900	4 793
July	4 927	1 578	0.32	0.84	1 740	10 201
August	19 075	4 128	0.22	−0.11	6 500	28 060
September	22 841	4 192	0.18	−0.03	12 200	32 468
October	15 633	3 284	0.21	0.53	7 540	27 140
November	8 201	2 025	0.25	0.41	4 120	13 300
December	5 519	1 254	0.23	0.88	2 830	10 120
Year	91 032	14 587	0.16	0.28	45 630	128 194

b_j is the regression coefficient of month j on month j-1 ($=\rho_j\sigma_j/\sigma_{j-1}$) and $\varepsilon_{t,j}$ are independently and identically distributed random variables with zero mean, unit variance and an assumed normal distribution.

Model 2: Seasonal lag-1 autoregressive model (log-normal distribution)
Model 2 uses the structure of model 1 with logarithmically transformed flows.

Model 3: Broken-Line model with disaggregation
This more complex model produces a simulated annual series generated as the sum of a number of independent broken-line processes (see second reference cited above). The line lengths and variances of the separate component processes are chosen to obtain a process with a given lag-1 correlation coefficient and a specified Hurst coefficient. A disaggregation model is then applied to obtain monthly flows (Valencia & Schaake, 1972). Disadvantages of this model are the lack of correlation between the last and first months of sequential years and an assumption of normality for the monthly flows.

Model 4: ARMA(1, 1) model with disaggregation
This first-order autoregressive moving average model which also preserves a measure of long-term persistence is given by

$$X_t = \phi X_{t-1} + \varepsilon_t - \theta\varepsilon_{t-1}$$

where X_t and X_{t-1} are sequential annual values, ϕ and θ are model parameters and ε_t is an independently distributed random variable. The disaggregation model used in Model 3 was applied to produce monthly flows.

Model 5: Seasonal multivariate autoregressive model with correlated noise (low-lag fitting)
This model seeks to combine the advantages of models 1 and 2 with short-term properties and models 3 and 4 with long-term properties without requiring disaggregation. The model is fitted at the monthly level but incorporates a direct measure of long-term dependence. The model may be written as

$$\mathbf{y}_t = \mathbf{A}\mathbf{y}_{t-1} + \boldsymbol{\eta}_t$$
$$\boldsymbol{\eta}_t = \mathbf{B}\mathbf{n}_{t-1} + \mathbf{C}\boldsymbol{\varepsilon}_t$$

with $\mathbf{y}_t$ an annual vector of 12 monthly zero mean flows and where $\boldsymbol{A}$, $\boldsymbol{B}$ and $\boldsymbol{C}$ are 12×12 matrices of parameters and $\boldsymbol{\varepsilon}_t$ is a 12×1 vector of independent random variables. The matrix $\boldsymbol{A}$ is estimated from the lag-k variance–covariance matrices of $\mathbf{y}_t$, k=0, 1, . . . in this model; a maximum lag of k=2 years is used. The model also preserves the monthly skewness by defining the appropriate skewness for $\boldsymbol{\varepsilon}_t$.

Model 6: Seasonal multivariate autoregressive model with correlated noise (high-lag fitting)
This is the same as model 5 but uses variance–covariance matrices up to a lag of k=20 years in fitting the model.

The parameters of all the models were obtained by fitting to the natural flows at

Aswan, July 1871–June 1976. Then each model was used to generate 1000 synthetic sequences of length 105 years. Except for those generated by model 3, each sequence was made conditional on the historic sequence using the last monthly flow value June 1976. A sample of the results for the driest and wettest months is given in Table 11.2.

Table 11.2 — Historic and model statistics

	Mean (M^3m)	Standard Deviation (Mm^3)	Skewness	Minimum (Mm^3)	Maximum (Mm^3)
May					
Historic	1 634	582	2.23	801	4 177
Model 1	1 647	566	0.09	265	3 090
Model 2	1 644	574	1.05	684	3 690
Model 3	1 637	561	0.09	286	3 104
Model 4	1 638	574	0.02	221	3 099
Model 5	1 623	590	1.19	522	4 027
Model 6	1 632	585	1.25	564	4 081
September					
Historic	22 841	4192	−0.03	12 200	32 468
Model 1	22 902	4031	−0.04	12 693	32 917
Model 2	22 878	4016	0.46	14 503	24 856
Model 3	22 886	3950	0.21	13 574	33 711
Model 4	22 895	4164	0.0	12 368	33 392
Model 5	22 794	4119	−0.02	12 065	33 343
Model 6	22 926	4450	−0.02	11 285	34 480

The model statistics are the means of the generated sequences statistics rounded to the nearest million cubic metre. All the models performed equally well in preserving means and standard deviations, and models 2, 5 and 6 performed best in reproducing the skewness, minimum and maximum flows. However, none of these statistics demonstrates the capability of the models to reproduce the long-term fluctuations in Nile flows so critical to the reliability of Lake Nasser supply. Rather than attempting to assess model performance through arbitrarily defined flow statistics, a model evaluation framework was created whereby the ability of each model to reproduce the observed historical behaviour in various reservoir performance measures could be quantified. This entailed developing a simulation model of Lake Nasser.

Simulation model

The Lake Nasser simulation model is a simple monthly water balance model:

$$\Delta V = I - O - E - S - \Delta B - T \tag{11.1}$$

where ΔV is the change in stored volume of water, I the inflow, O the released outflow from the Aswan High Dam, E the evaporation from the lake surface, S the regional throughflow, ΔB the change in bank storage and T the volume of water spilled to the Toshka depression. All the quantities are expressed in $10^9 m^3$ and the basic time unit is 1 month. The reservoir operating policy will specify upper and lower target levels for each month which put constraints on the model. The quantities E, S, ΔB and T are functions of water level and hence the water balance equation has to be solved by trial and error. A simple procedure is used which converges rapidly.

Calculation of I

The historic and synthetic naturalized inflows need to be modified to account for losses and changes in volumes in the reservoirs upstream and abstractions in the Sudan.

Reservoir inflow I is defined as

$$I = I_{nat} - \Delta\overline{V}_u - \overline{L}_{GA} - kA_{Sud} \quad (11.2)$$

where $\Delta\overline{V}_u$ is the lagged mean monthly change in upstream reservoir volumes, $\overline{L}_{GA}$ the mean monthly evaporation loss from Gebel Aulia, A_{Sud} the monthly Sudan abstractions and k the scaling factor equal to 1 under normal operating conditions or less than 1 when the reservoir drops below lower-rule curve.

Calculation of O

The released outflow depends on downstream water requirements within the limits of reservoir target levels. If the upper rule is violated, O is increased to a permitted maximum to prevent downstream degradation.

Calculation of E

Climatological data analyses demonstrated that there was little variation in the monthly evaporation pattern from year to year. The monthly evaporations were obtained from

$$E = 0.168(e_s - e_2)u_2 \quad (11.3)$$

where e_s is the saturated vapour pressure at surface water temperature, e_2 the vapour pressure at air temperature 2 m above surface and u_2 (m s^{-1}) the wind speed at 2 m above surface. Then the percentages of total annual evaporation for each month were obtained and the total annual evaporation was taken as a model parameter to be calibrated.

Calculation of S

The regional throughflow consisting of losses by recharge of the sandstone rocks below the reservoir is calculated from

$$S = \alpha(H - H_{0s}) \times 10^9 \text{ m}^3$$

where H is the water level and α and H_{0s} are parameters with values of 0.038 and 110 respectively, derived from associated groundwater studies carried out by the master water plan.

Calculation of ΔB

The change in bank storage differs in rising stages from falling stages. While Lake Nasser was filling, the following equation was used

$$\Delta B = a\,\Delta A(H - H_{0b})[1 + b\exp(-\tfrac{T}{c})] \tag{11.5}$$

where ΔA is the change in water surface area, H is the average water level over 1 month, H_{0b} is a fixed level, T months is the time from start of simulation period, and a, b, and c are parameters determined in calibration. As T becomes large the equation reduces to

$$\Delta B = a\,\Delta A(H - H_{0b}) \tag{11.6}$$

For the falling stage,

$$\Delta B = d\,\Delta A\,\Delta H \tag{11.7}$$

where ΔH is the change in water level and d a parameter to be determined.

Calculation of T

The simulation of diverted flows to the Toshka depression was calculated from

$$T = \beta(H - H_{0t})^{\gamma} \tag{11.8}$$

where H is the lake level, H_{0t} is the threshhold level, and β and γ are parameters characterizing the form of the diversion.

Calibration

The reservoir simulation model was calibrated from inflows for the period 1968–1977 at the upstream end of Lake Nasser and actual release data from the reservoir. Excluding T, the water balance equation was used to generate a series of monthly reservoir levels using calculated evaporation and initial estimates of the parameters a, b, c and d. Thus a series of residuals was generated:

$$\varepsilon_{i,H} = H_i - \hat{H}_i$$

where H_i and $\hat{H}_i$ are the observed and estimated levels. To ensure that the residuals summed to zero, a mean annual evaporation of 2.7 m was adopted. Using an optimization method, the bank storage parameters were obtained as $a = 0.060$, $b = 1.814$, $c = 46$, $d = -5.77$ and $H_{0b} = 126$ m. Further comparisons of observed and simulated volumes and levels produced test statistics of an acceptable level of accuracy.

Application of model evaluation framework

To apply the model evaluation framework, the following were required.

(1) Upper- and lower-rule curves defining an operating policy for the reservoir
(2) A monthly downstream demand scenario.
(3) Various parameter values within the simulation model.
(4) Reservoir reliability performance measures.

For the upper-rule curve, to ensure flood storage the reservoir has to be drawn down to 175 m AOD by 1 August each year. Target levels were specified for each month end to achieve this. To prevent downstream degradation, the maximum daily release is assumed to be 300 Mm^3. For the lower-rule curve, if the post-flood peak reservoir level is less than 168 m AOD, both Sudan abstractions and downstream releases are to be reduced according to a sliding scale. Appropriate monthly values for $\Delta\overline{V}_u$, $\overline{L}_{GA}$ and A_{Sud} in equation (11.2) were specified together with monthly values of downstream demand. For the Toshka spillway, the parameters of the stage–discharge relationship were taken as $H_{0t} = 178$ m, $\beta = 0.57$ and $\gamma = 1.6667$.

The following reservoir reliability measures (Klemes, 1973) were applied to Aswan release deficits, Aswan spills and Toshka spills.

(1) Occurrence-based measure of reliability

$$R_0 = \frac{T-r}{T} \quad 0 < R_0 < 1$$

where r is the number of failure years and T the number of years.

(2) Duration-based reliability

$$R_T = 1 - \Sigma \frac{\Delta T_i}{T} \quad 0 < R_T < 1$$

where ΔT_i is the duration of the ith of r failure periods.

(3) Average length of failure

$$\Delta \overline{T} = \Sigma \frac{\Delta T_i}{r} \quad \text{months.}$$

(4) Quantity-based measure of reliability

$$R_Q = 1 - \Sigma \frac{\Delta S_i}{D} \quad 0 < R_Q < 1$$

where D is the demand total and ΔS_i the shortfall during the ith failure period.

Since the planning horizon had been stipulated as 30 years, the model was run with three 30 year subsequences of the historic record and corresponding subsequences of the synthetic series for each model. The three sets of results for the historic record and the means of the synthetic flow model subsequences are given for the Aswan release deficits and the Aswan spills in Table 11.3. However, because of the large number of performance indices involved, a multivariate performance measure was computed for each synthetic sequence as

$$d_i = (\mathbf{g}_i - \hat{\boldsymbol{\mu}})^T \hat{\Sigma}^{-1} (\mathbf{g}_i - \hat{\boldsymbol{\mu}})$$

and for the historical sequence as

$$d_{hist} = (\mathbf{g}_{hist} - \hat{\boldsymbol{\mu}})^T \hat{\Sigma}^{-1} (\mathbf{g}_{hist} - \hat{\boldsymbol{\mu}})$$

where $\hat{\boldsymbol{\mu}}$ is the estimated mean vector and $\hat{\Sigma}$ is the estimated variance–covariance matrix of the performance measures $\mathbf{g}_i$. If for one model the proportion of d_i which are larger than d_{hist}, is small, say less than 0.1 or 0.05, then the performance of the model could be classified as poor, since d_{hist} would appear to come from a different population.

Table 11.3 — Sample reservoir simulation results

	Release deficits				Aswan spills			
	R_0	R_T	$\Delta\bar{T}$	R_Q	R_0	R_T	$\Delta\bar{T}$	R_Q
Historic 1	1.0	1.0	0	1.0	0.43	0.83	2.81	0.72
Historic 2	1.0	1.0	0	1.0	0.83	0.93	4.17	0.72
Historic 3	0.93	0.99	4	0.44	0.63	0.76	5.38	0.72
Model 1	0.97	1.0	0.78	0.77	0.54	0.81	3.70	0.74
Model 2	0.98	1.0	0.51	0.85	0.58	0.82	3.79	0.73
Model 3	0.95	0.99	0.64	0.84	0.56	0.80	3.74	0.75
Model 4	0.94	0.99	1.07	0.74	0.56	0.80	3.86	0.75
Model 5	0.92	0.99	1.21	0.68	0.63	0.83	3.83	0.75
Model 6	0.94	0.99	1.13	0.78	0.54	0.77	4.22	0.74

The results for the six models are summarized in Fig. 11.3 where standardized values of d_{hist} defined as

$$d^s_{hist} = \frac{d_{hist}}{\mu_d}$$

where μ_d is the mean of the distribution of d_i, are plotted for various combinations of performance measures for each model.

Conclusion

The results summarized in Fig. 11.3 indicate that models 4–6 perform better than models 1–3; these results were found to be heavily influenced by the ability of each model to preserve the long-term persistence. While model 6 appears best overall, it should be borne in mind that this model employs more parameters than does model 4 for example; hence the results need to be interpreted in this light. On balance, model 6 was adopted for further operational studies.

11.2 REGULATION OF THE RIVER SEVERN

LOCATION The River Severn, the longest river in the UK, rises in the Welsh Mountains, takes a circuitous easterly course and then flows south-westwards into the Bristol Channel.

SOURCES Goodhew, R.C. (1982) Performance of a major river regulation resource system under design conditions. *Exeter Symposium,* IAHS Publication No.135, International Association of Hydrological Sciences, pp. 337–355.

Severn–Trent Water Authority (1981) *Operating rules for regulation of the River Severn*, Internal Report, Directorate of Operations.

PROBLEM In the 1960s, the growing post-war demand for increased water supplies in the Midlands stimulated water engineers to reassess the potential resources of the River Severn. Lake Vyrnwy was a major source for Merseyside but the compensation water at least was available to the Severn. A scheme to provide a major storage for river regulation was implemented and Clywedog Reservoir designed on the basis of the 1949 drought with a capacity of 50 000 Ml (50 10^6 m^3) was brought into service in 1968. At a much later stage, in the 1980s the development of a major groundwater source in Shropshire added another storage to the system (Fig. 11.4).

The development of an operating strategy required assessment of the resources, freshwater and treated effluent and of the demands for domestic, agricultural and industrial supplies together with considerations of flood situations and environmental and amenity factors. The more economical use of the river as the water carrier to demand points instead of laying trunk mains from the storages means that transmission losses have to be moni-

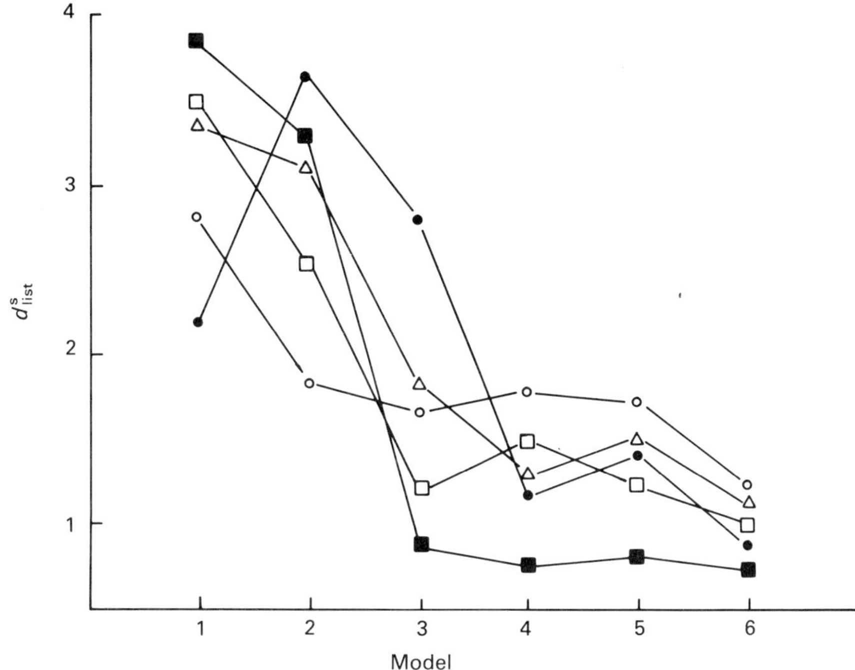

Fig. 11.3 — Standardized reservoir performance measures from the six models: ●, releases deficits; ○, Aswan spills; ■, Toshka spills; □, Aswan spills plus Toshka spills; △, releases deficits plus Aswan spill +, Toshka spills.

tored carefully. The evolution of the operating rules has taken place steadily over the years and continuous assessment of the procedures takes into account changes in the system.

Severn catchment

The river rises in the mountains of mid-Wales over 800 m AOD where the average annual rainfall exceeds 2000 mm. It has several mountainous tributaries before reaching the Montford gauging station near Shrewsbury. Downstream of Shrewsbury the river crosses the Shropshire Plain receiving steadily flowing streams from the Bunter sandstone area before reaching Bewdley gauging station at 17 m AOD (Fig. 11.4). The Bewdley gauging station has a continuous reliable flow record from 1921 and is the control point for the river regulation. The drainage area to this point is 4325 km^2 of which only 3% is reservoired.

Downstream of Bewdley, the Severn is joined by two major tributaries, on the right bank by the River Teme draining from the foothills of the Welsh Mountains and on the left bank by the River Avon draining from a lowland clay basin. The lowest gauging station is at Haw Bridge about 8 km upstream of Gloucester, the tidal limit.

Operating principles

The prime aim in river regulation is to arrange that there are sufficient water resources to meet the demands during periods of drought and to dispose of surpluses

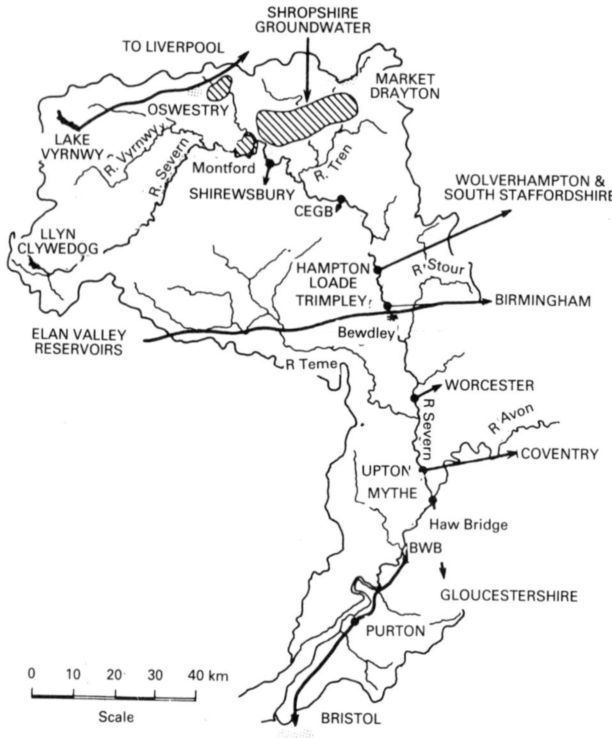

Fig. 11.4 — River Severn water resource system: ▧, proposed groundwater development areas; •⟶, major river intakes and aqueducts; BWB, British Waterways Board; CEGB Central Electricity Generating Board.

in times of plenty. The original assessment of the River Severn system was made in the 1960s when planning the storage required for Llyn Clywedog. At that time, a gross demand of 525 Ml day^{-1} reduced to a net rate of 336 Ml day^{-1} after allowing for effluent returns, pertained to the river upstream of the control point at Bewdley where the minimum flow had been 500 Ml day^{-1} in 1949. A minimum acceptable flow of 727 Ml day^{-1} at the control was proposed to serve the user needs down the whole length of the river provided that design abstractions were not exceeded and natural river flows including the downstream tributaries did not fall below the 1949 rates.

A more recent evaluation of the resources is given in Table 11.4 for key gauging points on the main river and for the main tributaries. The lower minimum flows, e.g. 280 Ml day^{-1} over a week at Bewdley were recorded in the more severe drought of 1975–1976 which was characterized by the unusual lack of winter rain to provide the annual replenishment of storages and aquifers. The operation of the storage resources in an average year would reflect the natural resources, being drawn down in winter to afford some flood storage, thereby mitigating floods downstream of the dams and making controlled releases to sustain abstractions and increased demands in the summer. In practice, extreme rainfall variations can occur at any season and thus operating rules are designed on a daily basis.

Table 11.4 — Water Resources, natural flows, 1965–1981

Station	Drainage area (km^2)	Average Annual rain (mm)	ADF (Ml day^{-1})		Minimum Weekly flow (Ml day^{-1})	
			Year	May–October	Average annual	Recorded
Clywedog Dam	49.0	1860	204	123	25	3
Vyrnwy Dam	94.3	1910	363	242	22	3
Montford	2025	1180	3700	2000	341	70
River Tern	852	729	636	438	219	92
Bewdley	4325	945	5340	2950	966	280
River Teme	1480	845	1600	790	249	66
River Avon	2210	683	1490	920	253	57
Haw Bridge	9895	776	9660	5400	1830	490

Particulars of the regulating storages are given in Table 11.5. The old Vyrnwy Reservoir built by Liverpool Corporation is still a major supplier to Merseyside and only 6000 Ml are available for regulation of the Severn at a net yield of 25 Ml day^{-1}. Llyn Clywedog is the most important component of the resources system but the newly developed groundwater resources in Shropshire are expected to provide a good back-up especially in dry summers, on average pumping 3 or 4 weeks 1 year in three.

A simplified list of abstractions and returning effluents is given in Table 11.6 (Fig.11.4). The major demands are for public water supply to the large conurbations of the Midlands and to Bristol. The supply to Bristol goes via the British Waterways Board canal from Gloucester. The principal industrial abstractor is the Central Electricity Generating Board taking cooling water for a power station but about 70% is returned to the river. A variable but increasing demand comes from spray irrigation by farmers and, while individual abstractors are licensed, the actual amounts of water taken are not readily available. For daily regulation of the river, estimates of such quantities and of transmission losses along the channel must be made. The crucial daily measurements of resources, rainfall, reservoir levels and river flows are recorded on site and relayed by telemetry to control headquarters. Pumping rates at abstraction points need to be reported regularly and as indicated in Table 11.6 when augmentation of the river is necessary, abstraction rates must be reduced accordingly.

Operating strategy

The aims of the river regulation may be considered separately in the two seasons: the refill season October–April when reservoir inflows normally exceed discharges and the drawdown season May–September when discharges exceed inflows.

Refill season objectives in priority order are as follows.

(1) To satisfy any river regulation demands.
(2) To provide adequate flood drawdown in the impounding reservoirs.
(3) To ensure that Llyn Clywedog is full by 1 May.
(4) To enable hydroelectric generation at Clywedog.
(5) To maintain reservoir levels for amenity.

Drawdown season objectives in priority order are as follows.

(1) To satisfy any river regulation demands.
(2) To identify when Llyn Clywedog alone cannot satisfy the demands.
(3) To select the required water quantities from Llyn Clywedog, Lake Vyrnwy and Shropshire groundwater.
(4) To enable hydroelectric generation.
(5) To manage reservoir levels for amenity.
(6) To maintain adequate flood drawdown.

A set of reservoir storage control curves for Llyn Clywedog is given in Fig. 11.5. In many years, Llyn Clywedog will be able to sustain the river abstractions without requiring help from the other sources. However, the curves indicate the reservoir levels when the other resources will be needed. Hydroelectric generation is possible

Table 11.5 — Storages

Reservoir	Commission date	Total capacity (Ml)	Regulation capacity (Ml)	Maximum release (Ml day^{-1})	Regulation yield (Ml day^{-1})
Llyn Clywedog	1968	49 900	49 900	500	500
Lake Vyrnwy	1890	59 700	6 000	405	25
Shropshire groundwater	1985				225

Table 11.6 — Abstractions (1981) and effluents (1976)

Intake	When not regulating		When regulating		Effluent (with respect to Bewdley) (Ml day^{-1})
	(Ml day^{-1})	Ml year^{-1})	For first 100 days (Ml)	when maximum Clywedog discharge (Ml day^{-1}	
Shrewsbury	38.6	14 104	3 860	38.6	31 (upstream)
Hampton Loade	220.0[a]	66 430	18 200	182.0	41 (downstream)
Trimpley	180.0[a]	22 000	6 000	60.0	0
Worcester	47.5	15 265	4 180	41.8	34 (downstream)
Upton	120.0	36 500	10 000	100.0	80 (downstream)
Mythe	125.0	39 820	10 900	109.0	44 (downstream)
British Waterways Board, Purton	185.0	62 000	17 000	170.0	0
Central Electricity Generating Board + spray irrigation	120.0			120.0	78 (upstream)

[a]Proportions may be varied but total not to be exceeded.

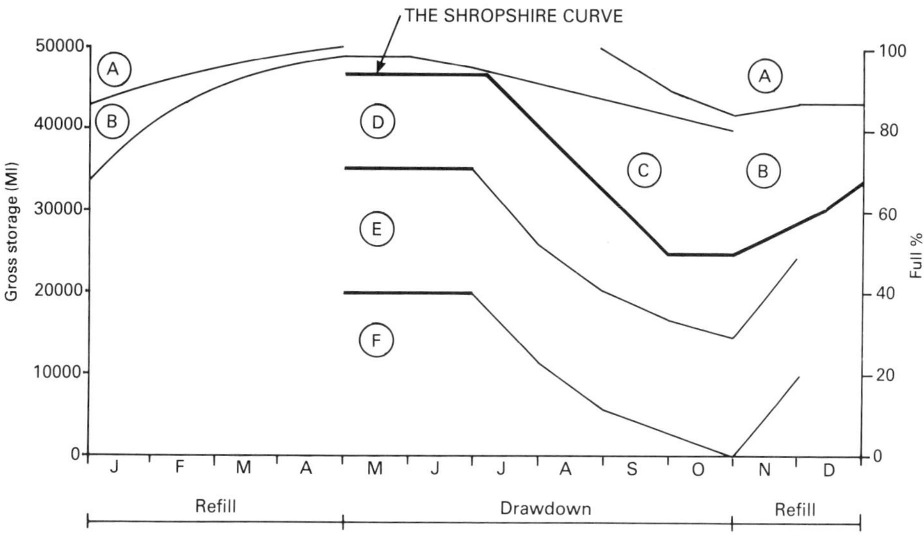

Fig. 11.5 — Llyn Clywedog control curves.

when Llyn Clywedog discharges exceed 130 Ml day^{-1} and therefore this rate of discharge for regulation will be provided first followed by discharges from Lake Vyrnwy.

The operational strategy with reference to Fig. 11.5 according to the current storage value is summarized as follows.

A	Discharge for flood retention provided that downstream river levels are low.
B	Discharge for hydroelectric power generation.
A, B	Discharge for river regulation when required; all water from Llyn Clywedog.
C	Discharge for river regulation when required; all water from Llyn Clywedog unless more than 500 Ml day^{-1}; then use Shropshire groundwater.
D, E, F	Discharge for river regulation when required with water from all storages.
E	Seek drought order.
F	Drought order powers in force.

If in the month of May the reservoir is less than 70% full, a serious situation arises whereby the water authority must obtain a statutory order with powers to restrict the use of water. The restrictions may continue even through the following refill season until supplies are assured.

There are very many details in the operation of the River Severn regulation scheme which cannot be covered in this account. The satisfying of any river regulation demands includes, in addition to the major water supply abstractors, the needs of farmers, fishermen and other sports activities. The maintenance of the river regime along the whole of its course has to be considered. The practical capacity of the channel downstream of the reservoir limits the maximum Clywedog discharge to 500 Ml day^{-1}. The regulation of the Severn was a successful operation in the 1975–1976 drought with the river flows at Bewdley being kept above 500 Ml day^{-1}; the estimated natural flow was below 300 Ml day^{-1} at the end of August 1976. The severe restrictions imposed on abstractors during August might not have been necessary if the Shropshire groundwater scheme had been in operation. This resource now is a safeguard for the future.

11.3 RIVER TORRENS FLOODPLAIN MANAGEMENT

LOCATION The River Torrens flows westwards from the Mount Lofty Range into the Gulf Saint Vincent via Adelaide, South Australia.

SOURCE Laing, I.E., Ockenden, A.P., & Botting, J.E. (1983) Floodplain management of the River Torrens. *Hydrology and Water Resources Symposium,* NCP 83/13. Institution of Engineers, Australia, pp. 306–310.

PROBLEM The city of Adelaide has seen extensive suburban development from the midtwentieth century. The urban extensions particularly to the west of the main centre have occurred on the flat land of the River Torrens floodplain (Fig. 11.6). The river channel is above the surrounding

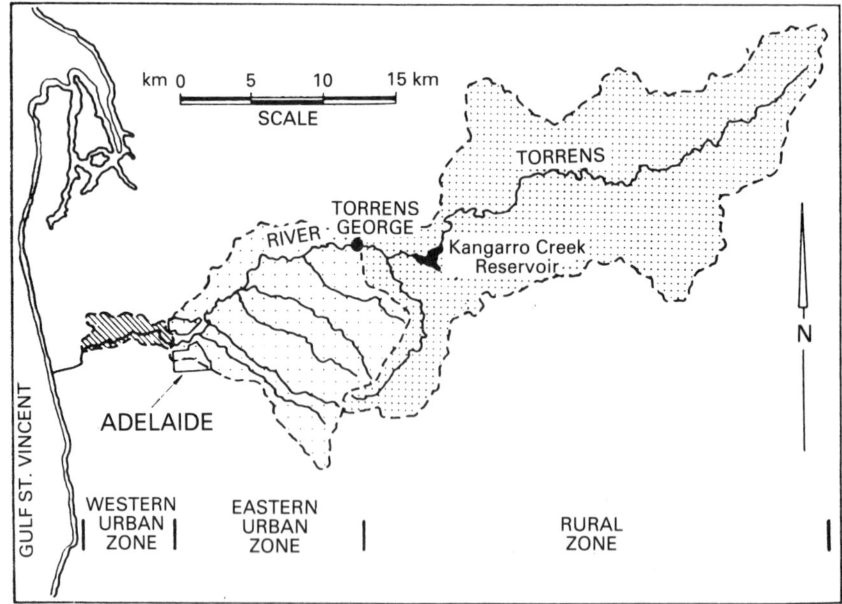

Fig. 11.6 — River Torrens catchment.

floodplain and, during floods, local drainage water and flood overspill cannot flow into the main stream. Before embarking upon a major flood mitigation scheme, it was decided to make an exhaustive cost–benefit analysis of the various options available for the effective management of the floodplain.

Floodplain damage

The extent of the floodplain was determined by the Snowy Mountains Engineering Corporation and a magnitude–frequency analysis provided information on the occurrences of flood flows. Most of the damage is caused in the western suburbs but an overall total of 6000 properties could be inundated by a 100 year return period flood (476 $m^3 s^{-1}$) and 13,500 properties by a 200 year flood (608 $m^3 s^{-1}$). Flood damage assessment was made by experienced valuers with series of sample surveys of residential, commercial and industrial properties. Estimates of expected flood damages were prepared for inundations to depths of 0.2, 1.0 and 2.0 m above floor level and a logarithmic stage–damage function evaluated. As an example, direct residential damage in Adelaide can rise (in Australian dollars ($A) per house (1973 values)) from 7000 at 0.2 m to 14 000 at 1 m and 16 000 at 2 m. Indirect damage due to communication and business disruption is very difficult to assess and it was considered appropriate to take the following proportions of the direct damages: residential, 15%; commercial, 55%; industrial, 70%. The total assessed potential average annual damage in the floodplain was $A3.3 million of which 80% was attributed to residential properties.

Management options

The structural measures that could be employed to control the flood flows included the following.

(1) Selective and/or extensive channel clearing.
(2) Channel realignment.
(3) Channel modification by lining.
(4) Construction of levee banks.
(5) Flood mitigation dams upstream in the Torrens Gorge.
(6) Modifications to Kangaroo Creek Dam.
(7) Modification of the catchment to change the runoff pattern.

Combinations of the structural measures were also considered. Non-structural measures of flood mitigation included land use control and acquisition of flood-prone land, flood warning, flood proofing of properties and flood insurance.

The hydrological, hydraulic and economic aspects of each option were studied and account was taken of ecological and social impacts. A comparison of all the options showed that most of the non-structural measures were inoperable owing to the extensive and varied floodplain development and the short warning time available in the relatively short steep catchment (rural catchment area, 350 km^2).

From economic considerations, the most feasible structural option in the rural zone of the catchment was the modification of Kangaroo Creek Dam, a water supply reservoir, to provide increased storage by raising the dam and/or lowering full supply water level and constructing additional outlet works. By analysing the effects of such modifications, the reduction in the peak flows of a range of floods was related to costs. For example, the 608 $m^3 s^{-1}$ of the 200 year flood could be reduced to 350 $m^3 s^{-1}$ at a cost of $A 2.66 million and the 476 $m^3 s^{-1}$ of the 100 year flood could be reduced to 250 $m^3 s^{-1}$ for an outlay of $A 2.48 million.

In the urban zones, structural options had to be compatible with a scheme to create a park along the urban reaches of the river and, hence, extensive channel clearing and lining and the construction a high levees could not be considered further. The options were confined to low levees, minor channel clearing, some channel widening and bridge modifications. The least-cost relationships in the western urban areas showed a rise in channel accommodation of an extra 100 $m^3 s^{-1}$ above 300 $m^3 s^{-1}$ for a cost of $A 1.5 million but beyond a discharge of 410 $m^3 s^{-1}$ the cost increased prohibitively. In comparison, the eastern suburban flood mitigation would be a minor cost.($A 0.2 million)

An optimization analysis of the several mitigation schemes was made to choose the best option by maximizing the economic net benefits. Taking the catchment as a whole, the least-cost options for each of the flood return periods were obtained by relating the cost of reducing the peak flows to the cost of providing the channel capacity in the urban areas to contain the reduced peaks. The comparable gross benefits were evaluated from the flood damage according to peak flows and overspill related to the average annual potential damage.

The relationship between costs and benefits for optimal flood mitigation is shown in Fig. 11.7. The net benefits are a maximum when the flood protection works are provided for a 200 a year return period flood. The economically best scheme consisted of the following

(1) Raising Kangaroo Creek Dam, lowering the full supply level and altering the spillway design.
(2) Increasing the capacity of the eastern reach by minor channel clearing, levee construction and minor channel enlargement.
(3) Modifying some bridges in the western reach and channel enlargement where necessary to increase the capacity to 410 $m^3 s^{-1}$.

The cost of this scheme amounted to $A 4.4 million (1980) and the worth of the net benefit was $A 14.9 million, giving a cost–to–benefit ratio of 1:3.4.

Conclusion

Protection against the 200 year return period flood represents a high standard but the study has justified its adoption. Indications of this result were seen when comparing the costs of modifying Kangaroo Creek Dam for the 100 year and 200 year floods. Absorbing the increased flood volume in an existing reservoir is very much cheaper than providing extra capacity in a restricted channel through an urban area. For high-value residential and industrial urban areas liable to river flooding, detailed economic investigations are recommended prior to designing mitigation schemes.

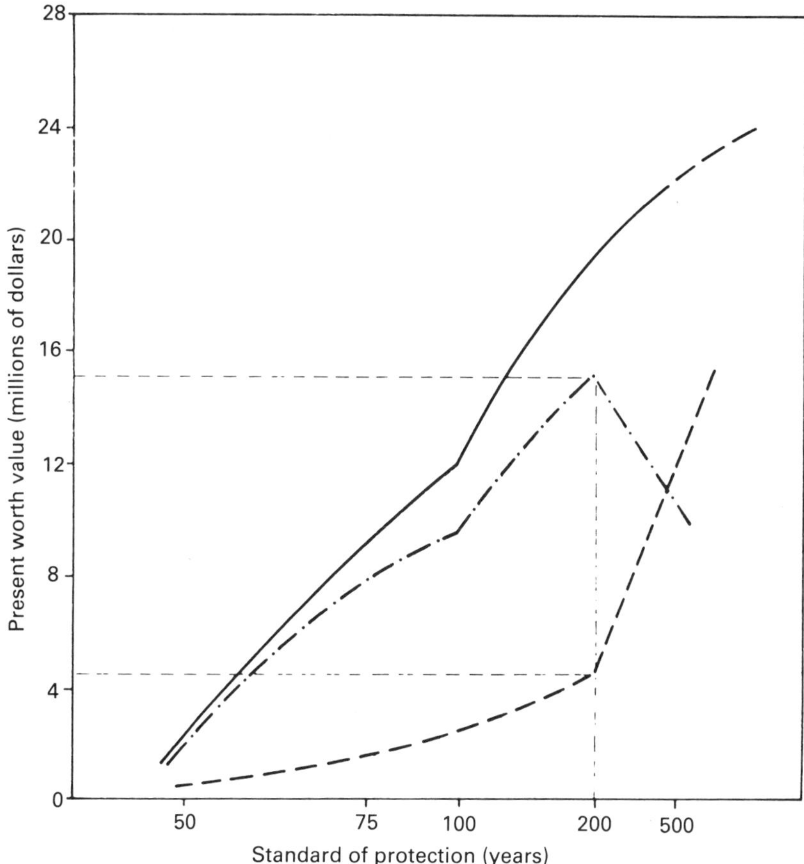

Fig. 11.7 — Benefit–cost functions for optimal mitigation: -··-, present worth net benefits; ——, worth gross benefits; - - -, present worth costs.

11.4 OPERATION OF THE KAUDULLA RICE IRRIGATION SCHEME

LOCATION Kaudulla is in east central Sri Lanka.

SOURCES Rees, D.H., & Hamlin, M.J. (1985) Optimal operation of a small-holder rice irrigation scheme in the humid tropics. *Scientific basis for water resources management*, IAHS Publication 153. International Association of Hydrological Sciences, pp. 295–310.

Rees, D.H., & Woodhead, T. (1986) The simulation of small-holder rice irrigation schemes in south east Asia. In: L.V. Tavares & J.E. da Silva (eds), *Systems analysis applied to water and related land resources,* IFAC Proceedings Series 1986/4. Pergamon, pp. 163–170.

PROBLEM The optimal management of water resources for irrigation in developing tropical countries has recently assumed importance to mitigate against the waste of water from expensive storage schemes. The UK Overseas Development Administration and the Natural Environment

Research Council have sponsored research into this problem and a study programme has been undertaken by the Department of Civil Engineering, The University of Birmingham, in collaboration with Hydraulics Research Ltd. The chosen area of study was the existing rice growing area of Kaudulla in the dry zone of Sri Lanka (Fig. 11.8). During the wet season, much

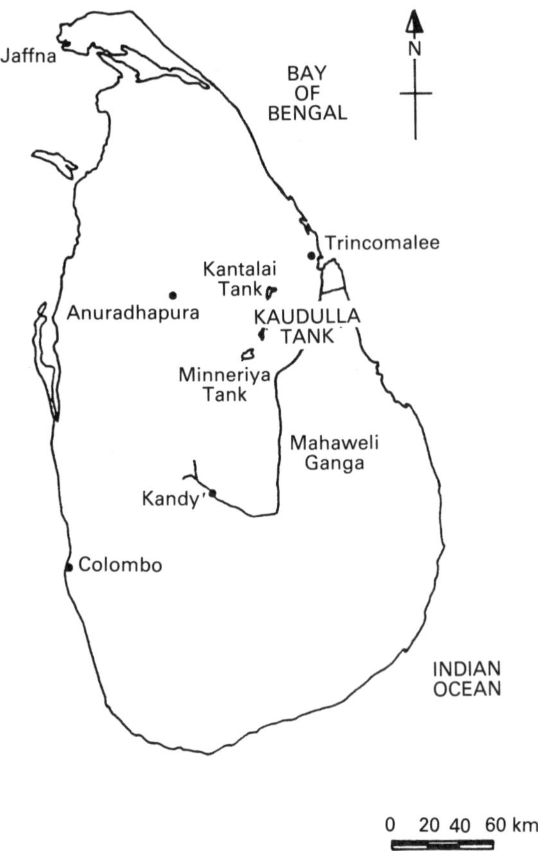

Fig. 11.8 — Location of Kaudulla scheme.

potential irrigation water could be wasted and yet the existing storage was insufficient to irrigate the dry-season crop. Extra supplies could be made available from the major Mahaweli interbasin transfer scheme. The main aim of the investigations was the development of a detailed simulation model to provide estimation of the water utilization in the whole system ranging from the catchment resources to the distribution of the crop irrigation needs and thereby to assess import requirements from Mahaweli.

Kaudulla irrigation system

A plan of the system is given in Fig. 11.9. The storage reservoir, Kaudulla Tank,

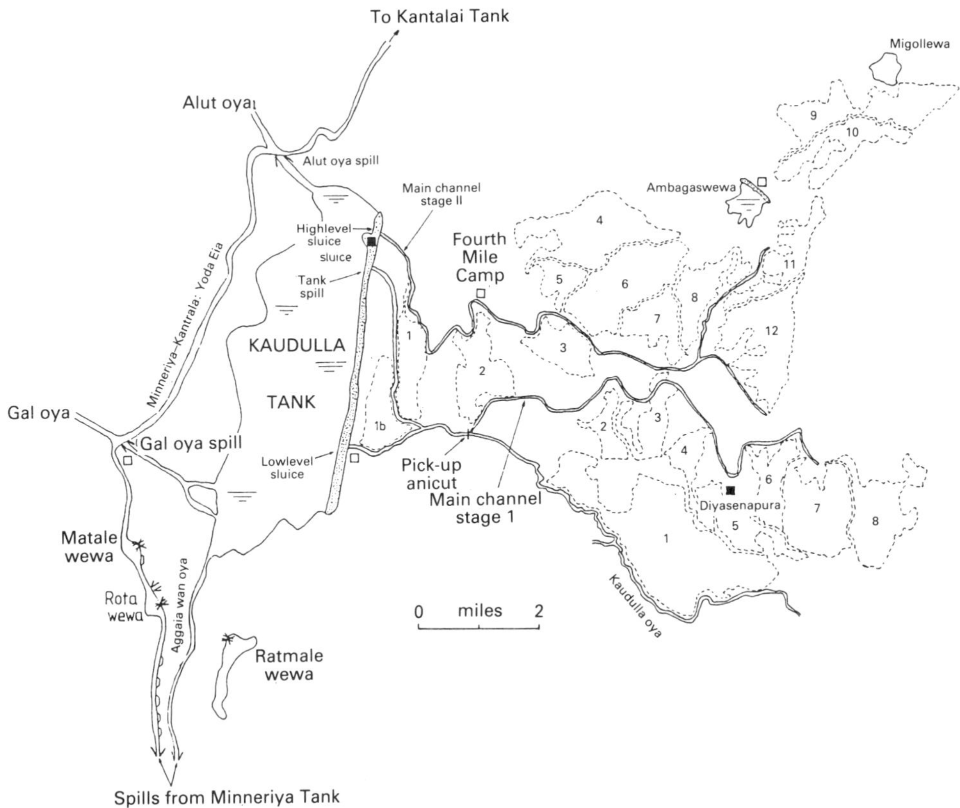

Fig. 11.9 — Kaudulla irrigation scheme: ----, irrigated paddy tracts; ★, tank sluices; spills with radial gates; causeway spill; □, daily rain gauges; ■, meterological stations.

occupies a shallow depression and its capacity of 128 Mm^3 is enhanced by a low earthen bund 9 km long. The top water level surface area is approximately 25 km^2 and the depth of live storage is less than 10 m. Two natural streams flowing to the tank from an area of 272 km^2 are intercepted by radial gates on a major canal, the Minneriya–Kantalai Yoda Ela which serves two neighbouring tanks. Spillage from the two streams and releases from other control points along the Minneriya–Kantalai Yoda Ela Canal drain to the Kaudulla Tank, thereby providing a complicated and uncertain input to the storage. Supplies from the Mahaweli system come from the Minneriya Tank via the Aggala Wan Oya.

Downstream, the 4250 ha of irrigation paddy is divided into two stages by firstly a low-level tank outlet 9.14 m below top supply level and secondly a high-level outlet at 7.62 m. The tank spillway leads down to a natural stream, the Kaudulla Oya, from which water is diverted into the first stage. Each stage has a main channel and a network of branch, distributory (secondary) and field (tertiary) channels in decreasing size, serving tracts of paddy fields 80–600 ha which are defined by natural streams acting as drainage channels.

Water resources

In the monsoonal climate of Sri Lanka, the winter northeast monsoon is the dominant wet season in this east central region. From the 32 year rainfall record at Minneriya, the neighbouring tank to the south, average rainfall in the Maha main growing season October–March is 1117 mm while the Yala dry season April–September only receives 336 mm. Four to five years of good-quality data in the Kaudulla area were available for rainfall, evaporation and irrigation releases.

The first analytical phase in producing a water resources model involved a study of the water balance of the Kaudulla Tank which was represented by the following relationship:

$$S_t = S_{t-1}+P_t-\mathrm{EO}_t+Q_t+\mathrm{MK}_t+\mathrm{AG}_t-\mathrm{SP}_t-I_t-G_t \qquad (11.9)$$

where S is the storage, P the rainfall on the tank surface (obtained using the area of S_{t-1}), EO the evaporation from the tank surface (also obtained using the area of S_{t-1}), Q the natural inflow, MK the diversions from Minneriya–Kantalai Yoda Ela, AG the direct transfers from Minneriya, SP the spillage when S_t exceeds 120 Mm^3, I the irrigation releases and G the groundwater seepage. All the variables are expressed in cubic metres per day with t in days. Of the nine components of this water balance model, only three were immeasurable: G the groundwater seepage from the bed of the reservoir, MK diversions from the Minneriya–Kantalai Yoda Ela canal and Q the natural river flows. For G, a satisfactory simple relationship relating the seepage to surface area A of the reservoir was found from the water balance using data when no inputs or outputs were discernible so that seepage was the only unknown term:

$$G\ (\mathrm{m}^3\ \mathrm{day}^{-1}) = \mathrm{b}A\ (\mathrm{m}^2) \qquad (11.10)$$

with the coefficient $b = 0.0045$ m day^{-1}. To find the combined MK and Q inputs, a deterministic lumped-parameter rainfall-runoff model normally calibrated on monthly data was modified to accept daily input data and to produce daily flows. Following repeated trials with manual calibration, a best fit of the model was finally obtained using an automatic optimization procedure. The model was found to give satisfactory results when generated tank storages using equation (11.9) and inserting seepage using equation (11.10) were compared with observed tank storages.

Distribution system

A model to evaluate rice irrigation water requirements developed by Hydraulics Research Ltd was applied to the distribution system. The model calculated the crop water needs and over a detailed representation of the distribution channel network aggregated the total water requirements. There are two main phases in the crop production: the land preparation and the period of crop growth which are considered separately.

During land preparation, irrigation IR to bring water level to operating depth OD is

$$IR = PL + FL + EO - ER$$

During the growing season,

$$IR = FL + ET_c - ER$$

where

$$ER = \min(RF, HB - OD).$$

The operating rule for the actual field requirement FIR is given by

$$FIR = 0 \text{ if } PWD > OD$$
$$= IR \text{ if } PWD < OD$$

where PL is the land preparation requirement, FL the field losses to deep percolation, EO the open-water evaporation, ER the effective rainfall, RF the total rainfall, ET_c the crop evapotranspiration, HB the height of bund spillage and PWD the paddy water depth. The variable units are all in millimetre depth. In operating the model, the calculations start at the furthest downstream point on the system and conveyance losses CL in the channels are accounted for as the flows required for a week to satisfy FIR are summed over the area. The order of magnitude of some of the water balance components is demonstrated in Table 11.7. The variables are as defined previously and with NR the net (effective) rainfall and RL the reservoir releases.

System operation

In the current operational practice, the existing storage is unable to support cultivation of the whole design area in the dry season and in the wet season the same release practice with much greater rainfall results in water being wasted.

The simulation model combining water resources and distribution was run with 32 years of historic data and assuming that the rainfall was fully utilized and in the wet season that the whole design area was cropped; the extent of the possible dry-season cropping area was varied in increments of 10%. The dry-season area was determined by the storage at the start of the dry season and the amount of Mahaweli water available.

Two operating rules were investigated.

Rule A

No water was to be imported until the crop cannot be supported from the local sources. This would result in no wastage of imports and maximum efficiency but would require accurate foreknowledge of requirement conditions. Two fixed storage thresholds were defined, the upper threshold, spillage level at 120 Mm^3, representing 100% live storage and the lower threshold when storage only contains 1 week's supply at 25.6 Mm^3 representing 12.6% of live storage. The import was the quantity required to restore storage to the lower threshold and its travel time was assumed to be zero.

Table 11.7 — Seasonal totals of water balance components

	ET (Mm^3)	FL (Mm^3)	RF (Mm^3)	NR (Mm^3)	FIR (Mm^3)	CL (Mm^3)	RL (Mm^3)
Dry season							
Maximum	29.8	23.4	13.5	12.6	60.1	8.0	68.1
Minimum	29.7	23.6	0.3	0.3	48.4	7.2	55.6
Average	29.7	23.5	4.6	4.5	56.1	7.8	64.0
Wet season							
Maximum	20.6	25.7	75.3	43.7	35.6	7.2	42.8
Minimum	20.4	26.5	18.6	18.6	13.8	3.3	17.1
Average	20.5	26.3	38.6	31.7	23.4	5.1	28.5

Rule B

Supplementary supplies would be imported when the reservoir storage was reduced to a pre-set target volume varying during the year. At the start of each calendar month, water would be imported if necessary to bring up the actual storage to the pre-set target volume. Again in the simulation program a zero travel time for imports was assumed. The optimum target volumes are given in Table 11.8. The highest levels in storage are naturally at the beginning of the dry season with the lowest at the peak of the wet season.

Results

Each iteration of the optimization procedure consisted of a number of runs of the simulation model with a given reservoir control rule. The objective of the procedure was to maximize the dry season irrigated area for the minimum import from the Mahaweli system.

Each run of the model covered the 32 year data period (using average daily evaporation, actual daily rainfall and synthesized natural inflows) and took a particular dry-season irrigated area and initial reservoir storage. The dry-season irrigated area was varied from 10 to 100% of the design area (4239 ha) in increments of 10%. The initial reservoir storage was varied between 12.6 and 100% of the live storage. The initial storage was reset during the run on 1 May of each year to calculate the probability distribution of annual Mahaweli imports for a given reservoir storage, irrigated area and control rule.

The numerical results of these multiple experimental simulations under the two operating rules are shown diagrammatically in Fig. 11.10. The probabilities of requiring imported water from Mahaweli (0–1.0) together with the average annual imports (0—70 Mm^3) have been plotted between the two ranges of variables.

Under rule A, no imported supplies would be needed if the initial storage and the dry-season cropped area fall above the 0 lines. In contrast, some additional supplies would always be needed with the variables below the 1.0 probability line corresponding to at least 20 Mm^3 if only 20% of the area was planted in the dry season.

Under rule B, the probability of requiring supplementary supplies is greater but it does not increase uniformly. Rule B is less efficient at lower initial storages and the corresponding greater quantities are due to the imports being made to bring up storage to the target levels and before it has been reduced to a critical level.

Fig. 11.11 shows a sample of the results throughout a year with varying initial storages when both dry- and wet-season crop areas are 100% of the design value. For the large initial storage of 90%, rule B requires the import in September when the target value is reduced to 32.6% whereas rule A allows the drawdown to 12.6% later in November before requiring further supplies. As the initial storage is reduced, e.g. to 10%, the import requirements of the two rules are not so dissimilar. The actual operation of rule B has been used to decide on the dry-season area to be planted given a fixed import allowance from Mahaweli. If thc initial storage is 60%, it is seen from Fig. 11.10 that the average level of imports would be A_1I_1, A_2I_2 and A_3I_3 for 80, 60 and 40% dry-season crop area respectively. These amount to 16.8, 9.8 and 4.2 Mm^3; so, if the Mahaweli authority could only supply a maximum of 15 Mm^3 that year, it would be prudent to plant only 60% of the area in that dry season. The probabilities of occurrence of imports have been assembled for those three crop

Table 11.8 — Target volumes as percentages of live storage for rule B

Dry season						
Month	April	May	June	July	August	September
Target volume (%)	42.6	42.6	40.1	37.6	35.1	32.6
Wet season						
Month	October	November	December	January	February	March
Target volume (%)	22.6	12.6	12.6	17.6	22.6	32.6

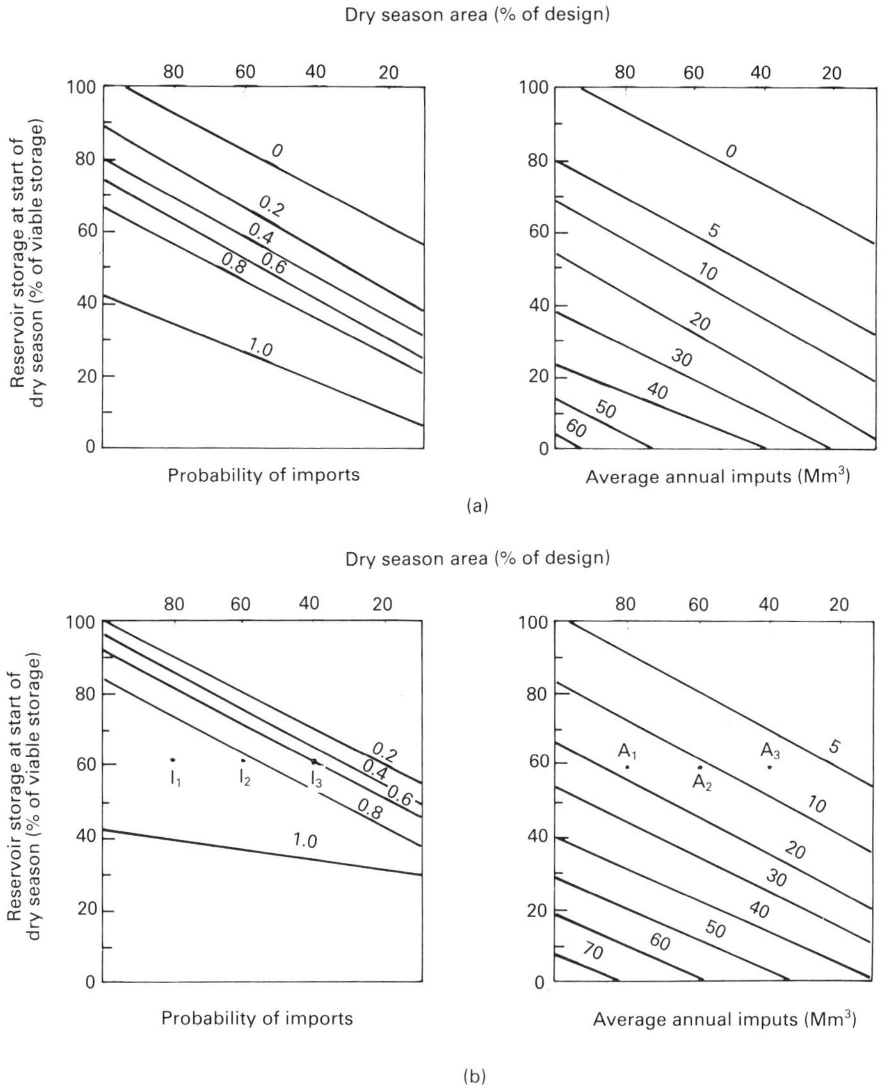

Fig. 11.10 — Probability of imports and average annual imports: (a) rule A; (b) rule B.

areas for the initial 60% storage and these are plotted in Fig. 11.12. From this, the chances of requiring an import of 15 Mm^3 are 66, 16 and 2% for the 80, 60 and 40% areas respectively. Thus given only 60% of initial storage at the beginning of the dry season it would be reasonably safe to plant 40% of the design area but 60% could be planted by incurring only a slightly greater risk.

Conclusions

The application of the simulation model to the operation of the Kaudulla rice irrigation scheme has shown how a limited period of good data can be used to

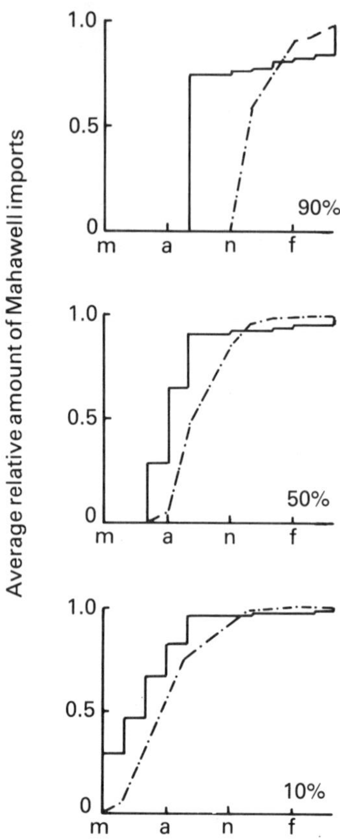

Fig. 11.11 — Temporal distribution of imports for selected percentages of initial storage on 1 May: —·—, rule A; —, rule B.

determine catchment hydrology and thereby to derive a seasonal operation rule to enable choice of dry-season cropping area knowing the initial reservoir storage and a probabilistic forecast of required supplementary water resources. Further work on rainfall forecasts and more detailed considerations of percolation losses in the rice fields have contributed refinements to the overall model.

11.5 DROUGHT MANAGEMENT IN THE THAMES BASIN

LOCATION The Thames Basin covers 13 100 km^2 of southern England.

SOURCES Moore, R.J., Jones, D.A., & Black, K.B. (1987) Risk assessment and drought management in the Thames Basin. *Seminar on Recent Developments and Perspectives in Systems Analysis in Water Resources Management, WARREDOC Meeting, Italy.*

Moore, R.J., Jones, D.A., & Black, K.B. (1987) A decision support

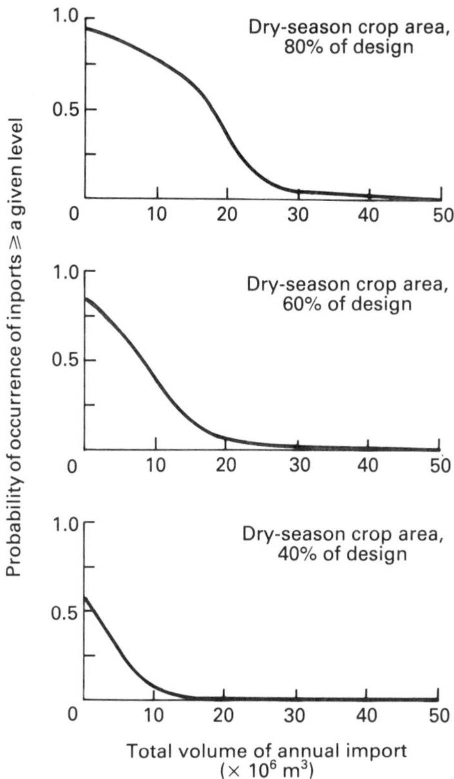

Fig. 11.12 — Probability of occurrence of imports with initial viable reservoir storage of 60% under rule B.

system for drought management in the Thames Basin. *National Hydrology Symposium, Hull.* British Hydrological Society, 34.

PROBLEM The occurrence of several seriously dry years, the increase in water demand resulting from rising standards of living and the reorganizations within the water industry have all contributed to the need for a new assessment of the management of the system of pumped storage reservoirs in the Thames Basin. These reservoirs provide about 58% of the water supply needs in the Thames Basin. They are replenished entirely by river abstractions apart from the net rainfall over the reservoired areas. Therefore, their ability to meet demands depends primarily on the variable river flows which are only available within the constraints of abstraction licences, reservoir and pump capacities, etc. The responsible authority, Thames Water, commissioned the Institute of Hydrology to develop and implement a computer-based procedure for forecasting the reliability of water supplies from these reservoirs during periods of drought as an aid to management.

Reservoir system

The location of the pumped storage reservoirs in Fig. 11.13 indicates that there are

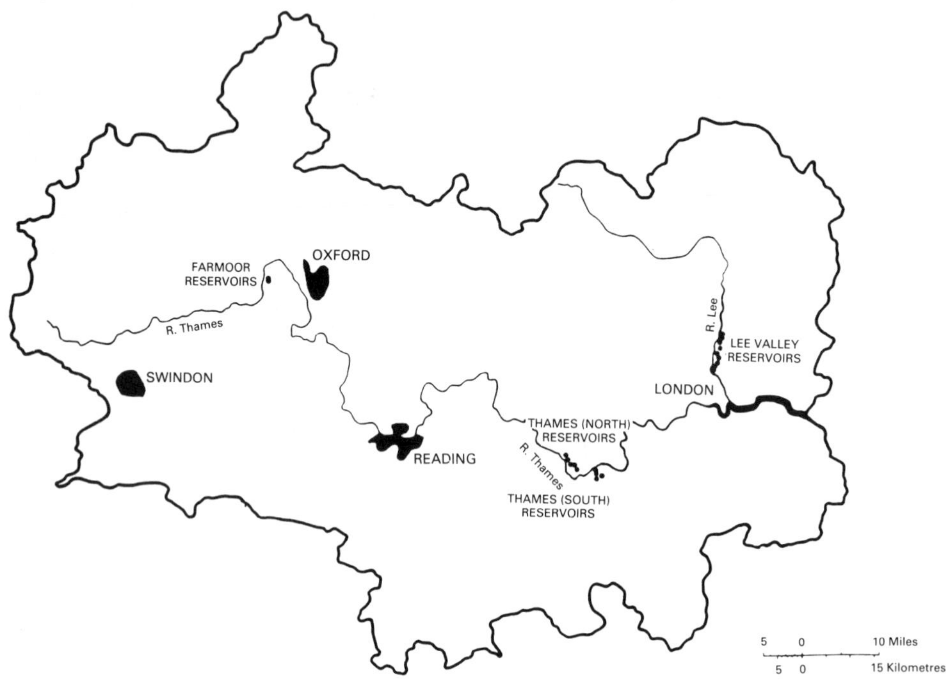

Fig. 11.13 — Location of pumped-storage reservoirs in the Thames Basin.

two distinctly separate systems; one is focussed on Farmoor Reservoir near Oxford serving parts of the Upper Thames valley and the second consists of the Lower Thames valley and Lee valley reservoirs serving the Greater London area. Simplified models of the two systems for operational purposes are shown in Fig. 11.14. Farmoor Reservoir has an operating capacity of 13 800 Ml supplying an average daily demand of 90 Ml. The complex dual London system has 18 reservoirs served by about 12 river intakes with a combined operating capacity of 206 400 Ml supplying an average daily demand of 1800 Ml. This system has been modelled by two reservoirs replenished at two notional abstraction points on the Thames and on the Lee.

For each system, there is a complex set of constraints and operating rules. The constraints include abstraction licences, residual flow requirements and pump and reservoir capacities. The operating rules take account of seasonal demands, demand restrictions during reservoir drawdown and the apportioning of water between the two London reservoir groups.

Risk assessment

The management of a water supply undertaking usually involves achieving compliance with a set of target levels of service during periods of drought. As the

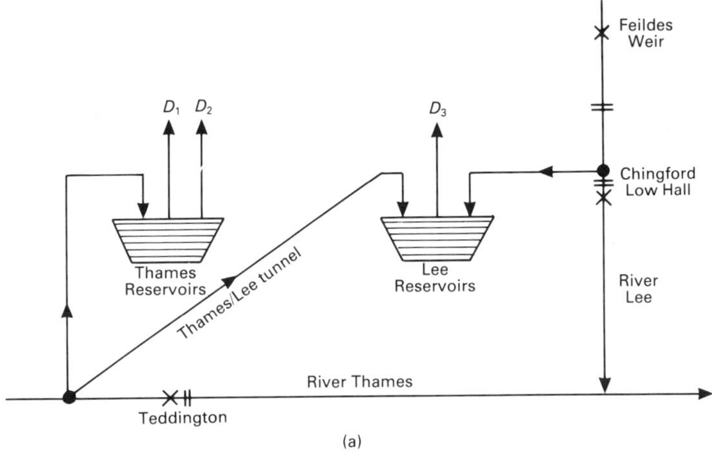

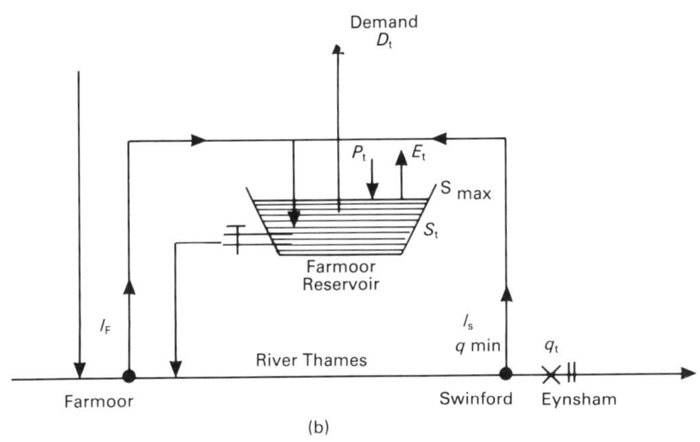

Fig. 11.14 — System diagrams for the Thames Basin reservoirs which consists of (a) the London reservoir system and (b) the Farmoor Reservoir system: ●, pumps; ||, minimum flow control; ×, gauging stations.

reservoir storage becomes depleted, a set of related demand restriction curves on a seasonal basis may be put into operation to maintain a desired level of service. For the London reservoir system, four demand restriction levels have been defined (Table 11.9). A method of risk assessment is required for the guidance of managers to indicate the likelihood that demand restrictions are imposed and their severity; seeking of a drought order to relax statutory limits on abstraction or of an alternative source of supply may be possible decision options.

The risk assessment procedure developed for the tactical management of the Thames Basin reservoirs is portrayed diagrammatically in Fig. 11.15. and its application may be considered in the following stages.

(1) The historical records of daily rainfall from 1890 for the Thames Basin are taken

Table 11.9 — Demand restriction levels and reliability standards

Level	Restriction	Maximum reduction in demand (%)	Maximum permitted risk of occurrence in any 1 year (%)
1	Hosepipe ban	7 (June–August)	17
2	Voluntary restriction; Reduction of pressure	17 (June–August)	5
3	Drought Act implemented; inessential uses banned; further pressure reductions	27 (June–August)	2
4	Major cuts in supply or use of stand-pipes	49 (September–October)	1

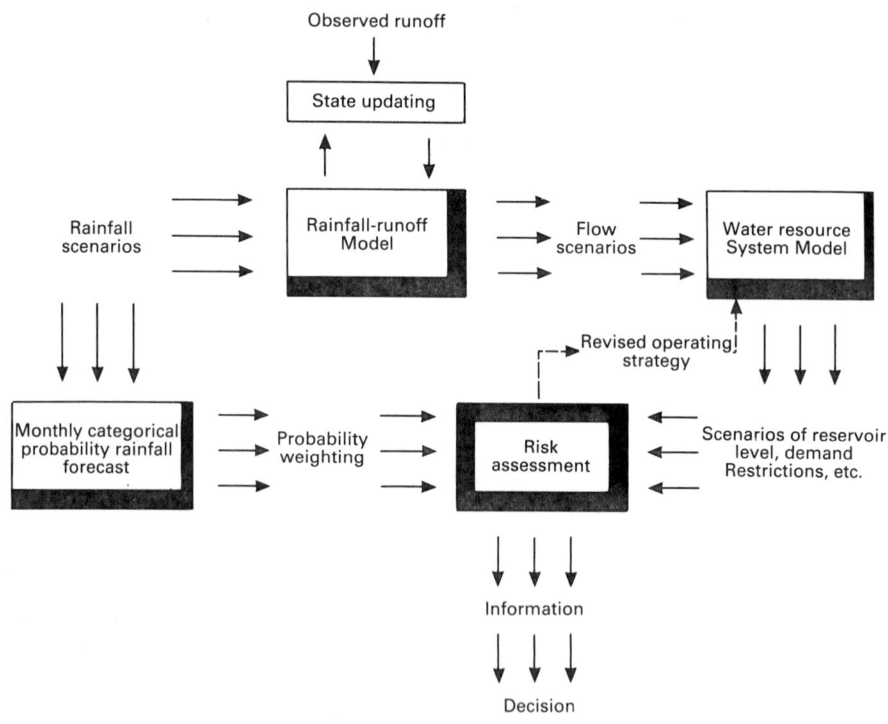

Fig. 11.15 — Risk assessment procedure for drought management.

to be representative of future rainfall sequences. Thus a set of rainfall sequences, corresponding to the forecast period, can be taken from each year of the historical record and used to represent equiprobable sequences of future rainfall.

(2) For a required forecast period of a month or 3 months ahead, rainfall-runoff models, updated to accord with current river flow records, produce a series of daily flow scenarios which are applied in turn to the water resource system models to give equiprobable sequences of daily reservoir levels, demand restrictions, etc. A statistical analysis of these data provides the probabilities of occurrence of any of these quantities falling below a given value.

(3) (a) Monthly rainfall forecasts provided by the UK Meteorological Office give the probability of rainfall amounts in each of three categories: below average; average; above average. The rainfall amounts defining the two category boundaries are determined from the historical record for the month concerned.

(b) Then, for each of the n years of record, the monthly rainfall category is identified and a weighting W_i is assigned where W_i is the forecast probability that rain falls in the category divided by the sum of the probabilities for every year of record. (A smoothing function between the categories has been developed to eradicate major differences across the boundaries).

(c) For a particular day of the forecast period, the n reservoir levels derived from each rainfall sequence are ranked, keeping the assigned weights attached.

(d) The weights are summed sequentially to obtain the probability p_r of exceedence of the rth ranked storage S_r.

$$p_r = \sum_{i=1}^{r} W_i$$

For $r = 1, 2, \ldots . n, p_r$ defines an empirical distribution function of storage for that particular day.

(e) The storage levels having specified risks of not being reached on that day can then be calculated from the empirical distribution function.

The sequence of steps (c)–(e) are repeated for each day of the forecast period to produce the final risk assessment information. Confidence limits for the risk estimates can also be calculated.

Risk assessments may also be made for the demand restrictions imposed during the depletion of the reservoir storage. A plot of the risk assessment of the four levels of demand restrictions made during the 1976 drought for the London reservoir system is seen in Fig. 11.16. This kind of information is clearly of utmost importance for tactical decision making in the management of the system.

Decision support system

The final product of the management study was a computer package for a PDP11/73 microcomputer to be operated as an interactive menu-and-form-driven decision support system. Colour graphical displays present the results to the user in an easily

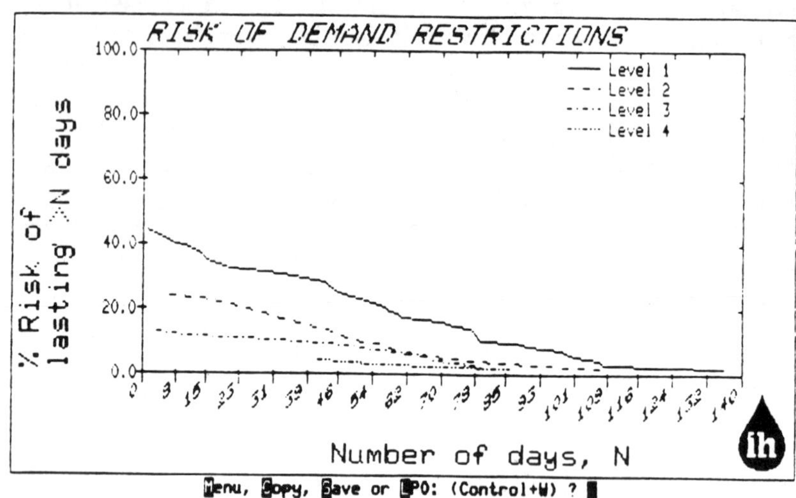

Fig. 11.16 — Risk of demand restrictions imposed at different levels of severity lasting for more than a given number of days during the 1976 drought for the London system.

assimilated form. The management of the large quantities of data involved is a special feature of the system. Provision is made for automatically updating the hydrometric data through a real-time communication link with a regional telemetry system.

In addition to providing guidance for week-to-week operations, risk assessments of water supply can be made for forecast periods up to a year ahead thereby giving indications of the longer-term viability of the water resource system. The decision support system serves as a valuable tool for management during periods of drought and is now operational within Thames Water.

11.6 KIELDER HYDROELECTRIC POWER GENERATION

LOCATION Kielder Water on the headwaters of the River North Tyne is in the Northumbrian Water authority area, UK.

SOURCE Johnson, P., & Sanderson, P.R. (1987) Control of Kielder hydroelectric power generation. *National Hydrology Symposium, Hull.* British Hydrological Society, 9.

PROBLEM Following the development of the Kielder water resources scheme between 1974 and 1981, the demand for water in northeast England declined owing to the recession in the area's formerly dominant heavy industry. It was decided to add to the water and amenity benefits from the reservoir by implementing the original scheme to generate hydroelectricity at the dam site. This was carried out between 1982 and 1984 by installing two turbines; a smaller turbine able to generate 0.5 MW from the continuous-compensation water releases and the larger turbine, 5.5 MW from controlled releases from the reservoir. The total power output was to be fed into the National Electricity Grid. This study by Northumbrian Water seeks to establish an optimum control procedure to maximize revenue from the hydroelectric power while continuing to fulfil the statutory demands on the system.

Kielder water system

A map of the water resource system is given in Fig. 11.17. The reservoir, impounding

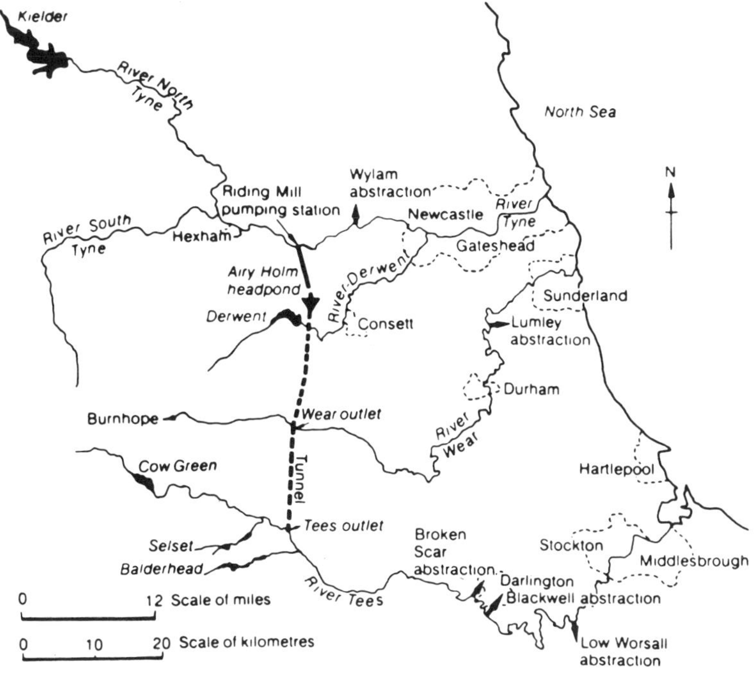

Fig. 11.17 — Kielder water scheme.

up to 200 Mm^3, has two main functions: to regulate the river flows in the Tyne from which large abstractions occur and to serve the water transfer system with up to 6 m^3 s^{-1} to supplement the Rivers Wear and Tees and the main from Derwent Reservoir.

Operational experience had demonstrated that the following criteria need consideration, in the following order of importance.

(1) Maintain compensation water into the North Tyne.
(2) Support minimum maintained flows in the Tyne, Wear and Tees prescribed by major abstraction licences.
(3) Support major abstractions to avoid violation of residual river flow conditions specified in the licences.
(4) Augment local resources temporarily limited by operational maintenance.

(5) Control Kielder levels with respect to dam maintenance and agreed low amenity levels.
(6) Support and encourage fish stocks.
(7) Assist in management of polluting spillages in the system.

With the addition of further controlled releases for the generation of hydroelectric power, which are required to vary diurnally, weekly and seasonally to match demand, specific constraints are needed to safeguard downstrean riparian and environmental interests. Of particular importance are the rates of change in the river flow. The maximum rise and fall in river level resulting from hydroelectric releases in dry-weather conditions is about 460 mm and to safeguard fish a stepping up of the releases over 2.5–3 h was found necessary; stepping down could be reduced to 1.5 h or less. An upstream constraint on amenity grounds had been agreed that the reservoir level should not be allowed to fall more than 7 m below the spillway crest for longer than 30 days in any 1 year.

Hydroelectric revenue

The revenue tariff paid by the Central Electricity Generating Board is renewed annually and has four varying rates.

The general tariff rate with a distinct diurnal pattern provides higher rates of return from 08.00 to 24.00 hours (3.2–3.8 p kW^{-1} h^{-1} in 1986–1987) with the lowest value about 1.4 p kW^{-1} h^{-1} from 01.00 to 06.00 hours. There are four different diurnal patterns according to day of the week and time of year.

A fuel cost adjustment factor is added to the general tariff rate to account for increases in coal prices.

A peak demand tariff of about 2.4 p kW^{-1} h^{-1} is added for hydroelectric power supplied during the 0.5 h before and the 1 h after the start of the peak energy demand period for the day except on weekdays during the summer months. The time is forecast each month by the Central Electricity Generating Board.

A basic capacity tariff is a special bonus for energy generated in the 3 months December, January and February from 08.00 to 24.00 hours each day, calculated from an effective average power output.

The gross revenue from the power generated each year is shared by the Central Electricity Generating Board and Northumbrian Water according to their relative capital investment in the hydroelectric scheme, with Northumbrian Water receiving about 40%.

Hydroelectric power release programme

From a number of operational considerations it was decided to plan the Kielder hydroelectric power releases on a weekly basis. Each week a 4 week programme is developed and in normal circumstances the first week is considered fixed and the remaining three are then forecast, to be modified according to changing weather conditions. The procedure to maximize revenue is carried out in two stages, considering first the diurnal profiles of releases and second the total weekly releases.

Diurnal Profile

The stepping up and down of the releases through the main turbine gives the time profile a trapezoidal shape. For a given weekly volume of release, a set of times for the profile exists for each day to give the maximum revenue from the specified tariffs. The determination of the times and the apportioning of the weekly volume to each hour of each day was made by the golden search method (Wilde, 1966).

Weekly Releases

To maximize revenue on a longer-term basis it was necessary to take into account the resources of the reservoir affected by the weather variabilities. A stochastic dynamic programming algorithm was developed to handle these uncertainties.

With the notation of Fig. 11.18 the power P generated in the week n to $n-1$ is

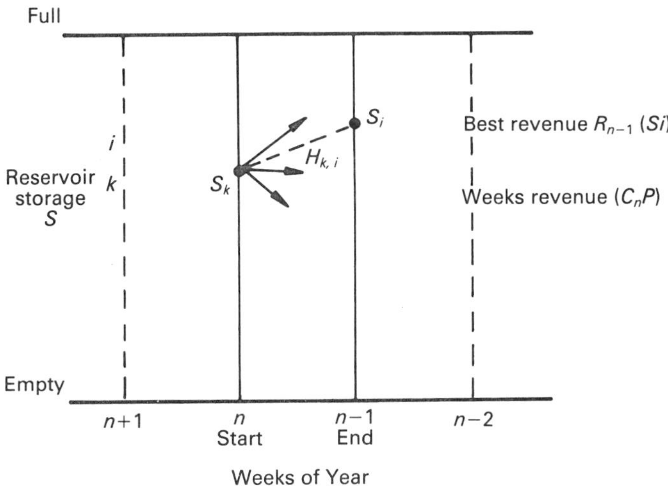

Fig. 11.18 — Reservoir level change in 1 week.

given by

$$P = \xi\gamma\overline{H}_{k,\,i}\,Q_n \quad \text{(kW)}$$

where γ is the specific weight of water, ξ is the overall power generation efficiency (0.84–0.87), $\overline{H}_{k,\,i}$ is the average head on the turbine as the reservoir level goes from k to i and Q_n is the chosen weekly volume of release at time n. The revenue R obtained in the week is

$$R = C_nP \quad (\pounds)$$

where C_n is the maximum tariff value of a kilowatt week of output. The chance of earning R is equal to the chance of a week's inflow I occurring to change the reservoir level from k to i since

$$I = S_i - S_k + Q_n \qquad 0 \leqslant S_i,\ S_k \leqslant \text{FULL},\ 0 \leqslant Q_n \leqslant Q_{\max}.$$

When the probability $p(I)$ of I is obtained from the river flow records, then the revenue contribution over many years from change in storage S_k to S_i with chosen Q_n is given by

$$p(I)C_nP$$

or

$$p(S_i - S_k + Q_n)\ C_nP$$

If in the long term the maximum expected revenue $R_{n-1}(S_1)$ starting with storage S_i is known, then the total expected revenue at $n-1$ by the previous week's decision of Q_n release to cause level transition k to i is given by

$$p(I)\ [C_nP + R_{n-1}\ (S_i)].$$

Taking a range of values of I leading to a range of reservoir levels, the total expected revenue for this 1 week with release decision Q_n is therefore the sum of all such expected contributions given by

$$\sum^{i} p(I)\ [C_nP + R_{n-1}(S_i)]$$

i.e.

$$\sum^{i} p(S_i - S_k + Q_n)\ [C_n\ \xi\gamma H_{k,i}\ Q_n + R_{n-1}(S_i)].$$

It is required to choose a Q_n which gives a maximum revenue $R_n(S_k)$. A computer program was run to search and examine for each starting level S_k giving the value of Q_n which would yield maximum return and satisfy the equation

$$R_n(S_k) = \begin{pmatrix} \text{Max} \\ Q_n \end{pmatrix} \sum^{i} p(S_i - S_k + Q_n)\ [C_nP + R_{n-1}(S_i)].$$

The $R_{n-1}(S_i)$ values become specified in the analysis. A time-dependent matrix of maximum future expected revenues for each week of the year and for each starting level of reservoir storage is produced. A corresponding decision matrix is evolved to give the Q_n to yield the maximum expected revenue while allowing for the uncertainty of inflows.

Model application

The algorithm was programmed for a mainframe computer and the reservoir storages and inflow and outflow volumes were entered in units of 0.4 Mm^3 to reduce data storage and to speed up computation. The probability of the weekly inflows was based initially on 19 years (since increased to 30 years) of flow records on the North

Tyne. The year was divided into 13 blocks of 4 weeks. For each week of a 4 week period over the 19 years, the flows were assumed to come from a common frequency population and thus 76 points were obtained to assess the 13 sets of probabilities.

For the turbine releases, 0.4 Mm^3 was approximately equivalent to 1 unit, the minimum prescribed compensation water discharge, and 24 units was the maximum weekly release. For each of these possible releases, golden search maximum revenues were pre-calculated for each of the distinct tariff periods.

Operational management uses the decision matrix of Q_n, the weekly turbine discharges for each calendar month and for each 0.2 m storage level in the reservoir, to choose 4 weeks of scheduled reservoir releases. Each week is subjected to the golden search program to give the times and durations for discharges each day. A comprehensive schedule of power generation is thus provided and published each week by Northumbrian Water.

With the extension of available river flow records, the flow probabilities have been updated and the model has proved invaluable in solving operational problems resulting from reservoir level constraints.

11.7 OPERATION OF ROADFORD RESERVOIR

LOCATION Roadford Reservoir is a new water resource in the Tamar valley of the South West Water authority area, England.

SOURCE Whiter, N.E., (1987) A hydrological and operational model of the Roadford Reservoir system. *National Hydrology Symposium, Hull.* British Hydrological Society, 10.

PROBLEM The growing demands for water in the South West Water area, particularly the seasonal needs of the tourist industry, are to be met by the construction of a third large regulating storage reservoir. The first two are Wimbleball on Exmoor which regulates the River Exe, thus serving the eastern parts of the area, and Colliford on Bodmin Moor which regulates the River Fowey and serves Cornwall. After a long period of planning and public inquiries, the Roadford Reservoir scheme was approved in 1986 and completion is expected in 1990. The reservoir (capacity, 36.9 Mm^3) is to be on the River Wolf, a headwater of the River Tamar. In its central location shown in Fig. 11.19, it is to serve north Devon and the Plymouth area and, by relieving the demands on Burrator Reservoir, will allow increased supplies to Torbay.

A comprehensive hydrological model of the Roadford Reservoir system has been developed to optimize the allocation of the water resources and to answer resultant environmental questions. It will also provide information for the water quality chemists and the fisheries biologists. At this stage in the progression of the scheme, the operation of the model is not to be carried out in a real-time sense.

The model

The construction of the numerical model to simulate the working of the Roadford Reservoir system used a basic pragmatic approach with many trials and consider-

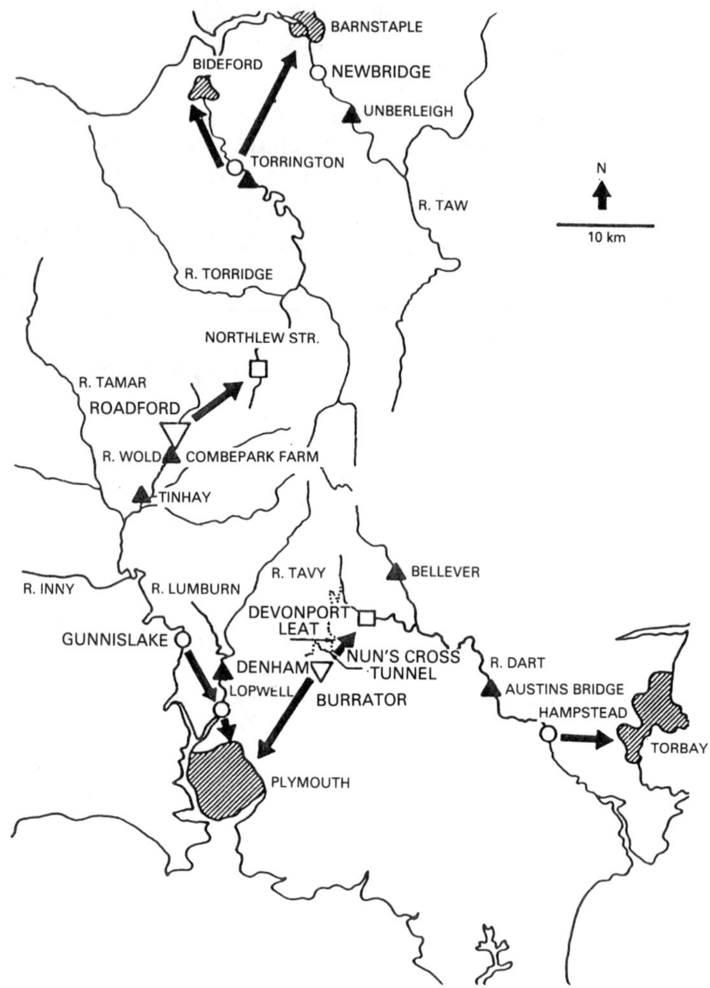

Fig. 11.19 — Roadford Reservoir system: □, discharge points; ○, intakes; ▲, gauging stations.

ations of varying time increments and boundary conditions. The purpose of the model was to find optimal methods of operating the system under historic hydrological conditions to serve future demands. A schematic plan of the hydrological and operational model for evaluating Roadford (HOMER) model is shown in Fig. 11.20. It incorporates the principal reservoir sources with their interconnections via rivers and pipelines and the main demand centres with their treatment works and service reservoirs.

Input data

River flow

The basic data measuring the water availability are the daily mean flows from nine river gauging stations itemized in Table 11.10. In order to have comparable records,

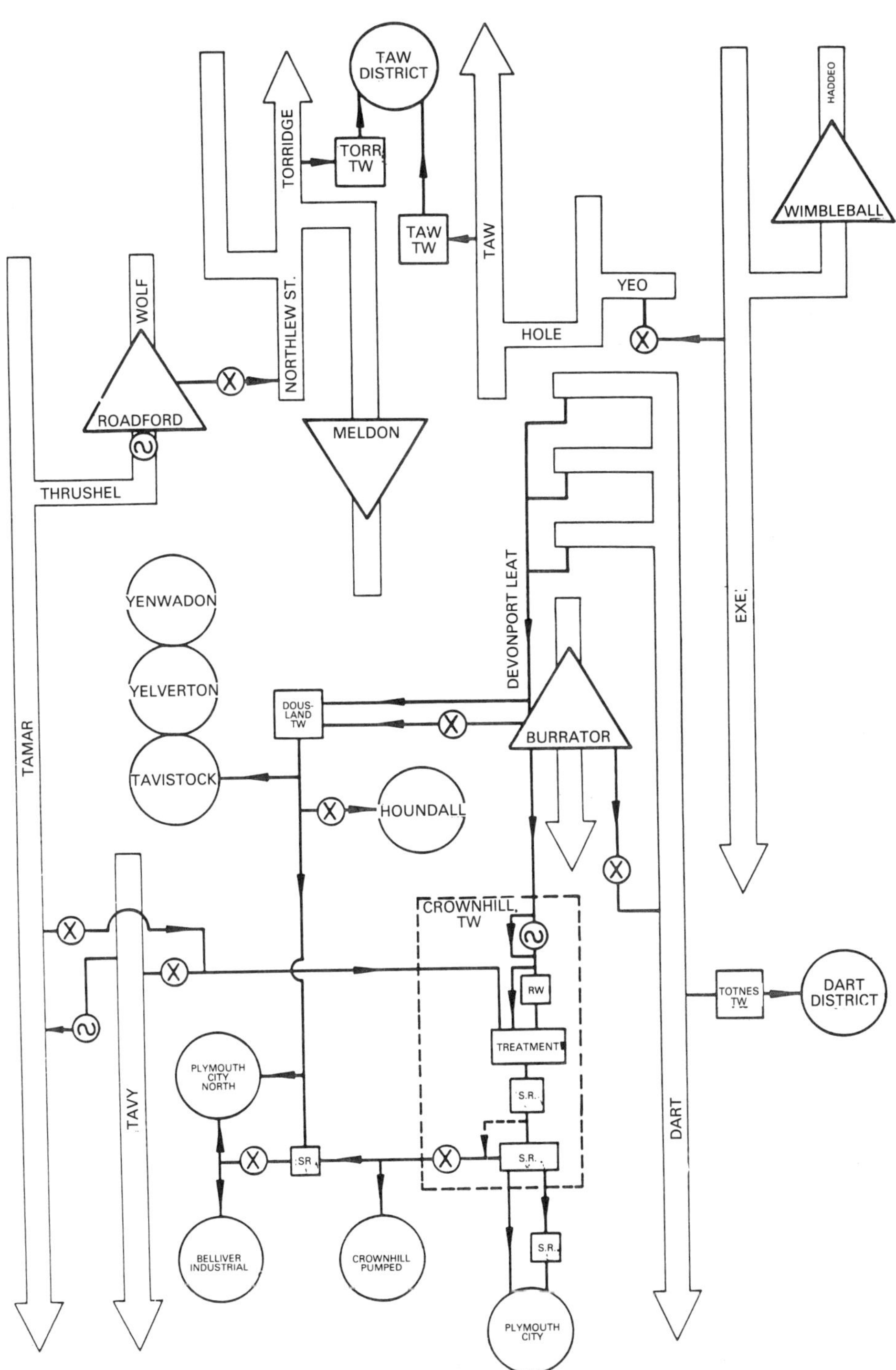

Fig. 11.20 — HOMER: ⓝ, hydrpower; ⊗, pumping; S.R., service reservoirs; TW, treatment works.

Table 11.10 — River gauging stations

Station	River	Start of Record	Catchment area (km^2)	ADF ($m^3 s^{-1}$)
Gunnislake	Tamar	1957	917	21.16
Tinhay	Thrushel	1970	113	2.41
Combepark Farm	Wolf	1978	31	0.71
Denham	Tavy	1978	197	6.46
Torrington	Torridge	1963	663	14.88
Bellever	East Dart	1965	22	1.20
Austins Bridge	Dart	1959	248	10.10
Nuns Cross	Devonport Leat	1974	25	—
Umberleigh	Taw	1959	826	18.33

30 year series from 1957 were compiled by generating data to fill in the missing years. Various methods of data generation were tried but the following simple equation was adopted:

$$\text{Pentad}(y) = C(\text{Pentad}(x) + K)^p$$

where Pentad is the average 5 day flow ($m^3 s^{-1}$) and C, K and p are constants.

The Gunnislake record was used for all the early years; then the nearest station to an unknown station was used as the records started. For each correlation, values of C, K and p were determined and these constants used to generate mean daily flows as follows:

$$\text{MDF}(y) = C[\text{MDF}(x) + K]^p$$

The results were tested by examining the cross-correlation coefficients and comparing sample hydrographs of recorded and generated flows. The generation of the 30 year daily flow sequences was preferred to the derivation of theoretical droughts with related problems of probability, since the period included the major droughts of 1959, 1975–1976 and 1984 to which the local people can relate experienced river conditions. This is very important in such a noted fishing area.

Demand data

The annual average demands predicted for the years 1991–2011 for the 10 demand centres and a seasonal demand pattern on a weekly basis for each centre formed the demand input data. The selection of the weekly demand pattern throughout a year needed careful consideration since an average set of values over several years could even out important salient features and yet the choice of a single year might be too closely related to extreme weather conditions. A compromise average over three typical years was taken.

In addition to the bulk public water supplies, details of the private licensed abstractors, their prescribed flow and percentage take together with their intake pump particulars must also be included.

Reservoir particulars

The gross and net storages of the reservoirs, Roadford and Burrator, together with details of compensation releases and the storage–area relationships with historic average monthly evaporation for the calculation of evaporation losses, are all needed for the operation of the model. The storage–depth relationships were also included to provide information for the possible generation of hydroelectric power. Estimates of times of travel of water releases and percentage losses are also required and these have a direct bearing on the question of the timing of operational rules.

Control curves for Burrator Reservoir, the main source of water for Plymouth since 1898, were developed for the conjunctive use of Burrator and the Tavy River intake at Lopwell. The resources of Burrator will be partly redeployed when Roadford is completed and supplementary supplies pumped to the River Dart. Thus new operational rules for this reservoir have had to be derived.

Model output

The model of the Roadford Reservoir system is a valuable tool in the management of the water resources. By varying its operation with differing demands drawing on the several sources, the effects on a whole range of consequences, such as on the costs of the operation or on the desirable river regimes for the fisheries, can be determined.

The choice of objectives in the operation must be made by the managers. The output of the model to provide the necessary information on which to base decisions, is as follows.

(1) Daily or weekly hydrographs and duration curves, both pre-reservoir and post-reservoir, at any location.
(2) The proportion of natural river water to Roadford augmentation water at any point.
(3) The effect of the stepping, up or down, of releases on the river flows.
(4) Volume of pumped water and pumping costs.

It is expected that pumping costs will be high in the operation of the Roadford Reservoir system but it may be possible to minimize the costs and there may be some helpful returns from hydropower generation.

Conclusion

The model HOMER is a flexible tool which can be continually modified or refined. In the future it could be made the basis of a real-time operational model. It should certainly help to resolve inevitable conflicts in the operation of such a complex system. Following the necessary compromises, the consequent optimization of the system becomes subjective rather than objective.

References

Allard, W., Glasspoole, J., & Wolf, P. O. (1960) Floods in the British Isles. *Proc. Inst. Civ. Eng.* **15**, 119–144.

ASCE Task Committee of Hydrometeorology of the Hydraulics Division (1973) Reevaluating spillway adequacy of existing dams. *J. Hydraul. Div. Am. Soc. Civ. Eng.* **99**, 337.

Bell, F. C. (1969) Generalised rainfall–duration–frequency relationships. *J. Hydraul. Div., Am. Soc. Civ. Eng.* **95**, 311–327.

Benson, M. A. (1968) Uniform flood–frequency estimating methods for federal agencies. *Water Resour. Res.* **4**, No. 5, 891–908.

Brandon, T. W. (1987) *River engineering*, Part I, *Design principles. Water practice manuals*. Institution of Water Engineers and Scientists for Institution of Water and Evironmental Management.

Chidley, T. R. E., & Lloyd, J. W. (1977) A mathematical model study of fresh water lenses. *Ground Water* **15** No. 3, 215–222.

Chow, V. T. (1959) *Open channel hydraulics*. McGraw-Hill, pp. 146–148.

Creager, W. P., Justin, J. D., & Hinds, J. (1964) *Engineering for dams*, Vol. I. John Wiley.

Drayton, R. S., Kidd, C. R. H., Mandeville, A. N., & Miller, J. B. (1980) *A regional analysis of river floods and low flows in Malawi*, IH Report, No. 72. Institute of Hydrology.

Fiddes, D. (1977) Flood estimation for small East African rural catchments. *Proc. Inst. Civ. Eng.* Part 2, **63**, 21–34.

FAO (1977) Crop water requirements. Irrigation and Drainage, Paper 24. Food and Agricultural Organization.

Grindley, J. (1967) Estimation of soil moisture deficits. *Meteorol. Mag.* **96**, 97–108.

Hall, M. J. (1984) *Urban hydrology*. Elsevier.

Hall, M. J., & Hockin, D. L. (1980) *Guide to the design of storage ponds for flood control in partly urbanised catchment areas*. CIRIA Technical Note No. 100. Construction Industry Research and Information Association.

ICE (1960) *Floods in relation to reservoir practice*, Interim Report 1933, reprint. Institution of Civil Engineers.

ICE (1978) *Floods and reservoir safety: an engineering guide.* Institution of Civil Engineers.

IH (1979) *Design flood estimation in catchments subject to urbanisation*, The Flood Studies Supplementary Report No.5. Institute of Hydrology.

Interagency Advisory Committee on Water Data (1982) *Guidelines for determining flood flow frequency*, Hydrology Subcommittee Office of Water Bulletin 17B. Data Coordination, Geological Survey, US Department of the Interior.

Kibler, D. F., & Woolhiser, D. A. (1970) *The kinematic cascade as a hydrological model*, Hydrology Paper 39. Colorado State University.

Klemes, V. (1973) *Applications of hydrology to water resources management*, Operational Hydrology Report No.4. World Meteorological Organization.

Lencastre, A. (1987) *Handbook of hydraulic engineering.* Ellis Horwood.

Massey, B. S. (1986) *Measures in science and engineering.* Ellis Horwood.

McMahon, T. A., & Mein, R. G. (1978) *Reservoir capacity and yield.* Elsevier.

NERC (1975) *The Flood Studies Report, Vols I–V.* Natural Environment Research Council.

NWC–DOE (1981) *Design and analysis of urban storm drainage. The Wallingford procedure*, 5 vols. National Water Council–Department of Environment.

Office of Water Data Coordination (1986) *Feasibility of assigning a probability to the probable maximum flood.* Hydrology Subcommittee of the Interagency Advisory Committee on Water Data, Washington,USA.

Packman, J. C. (1981) Effects of catchment urbanisation on flood flows, *The Flood Studies Report — Five Years On.* Institution of Civil Engineers, pp. 121–129.

Parks, Y. P., & Gustard, A. (1982) A reservoir storage yield analysis for arid and semiarid climates. *Exeter Symposium.* IAHS Publication No. 135, 49–57. International Association of Hydrological Sciences.

Penning-Rowsell, E. C., & Chatterton, J. B. (1977) *The benefits of flood alleviation : a manual of assessment techniques.* Saxon House, UK.

Pilgrim, D. H. (ed.) (1987) *Australian rainfall and runoff, a guide to flood estimation.* Institution of Engineers, Australia.

Pitman, W. V. (1973) *A mathematical model for generating monthly river flows from meteorological data in South Africa.* Report No. 2/73. Hydrological Research Unit, University of Witwatersrand.

Ponce, V. M., & Yevjevich, V (1978) Muskingum-Cunge Method with variable parameters. *J. Hydraul. Div., Am. Soc. Civ. Eng.* **104**, 1663–1667.

Price, R. K. (1980) *FLOUT — a river catchment flood model.* Report No. IT 168, revised. Hydraulics Research Ltd.

Richards, B. D. (1955) *Flood estimation and control*, 3rd edn. Chapman & Hall.

Rodda, J. C. (1969) *Hydrological network design — needs, problems and approaches*, WMO/IHD Projects Report No. 12. World Meteorological Organization.

Rodier, J. A., & Auvray, C. (1965) Preliminary general studies of floods on experimental and representative catchment areas in Tropical Africa. *Budapest Symposium*, Vol. I, International Association of Hydrological Sciences, pp. 22–28.

Rogers, P. P., & Fiering, M. B. (1986) Use of systems analysis in water management. *Water Resour. Res.* **22**, No. 9, 146S–158S.

Rushton, K. R., & Ward, C. (1979) The estimation of groundwater recharge. *J. Hydrol.* **41**, 345–361.

Shaw, E. M. (1988) *Hydrology in practice*, 2nd edn. Van Nostrand Reinhold.

Simpson, R. W., & Thorpe, G. R. (1982) Use of a generalised computer program for resource systems optimisation in developing countries. *Exeter Symposium*, IAHS Publication No. 135, International Association of Hydrological Sciences, pp. 285–298.

Sutcliffe, J. V. (1978) *Methods of flood estimation — a guide to The Flood Studies Report*, IH Report No. 49. Institute of Hydrology.

Thomas, H. A., & Fiering, M. B. (1962) Mathematical synthesis of streamflow sequences for the analysis of river basins by simulation. In: A. Maas *et al.* (eds.), *Design of water resource systems*. Harvard University Press, Chapter 12.

Twort, A. C., Hoather, R. C., & Law, F. M. (1974) *Water supply*, 2nd edn. Edward Arnold.

UN–WMO (1960) *Hydrologic networks and methods*, Flood Control Series No.15. United Nations–World Meteorological Organization.

US Army Corps of Engineers (1971) *HEC-4 monthly streamflow simulation generalised computer program*. Hydraulic Engineering Centre.

USBR (1973) *Design of small dams*, Water Resources Technical Publication 2nd edn. US Bureau of Reclamation, US Department of Interior.

USDA (1967) *Irrigation water requirements*, Technical Publication No. TP 21. Soil Conservation Service, US Department of Agriculture.

USDA (1972) *National Engineering Handbook*, Section 4, *Hydrology*. Soil Conservation Service, US Department of Agriculture.

Valencia, D., & Schaake, J. C. (1972) Disaggregation processes in stochastic hydrology. *Water Resour. Res.* **9** No. 3, 580–585.

WAA–WRc (1986) *Sewerage rehabilitation manual*, 2nd edn. Water Authorities Association–Water Research Centre.

Wilde, D. J. (1966) *Optimum seeking methods*. Prentice-Hall.

WMO (1958) M. A. Kohler (ed.), *Design of Hydrological Networks*, Technical Note No. 25; R. K. Linsley (ed.), *Techniques for surveying surface water resources*, Technical Note No. 26. World Meteorological Organization.

WMO (1972) *Casebook on Hydrological network design practice*, WMO Publications No. 324. World Meteorological Organization.

WMO (1973) *Manual for estimation of probable maximum precipitation*, Operational Hydrology Report No.1. World Meteorological Organization.

WMO (1976) *Hydrological network design and information transfer*, WMO Publication No. 433. World Meteorological Organization.

Author and geographical index

Hydrological subject and technique index